TOURISM AND SUSTAINABILITY

TOURISM AND SUSTAINABILITY

Principles to Practice

Edited by

M.J. Stabler

Department of Economics
University of Reading
UK

CAB INTERNATIONAL

CAB INTERNATIONAL
Wallingford
Oxon OX10 8DE
UK

Tel: +44 (0)1491 832111
Fax: +44 (0)1491 833508
E-mail: cabi@cabi.org

CAB INTERNATIONAL
198 Madison Avenue
New York, NY 10016-4314
USA

Tel: +1 212 726 6490
Fax: +1 212 686 7993
E-mail: cabi-nao@cabi.org

A catalogue record for this book is available from the British Library, London, UK.
Tourism and sustainability : from principles to practice / edited by
  M.J. Stabler.
        p.   cm.
     Includes index.
     ISBN 0-85199-184-X (alk. paper)
     1. Tourist trade—Congresses.   2. Sustainable development-
  -Congresses.   I. Stabler, Mike.
  G154.9.T6844   1997   97-5265
  338.4'791—dc21   CIP

ISBN 0 85199 184 X

Typeset by Solidus (Bristol) Ltd
Printed and bound in the UK by Biddles Ltd, Guildford and King's Lynn

Contents

Contributors

Rosemary Burton is Senior Lecturer in Environmental Management and Geography at the University of the West of England, Bristol. Her main teaching and research interests are leisure, recreation and tourism planning and management, and specifically the geography of tourism. She is the author of *Travel Geography* and currently researching ecotourism and the management of tourism in Australian National Parks.

Jim Butcher lives in Birmingham, and is a lecturer at the Birmingham College of Food, Tourism and Creative Studies. He is currently conducting research at the Institute for German Studies, University of Birmingham. He has a first degree in Economics, and a Master's degree in European Studies from Coventry University. His main academic interests are the politics and sociology of contemporary society.

Stroma Cole is a lecturer of Tourism at Buckinghamshire College, a college of Brunel University. Until 1995 she lived in Indonesia, working for 12 years first as an anthropologist and then as a teacher. In 1989, after completing her Master's degree at SOAS, London University, she set up a tour company specializing in culture and nature tourism in the remoter regions of Indonesia. Stroma has worked as a consultant on ecotourism projects in Indonesia. She writes from the perspective of both an academic and a practitioner in the field.

Diane Davis is Head of the Department of Hospitality and Tourism at the University of Central Lancashire. She is particularly interested in organizational behaviour and the problem of managing human resources in international hotel chains. Her recent research, in Greece and in Israel, concerns the management of the service encounter.

Ian Eastwood is Head of the Department of Environmental and Leisure Studies, Crewe & Alsager Faculty, Manchester Metropolitan University, UK. He gained a Bachelor's degree in Environmental Studies from Sheffield City Polytechnic, a Grad. Cert. Ed. from Huddersfield Polytechnic and was awarded a PhD by Sheffield City Polytechnic for research into pollution of soils by lead and its uptake and pathway in the ecosystem. He is author of published articles and conference papers on aspects of heavy metals in the environment and tourism and the environment. His research interests include environmental stress resulting from tourism and recreation activity; degradation of soil systems by human activity; soil contamination resulting from historical heavy metal mining and processing activity.

Brian Eaton is Senior Lecturer at Coventry Business School, where he specializes in marketing research for leisure industries. His publications cover a variety of leisure subjects including visitor management in tourist areas, rural leisure and tourist transport. He has carried out work for the Countryside Commission and is currently continuing research into the effects of privatization of Britain's railways. His most recent book was *European Leisure Businesses: Strategies for the Future*, published by ELM Publications, Huntingdon. He lives at Brassenthwaite Lake in Cumbria.

Ann Edington is a microbiologist and formerly lecturer in the Departments of Microbiology and Zoology at University of Wales, Cardiff. She is now Visiting Lecturer in Environmental Engineering at Cardiff School of Engineering. She also works with Dr John Edington in carrying out environmental appraisals of development projects.

John Edington is a zoologist and formerly Director of Environmental Studies at University of Wales, Cardiff. He now works as a consultant in curriculum development and environmental assessment. He has a particular interest in environmentally-sensitive tourism development.

Helen Farr is Rural Development Programme Officer for Cornwall and Isles of Scilly. The Rural Development Programme has an annual financial allocation from the Rural Development Commission which is available to grant aid to economic and community development projects in the Rural Development Area. Prior to moving back to Cornwall, Helen was part of the research team in the Department of Agriculture, University of Aberdeen, investigating the economic impact of rural tourism on rural economies.

Alan Fyall is a senior lecturer in Tourism Marketing at Southampton Business School, Southampton Institute. He completed his MPhil in 1993 on the development of combined leisure and shopping centres and has consultancy experience with a number of major leisure companies in the UK.

Brian Garrod is a senior lecturer in Environmental Economics in the Faculty of Economics and Social Science, University of the West of England. He is also a Fellow of the University's Centre for Social and Economic Research. He completed his PhD in 1993 and has published a number of articles on environmental economics.

Brian Goodall is Joint Director of the Tourism Research and Policy Unit, University of Reading, and Professor and Head of the Department of Geography and Consultant Director of the NERC Unit for Thematic Information Systems. He was formerly Dean of Urban and Regional Studies. Professor Goodall has been researching tourism and recreation for over 20 years; his interests centre on the structure of the tourism industry, the evolution and planning of destination regions, and environmental assessment for tourism. He has published widely in books and journals.

Amran Hamzah is a chartered town planner and lecturer in the Department of Urban and Regional Planning at the Universiti Teknologi Malaysia, Johor Bahru. His research interests and consultancy are in the field of environmental and resource management and planning.

Paul Harper is an ADAS (Agricultural Development Advisory Service) tourism and rural economy consultant. He has had 18 years' experience in the advisory field and in particular, while working in Cumbria with farm tourism operators, has spent three years as project manager of the Cumbria Farm Tourism Initiative.

Joanna House is a lecturer in the Department of Social Science at the University College of St Mark and St John, Plymouth. She has gained an MA in Tourism and Social Responsibility and is currently a doctoral student specializing in the study of alternative cultures. She has lately been undertaking ethnographic and qualitative field research in several areas in Devon and Cornwall in the UK.

Howard Hughes is Professor of Tourism Management at Manchester Metropolitan University. He gained Bachelor's and Master's degrees in Economics from the University of Wales and Leeds University respectively and was awarded his PhD by the City University, London, for research into the tourism dimensions of a UK opera company. He is author of a specialist economics textbook *Economics for Hotel and Catering Students*, published in 1986 and now in its second edition. He has over 30 published papers on aspects of the hotel and tourism industries to his name. He has presented papers at conferences and seminars in overseas countries including China, Israel, the Netherlands, USA and Canada. Currently he is engaged in a research project examining the entertainment–vacation link. Other research interests include urban tourism and hallmark events; the government–tourism relationship; holidays for disadvantaged groups.

Philippa Hunter-Jones is a lecturer in Tourism and Leisure Studies at Manchester Metropolitan University, UK. She gained a first degree in hospitality management from Leeds Polytechnic. She has presented papers at conferences and seminars on aspects of tourism and the environment with particular reference to Spain. She is currently engaged in a research project examining community attitudes towards tourism and the environment in Spain. Other research interests include pilgrimage tourism in Spain; tourism and health issues; tourism and developing countries.

Michael Ireland is a senior lecturer in Tourism and Course Leader on the MA in Tourism and Social Responsibility at the University College of St Mark and St John, Plymouth. His interests include rural and maritime tourism.

Terry McCormick, as a Leverhulme Research Fellow, was responsible for coordinating the development of the Wordsworth Museum in the early 1980s and was subsequently curator for the Wordsworth Trust for 11 years. In 1996 he became a freelance heritage and tourism consultant, recently being concerned with museum networking, and project design and business planning for visitor centres.

Fraser McLeay holds a first degree in Agricultural Science and completed a PhD at Lincoln, New Zealand. His main areas of research are marketing management and business strategy in small- and medium-sized enterprises. He is currently a lecturer at the University of Newcastle upon Tyne.

Rory MacLellan has a first degree in Geography from the University of Aberdeen and an MSc in Tourism from the University of Strathclyde. After a period with the Welsh Tourist Board he lectured in tourism in Portsmouth, Cardiff and Swansea before taking his current position at the University of Strathclyde. His teaching and research interests include public sector tourism organizations, rural tourism, outdoor recreation, and tourism and the environment. He has published on aspects of tourism and the Scottish natural environment and is involved in consultancy with the Scottish Tourism Research Unit, mainly for tourism and economic development agencies in Scotland.

Victor Middleton, a graduate of the London School of Economics, is an independent consultant with wide-ranging interests in the travel and tourism industry. He is Visiting Professor at Oxford Brookes University where he was also first Director of the internationally sponsored World Travel and Tourism Environmental Research Centre. He is a founder fellow and recent past chairman of the UK Tourism Society. He has published a very large number of research reports, journal articles and several books over the last two decades.

Alan Morrison is a principal lecturer in Environmental Science at Manchester Metropolitan University, UK. His interests are in the field of environmental degradation related to people's activities, including tourism. His more specific interests are focused around woodland, forests and water quality issues. He has presented papers at conferences and seminars on factors affecting sedimentation in lake reservoirs and is currently engaged in a research project examining the effects of tree shelter belts in preventing pollution transmission from major roadways.

Andrew Moxey, after studying agricultural economics at the University of Newcastle upon Tyne, now holds a post as lecturer there. He is principally interested in applying quantitative analysis to marketing and economic issues within the rural environment.

John Parsler is a lecturer in Environmental Management at the University of Central Lancashire, previously having worked in the voluntary sector for the Council for the Protection of Rural England and the Marine Conservation Society. His research interests are biodiversity conservation and eco-tourism with particular reference to Madagascar.

Eleftheria Prinianaki-Tzorakoleftheraki, a postgraduate of the University of Surrey, is an assistant professor at the Department of Tourism Industries of the Technological Educational Institute of Heraklion, Crete, Greece. She is a member of the Greek and the World Federation of Travel Journalists and Writers as well as of the Economic Chamber of Greece.

Julia Sharpley is a senior lecturer in Travel and Tourism at the University of Hertfordshire. Her current research focuses primarily upon the potential for tourism development in urban areas in the UK and she has also undertaken research into rural tourism, sustainable tourism development and, in particular, tourism development issues in The Gambia.

Richard Sharpley is Senior Research Fellow in the Department of Tourism and Leisure at the University of Luton. He lectures mainly on rural tourism and the sociology of tourism and has published textbooks in both areas. His current research is primarily concerned with tourism development in The Gambia and the implications for sustainable tourism development of socio-cultural change and the consumption of tourism in modern society.

Bill Slee is Senior Lecturer in Rural Economics in the Department of Agriculture at the University of Aberdeen. His research interests include rural tourism, farm and rural economic diversification and agri-environmental policy. He recently coordinated a study on the economics of rural tourism in three European Union countries and has worked for a variety of organizations including the World Bank, the European Union, the Forestry Commission, the Scottish Office and Scottish Natural Heritage.

Patrick Snowdon is a teaching assistant in the Department of Agriculture at the University of Aberdeen. His research interests focus on diversification in the rural economy and on the role of European Union policy in rural development. He has recently engaged in a major European Union-funded study on the economic impact of rural tourism and in research for the Forestry Commission investigating the potential for community participation in forest management in remote rural areas. He has a BA in Geography and French from the University of Sussex, and MSc in Recreation and Conservation Management from the University of Hertfordshire.

Michael Stabler worked in local government and commercial management and gained a number of business qualifications, before becoming a lecturer at a company staff college. He was lately a senior lecturer in Economics in the Centre for Spatial and Real Estate Economics at the University of Reading. He is currently Joint Director of the Tourism Research and Policy Unit and a member of the School of Planning Studies at Reading. In addition to his long-term interest in environmental, leisure and land-use policy issues, he has investigated the economics of tourism, including its economic, environmental and socio-cultural impact. At present he is considering the role of land-use planning systems, business organizations and local institutions in achieving sustainable tourism. His research on these topics has been published in books, international journals, conference proceedings and research reports.

George Taylor is a senior lecturer in Tourism at the University of Central Lancashire specializing in international tourism development and socio-cultural impacts. His research and publications have been largely in the areas of resident responsiveness to tourism development and the transfer of corporate culture in international hotel chains. His current research concerns the nature of the relationship between traditional and commercial hospitality in Crete.

Angela Tregear is a lecturer in the Department of Agricultural Economics at the University of Newcastle upon Tyne. She studied marketing and modern languages and her research interests focus on theories of marketing and the application of these to aspects of the rural environment, particularly food and tourism.

Brian Wheeller is Senior Lecturer in Tourism at the University of Birmingham. He is Course Director of the Tourism Policy and Management M. Soc. Sci. programme. His research interests include a critique of eco/ego/sustainable tourism and the relationship between travel, tourism and popular culture.

David Wilson is a lecturer in Social Anthropology at the Queen's University of Belfast. He is the author of a number of articles on tourism in Goa, the Seychelles, the Caribbean, Northern Ireland and the North of England. His research interests include the socio-cultural impacts of tourism, heritage, images of place, public policy and sustainable development.

Foreword

As well as having its own intrinsic value, the beauty of the environment is also a key factor in attracting visitors. A large part of the tourist industry exists chiefly because of the attractiveness and quality of the environment and it is therefore vital for that reason alone, quite apart from other considerations, that tourism does not impair the very thing that the industry needs for its future success. Tourism and the environment must develop in harmony and not at each other's expense. It is, after all, right that future generations should be able to enjoy the heritage, culture and countryside as much as the present one does.

One cannot stop people wanting to travel. When they become tourists, people have requirements that need to be met, ranging from parking to accommodation. Unfortunately, the most attractive features of an environment are often those most sensitive to visitor pressures. Sustainable tourism is about managing tourism development in an acceptable way to ensure changes to the environment are benign. It is not about denying any development at all.

The tourist industry makes an enormous contribution to the prosperity of a country: for example in the UK it generates over £35 billion a year in income and provides one and a half million jobs. With the world tourism market expected to grow by 4% a year over the next decade, and tourism set to become globally the largest industry by the year 2000, the potential for growth in the tourism market is there. The aim in any destination should be to create the conditions which will encourage inward and domestic tourism so that the industry can make its full contribution to the economy, increase opportunities for access to culture and heritage and provide resources for them. The need for tourism to prosper and grow, the need to preserve and protect areas of natural beauty and the need to recognize the role played by a primary industry, such as agriculture, in this process are aims that it is not always easy to reconcile. It is a question of balancing the needs of the visitor with the needs of other interests and the environment. These are by no means mutually incompatible, especially as today's more selective and discerning tourist will expect to see a number of economic activities in the countryside and that the environment is well cared for. The funding of this balance is not a desirable optional extra, it is an essential component of any tourism strategy and essential for safeguarding the future.

In general, problems caused by visitor pressure are not widespread or chronic. Where they do occur, they are often highly localized in time and place, and in many cases can be addressed through improved management. It should be recognized that there is the need for greater dissemination of good practice, more assessment of the effects of tourism, and more innovation, coordination and collaboration in tackling this issue at a local level.

The growth in the tourism market offers the industry immense opportunities, but also places on it great responsibilities. The need for visitor management is a key message. A need for practical measures to manage visitors will continue to grow. Properly managed tourism to a heritage site, to the countryside, or to a historic town can increase the public's appreciation, contribute to the maintenance and sensitive development of a place, and thus make a very positive contribution to the environment.

It is sometimes too easy to lose sight of the positive impact that tourism can have. It can bring environmental benefits: tourism can offer new life to redundant buildings or despoiled lands, can help to fund the maintenance of the built and natural heritage, and can stimulate public awareness of the importance of protecting the environment, including some of the best-loved features of a country.

Tourism can also bring indirect economic benefits to communities. Non-tourism enterprises benefit from visitor spending, and whole communities, particularly in rural areas, benefit from visitor support of local services. Furthermore, the continuing growth in tourism employment is playing a significant part, certainly in developed countries, in offsetting the loss of jobs in agriculture and other declining rural sources of work. Indeed, the importance of tourism to the rural economy is increasing.

The contribution made by tourism in rural settlements often goes unrecognized, because the damaging impacts, such as traffic congestion, is more visible and emphasized more strongly. It is, however, possible to increase the scale of benefits and change negative views among residents, particularly through employing visitor management techniques and through the increased involvement of the local community.

An environment which is well maintained and cared for is part of that 'visitor's experience' and something which, increasingly, he/she is coming to expect. The kind of tourism wanted can only flourish in a high-quality environment. Damage to the environment, will, in the long run, mean damage to the tourism industry, as well as to a country's own heritage, and will limit its successful growth. The heritage, culture and countryside over the globe have given pleasure to millions through the years and it is imperative that they will continue to do so for future generations.

Lord Inglewood
Under Secretary of State
Department of Natural Heritage, UK

Preface

This book contains a selection of edited and revised papers which were presented at a conference mounted in April 1996 in part celebration of the centenary of the founding of Newton Rigg College, Cumbria, UK, organized in conjunction with the University of Central Lancashire. The chosen theme of 'Sustainable Tourism: Ethics, Economics and the Environment' was apt, given the location of the college bordering one of the most attractive but intensively used tourism destinations in the UK, and the recognition that sustainable development has become a central issue in academic institutions and organizations concerned with rural economies, particularly as tourism development is seen as the salvation of more remote but high-value scenic areas as traditional economic activities decline. The conference was organized to provide a forum to discuss the environmental and cultural impact of tourism and the strategies for sustaining and managing it.

In the invitation to participate in the conference, the World Tourism Organization's reinterpretation of the widely accepted definition of sustainable development was cited as the starting point for the realization of sustainable tourism: 'Tourism which meets the needs of present tourists and host regions while protecting and enhancing opportunity for the future.'

The question posed was that the translation of this principle into practice would require a huge shift in the attitudes to and actions of investors, managers and destination residents, as well as tourists, in the face of a primary objective of increasing tourism activity. The intention of the organizing committee of the conference was to attract managers, planners and academics to debate the implementation of sustainability principles in both theory and practice.

In making the selection of papers, the editor has endeavoured to adhere to the spirit of the conference and debate the organizers wished to engender. It is hoped that the chapters will reflect the:

- flavour of the commercial economic, ecological, cultural, political and social issues which sustainability raises.
- different concepts and interpretations given to sustainable tourism.
- difficulties of translating principles of sustainable tourism into policies and implementing them.
- opening up the debate through consideration of cases within both developing and developed countries where issues are sharply focused.

Inevitably academics tend to dominate conferences, the themes of which are strongly conceptual and methodological. Nevertheless, strenuous attempts have been made to strike a balance by including as many contributions by practitioners as it was feasible to do. Moreover, due consideration was accorded to the work of young researchers particularly where new light was thrown on sustainable tourism issues.

The temptation merely to juxtapose contributions which echoed the three themes of the conference was, in the event, not strong. It became readily apparent that the range and diversity of the papers were quite wide and that issues concerning ethics, economics and sustainability were much more subtly incorporated, often being implicit rather than explicit. This made it more difficult to group chapters with related themes to give cohesion to the text. Given the key issues emerging from the sustainable tourism debate relating to: the concept and its interpretation; the likelihood of self-regulation by the tourism industry; the increasing need for empirical evidence on the impact of tourism development; and moves towards making it sustainable and implementation of the necessary policies, each contribution was assigned to one of four parts of the book which largely accord with these four issue areas.

The editor wishes to acknowledge the support of Andrew Humphries and Stephen Oliver-Watts of Newton Rigg College, who acted as editorial advisers on the selection of papers for publication in this book. Grateful thanks are, however, due to the contributing authors, who responded most readily and positively to suggested changes and required revisions.

An Overview of the Sustainable Tourism Debate and the Scope and Content of the Book

M.J. Stabler

Department of Economics, Faculty of Urban and Regional Studies, University of Reading, Whiteknights, PO Box 219, Reading RG6 6AW, UK

It is perfectly legitimate to raise the question as to the relationship between ethics, economics, the environment and sustainable tourism. The choice of the conference title presupposes that the notion of sustainable tourism involves ethical and economic issues. Indeed, it infers that in order to attain sustainable tourism they are both conceptual and policy imperatives. In essence they are inextricable elements in the process of that attainment. However, what do the terms ethics and economics in particular signify in the context of sustainability? They are each open to several interpretations and perspectives as to their significance, and, in the case of ethics, as to its meaning.

Ethics

At the theoretical, perhaps academic, level the researcher should, for example, consider such matters as the ethical implications of pursuing sustainability as a goal *per se* or for destination communities and environments, or for current and future generations. Conversely, the ethical significance of not attempting to attain sustainability should be examined. In a practical context, ethical issues arise in the conduct of tourism business and certainly, very often on income and wealth distributional grounds, with respect to government policies involving regulation or taxation.

To date in the tourism literature, and indeed in a wider context of economic development in both developing and developed countries, there is a supposition that sustainability is universally acceptable on ethical grounds.

In short, there is simply no argument. It is ethical. Consequently, discussion tends to be suppressed, with any dissenting view being perceived as bizarre, if not heretical, so that what ethics means has not been established, or at best understanding of it and its significance is confused. The dictionary definition of ethics is that it is the philosophical study of the moral value of human conduct and of the rules and principles that ought to govern it. It is also defined as a social, religious or civil code of behaviour that is considered correct, especially of individuals, professions or specific groups, leading to questions as to the moral fitness of a course of action, effectively as to whether that action is 'ethical'. A code of behaviour or set of moral values held by an individual or groups can be designated an 'ethic', such as the 'Puritan Ethic'. Therefore, it is possible to conceive of there being a 'sustainability ethic' which is not the same as considering whether, as a goal, it is ethical to endeavour to attain it, or that the means of doing so are ethical; somewhat along the lines of the aphorism of the ends justifying the means.

What appears to be occurring in the sustainability debate is that the 'ethic' and the 'ethics' (that is whether it is ethical) have got muddled up together. Indeed, one can take this point further because both the ethic and the ethical ramifications of the whole issue of sustainability have also got bound up with the meta ethics of it, by which is meant the questions of the nature of ethical judgements as distinct from those of normative ethics, for instance, do ethical judgements state facts (effectively are they objective) or do they merely reflect attitudes and opinions? This in turn raises the issue as to whether there are objective standards of morality and how moral judgements are made. Identifying such questions is not merely indulging in semantics, in the linguistic sense of trying to make fine distinctions between the meaning of terms or words, it is an attempt to clarify the issues which surround the whole concept of sustainability. An illustration might serve to justify why the matter has been raised here.

There is widespread acceptance of the World Commission on Environment and Development (1987) Brundtland report, which has subsequently been reinforced by governments' and environmental bodies' utterances, of the principle of sustainability, particularly the intergenerational responsibility placed on the current generation. This is the ethic which might be questioned on the grounds that it is anthropocentric. (It is acknowledged, however, that there are other ethics, for example ecocentrism.) This principle is being translated into action, certainly within such initiatives as Agenda 21 (United Nations Conference on Environment and Development, 1992) and those in the European Union (Commission of the European Communities, 1993), and there are different stances on how this should be accomplished, the extremes of which, for instance, might be unfettered market forces at the zero end of a spectrum of intervention and command and control at the other. The former may take a very long time before scarcity of resources and pollution induce initiatives to secure sustainability via the price mechanism while the latter may be effective in the short run but prove to be inefficient in the long run. Either approach has an impact on the allocation of resources and distribution of income and wealth, in both a vertical (within a given society) and horizontal (spatially, say between developed and developing countries) sense. In essence, the means of achieving sustainability may be considered unethical because it benefits some, perhaps higher income groups or those in richer countries, while the costs are borne by others, possibly the majority, who are in lower income groups or in poorer nations. Both the 'ethic' and the 'ethical' considerations have meta ethical connotations because the fundamental issue is raised as to the nature of the decision, and the moral position of those making it, on the goal (ethic) and the means (ethical acceptability) of attaining it. For example, it has been suggested that sustainability is an ethic which has been devised by those in the developed world and foisted on those in developing countries, who have also been instructed as to how it should be attained, very probably using means (for example, equipment and expertise) supplied by industries in the former, which make profits benefiting its residents (unethical behaviour). This illustration can be considered as an ethnocentric sustainability ethic.

In relation to the theme of the conference and the extent to which the chapters in this book consider 'ethics', used in a generic sense to encompass all three aspects identified here, it is not too much of an injustice to suggest that they do so implicitly and largely as to whether the means of pursuing sustainability are ethical or not. No author fundamentally questions the consensus on sustainability and only one or two direct somewhat sidelong glances at the meta ethical issues. However, some evidence of what might be construed as ethical issues emerges. The chapters which touch on such matters, particularly in the case studies, refer to the attitudes and actions of governments and bodies concerned with economic and tourism development, but the most frequently quoted examples are of the activities of tourism businesses. An interesting feature of this, in some developing countries, is that many commercial activities are ostensibly illegal in that there are environmental laws and regulations forbidding them but they are condoned by the authorities. In the eyes of the operators such behaviour is not perceived as unethical, since after all a service is rendered to tourists which would not be available if there was strict enforcement. Does this perhaps suggest that it is really tourists who are acting unethically or can it be argued that this is so only if they are aware that the supplier is operating illegally? Does this place an obligation on governments, bodies and businesses in destination areas and tour operators in origin areas to bring such violations of legal and regulatory policies to the attention of would-be tourists? Moreover, can international and national bodies and businesses, which offer aid on the understanding that it will be applied to environmental protection or enhancement, be indicted for being negligent if they fail to ensure that it is used for what it was intended? There are also many apocryphal accounts of governments, as well as destination communities, playing lip-service to sustainable tourism in order to gain funding with a view to applying it to economic development through tourism. It is little wonder, therefore, that the term sustainable tourism is derided as a monumental oxymoron. These are surely ethical issues, as well as often being legal ones. With a few exceptions, such questions have hardly been raised, let alone debated, in the tourism literature. In the contributions in this book, the connections between sustainable tourism and ethics are beginning to be made, particularly in the earlier chapters concerned with more general issues.

Economics

Starting from the premise that tourism is a market-based activity and that setting a goal of sustainability involves environmental protection and enhancement, which impinge on the industry and the market, then economic considerations are inescapable and apparently clear. Unfortunately, this is not so, for the term economics can embrace what might be more appropriately viewed as business practice and management. In tourism studies, it is often treated in this way in taking an industry perspective on environmental and sustainability issues. More correctly, an economic perspective should be much wider to encompass analysis of the issues of the principles of sustainable tourism and the policies to implement it. Indeed, environmental economics, which underpins the subject's approach to sustainability, embodies many aspects which are pertinent but what is not always appreciated by those outside the discipline is that it is a generic term which covers more specific areas of analysis.

The longest-standing foundation of environmental economics is conservation economics, which tends to emphasize the impact of economic activity on the demand for productive and energy resources. This branch of economics considers their optimal use over time, the strategies advocated being determined by whether resources are exhaustible or renewable. It is only in recent years that conservation economics has broadened its horizons to include natural and man-made environments, especially in examining the role they play as inputs in an activity such as tourism. These, as the resource base of tourism, are therefore treated in the same way as productive inputs, such as minerals or energy; some are exhaustible, or perhaps more appropriately non-renewable, such as unique environments like the Grand Canyon, the Antarctic and the Himalayas, while others are renewable if properly managed, for example flora and fauna and many man-made landscapes. A complication which arises, however, is the non-priced nature of many such resources and this creates allocation problems where there is competition for their use which manifests itself in the market. For instance, agriculture and forestry are productive market-based uses of rural land which are likely to outbid any amenity demand for the land for informal recreation and tourism. The methods which economists have proposed to overcome such problems are considered later.

The area of interest which has grown out of conservation economics, where the environment is used as a free good, is pollution abatement. This is essentially the core of environmental economics and most attention has been paid to the costs of emissions into the atmosphere, discharges of effluents into water courses and solid waste disposal in the form of landfill or dumping at sea. Lately, noise and visual pollution have been conceived as being significant. It is the identification, measurement and evaluation of pollution, the means of abating it and the effects of the ameliorative action, on which economists have concentrated. It is the concept of social cost (a detrimental externality) which economics has devised to show that the perpetrators of pollution, most often the producers of goods and services, do not bear the full costs, these falling on the community at large, that has had the greatest impact on both the direction of environmental economics and policy initiatives. It has also largely determined the course of the sustainability debate and the identification of its ethic, particularly with regard to the intergenerational issue of the state of the environment which is likely to be handed on by the present to future generations.

Tourism as a major industry can, and should, be viewed in the same light as manufacturing industries with respect to its pollution of the environment and generation of externalities. It consumes materials and energy and produces wastes, in origin, transit and destination areas. However, in some respects its impact is more marked for in using natural and man-made environments as its basic input, indeed as a free good, it exacerbates the problem. This occurs because of what economists term 'market failure', the two most important features of which are: (i) the problem of open access resources (collective consumption or public goods) whereby because users cannot be excluded no price can be charged; and (ii) the generation of externalities (social benefits or social costs). Typical cases of collective consumption goods are resources, such as beaches, mountains, lakes, forests and heritage artefacts (for instance monuments and historic buildings), which become heavily used and can possibly be degraded. There are no mechanisms for restricting use, for example, the price system, which could be used to choke off excess demand, or physical exclusion, which is not possible if the resources carry no property rights or even if they did exist would be too costly to exercise. Therefore, to a large extent, the tourist and tourism industry are free riders because they cannot be compelled to pay either to use the resources or be charged for the externalities generated, whether beneficial or detrimental.

In effect, the economics of conservation and pollution abatement constitute a conceptual contribution by the discipline to understanding and explaining tourism environmental issues, but it is the methods associated with these two analytical areas which are of more significance. Three are central to environmental economics and of relevance to sustainable tourism. First, economics has developed cost–benefit analysis (CBA) over many years by which to appraise projects, many of an environmental nature. More recently it has been refined (Lichfield *et al.*, 1993) to accommodate more fully distributional issues (particularly intra- and intergenerational). Such a method, which endeavours to incorporate both monetary and non-monetary benefits and costs, takes a long time horizon and adopts discounting techniques which do not discriminate unduly against pay-offs well into the future, is essential in considering sustainable tourism. For instance, decisions on investment in tourism developments in destinations which will have long term impacts on the cultural and natural environment, as well as on the national, regional and local economy, need to be made within a comprehensive framework. Without explicitly stating it as such, Pearce (1989) effectively sets out a procedure for appraising tourism projects.

The second important methodological contribution economics has made is in the derivation of a number of methods for estimating non-monetary values. The principal approaches have been applied mostly to environmental pollution abatement and rural recreation (see Hanley and Spash, 1993, for a clear and concise review). From an economic perspective, the prime purpose of valuing non-market resources is to facilitate decision-making on their allocation. By expressing their value in monetary terms, the amenity demand for non-priced resources can be assessed on a like-for-like basis with market-based activities, such as agriculture and forestry. Increasingly, however, the methods are being employed in urban contexts, for example the valuation of heritage buildings (Willis *et al.*, 1993; Stabler, 1995; Allison *et al.*, 1996), which relates them much more closely to tourism.

The main methods which have been developed within economics to place a value on the environment and which can be applied to the valuation of the non-priced tourism base are: the hedonic pricing; travel cost and contingent valuation methods, or combinations of these. Other economically-based approaches, particularly where detrimental impacts occur, are those founded on household production analysis, for example avoided cost or replacement cost. In addition there are methods which, although they do not attach monetary values, can quantify the environmental impact of tourism. One such technique is Delphi analysis whereby a panel of experts, on a repeated and convergent assessment basis, estimate the effect on the environment of one or more scenarios.

Economists prefer methods which are based on demand analysis for they reflect consumers' willingness to pay, or preferences, for goods and services and the three main methods are acceptable because they possess such an attribute. There are two ways in which preferences might be ascertained. The hedonic price method (HPM) and travel cost method (TCM) are termed 'revealed preference' models for they are indirect approaches, by which is meant they use a surrogate or proxy market as a means of estimating the demand for non-priced resources. For example, the TCM uses the distance travelled to, say, a rural historic site, as the basic information for estimating travel costs and the construction of a demand curve. The contingent valuation method (CVM) is considered as a direct approach for it seeks to elicit 'expressed preferences', that is consumers are directly asked to indicate their willingness to pay or accept (explained below).

The hedonic price method (HPM)

The HPM stems from a new approach to consumer theory by Lancaster (1966) to ascertain the value of unpriced characteristics of goods and services. It takes as its starting point the idea that any differentiated product can be viewed as a bundle of characteristics each with an implicit price. For example, with respect to the built environment, such as an historic property, the characteristics may be structural in terms of style, age, size and so on, or environmental, for instance the location, the features of the surrounding properties and the landscape setting. This example is relevant to tourism in that the capability of historic buildings or areas to attract visitors can have a marked impact on local economies, in both a beneficial and detrimental way, as well as on the local community, for instance, in terms of enhancement of property values. Indeed, it is possible that the designation of conservation areas or renovation of run-down historic buildings will trigger, through tourism, regeneration of a city or town. The HPM is also useful in relation to natural environments and tourism in that their attractiveness, for example, the presence of forests or watercourses, draws visitors to an area and enhances property values. Willis and Garrod (1993) found that the proximity of properties to waterways added 5% to their values. Likewise, Garrod and Willis (1991) estimated that woodland enhanced values by 7%. A related area of study in which HPM has been used is in the pricing of package holidays where the natural tourism base is a key element. The price competitiveness of such holidays cannot be compared directly because of differences in their characteristics but HPM can estimate the price differences which arise from these. For instance, Clewer et al. (1992) compared the price competitiveness of city package holidays using the HPM.

Travel cost method (TCM)

Sometimes known as the Clawson method because of its development by him (Clawson and Knetsch, 1966), the approach is based on the idea that the cost of travel to a recreational site or tourist destination can be taken as a measure of the willingness to pay and, if aggregation of individual travel costs by the number of visitors is conducted, a valuation of each site is obtained. The method has been applied widely in the rural recreation field with respect to fishing, forest visits and hunting (for example Loomis *et al.*, 1991; Smith *et al.*, 1991; Hanley and Ruffel, 1992). Even if visitors do not have to pay an entry fee, they have incurred expenditure either explicitly or implicitly in travelling to it. Thus the cost of the use of a car or public transport is an explicit cost, while even if access to a site is gained on foot, involving no expenditure, there is an implicit time cost. Whether on-site and travel time should be included into the estimate of total cost is a point of debate. Very few studies of visits to urban sites using the TCM have been undertaken to evaluate urban resources because of the problems of separately identifying the benefits generated by specific resources and the often multi-purpose nature of journeys to cities and towns. The method has not been applied to tourism travel, certainly not for international trips.

Contingent valuation method (CVM)

This method, unlike the two indirect approaches outlined, directly questions consumers on their stated willingness to pay (WTP) for, say, an environmental improvement, or their willingness to accept (WTA) a fall in its quality. In the context of tourism, CVM can relate to a WTP for both an environmental quality improvement and access to resources, or conversely to a WTA degradation or foregoing access. The method also enables non-use option and existence valuations to be estimated, for example to preserve an informal rural recreational site or a wetland of which respondents are not users.

The CVM is a survey-based method with responses being obtained either by self-completed questionnaire or by interview. A number of stages are necessary: first setting up a hypothetical reason for payment or compensation; then the cost of the action and means of payment or compensation are indicated and subsequently respondents are asked for their bids, which normally include a maximum willingness to pay or minimum compensation of the scheme proceeds. The final stage is the estimation of a bid curve. Despite a number of misgivings concerning the reliability and validity of CVM, it has grown rapidly as an important valuation method because of its potential ability to yield use and non-use benefit estimation, its almost universal coverage of the population if required and its political acceptability, being seen as a democratic means of gaining society's evaluation of environmental and other issues.

In environmental and other tourism related fields, the CVM has been extensively applied with regard to landscape preservation (Bateman *et al.*, 1994), improved park facilities (Combs *et al.*, 1993), the value of wildlife (Brown and Henry, 1989) and the retention and provision of national parks (Lockwood *et al.*, 1993).

The three methods do not constitute alternative means for valuing non-priced benefits because they focus on different sectors of consumers and types of values. The HPM takes account of the value that consumers place on environmental attributes only insofar as they enter into product prices or are capitalized into land and property values. In short, it captures only internalized benefits. Therefore, its validity depends on the extent to which the tourism product price is determined by the attributes of its environmental elements, for example the quality of skiing pistes or the culture of a particular city, say, Venice or Paris. The method cannot embody non-use values and there are also some technical drawbacks, fundamental to which are its restrictive assumptions, data requirements and the danger of omitted variables. The TCM is most effective when visitors live a long way from the site or destination being studied but it does not cover non-user demand, especially that by residents, and it is difficult to aggregate values obtained from individual sites. The disadvantages of the HPM and TCM tend to suggest that the CVM is likely to gain ascendancy as the preferred approach, although it is possible to consider the three methods as complementary and possibly additive, but such developments are still in their infancy. Likewise, combining these, as measures of static benefits, with methods which ascertain dynamic benefits, such as tourism multiplier and input–output analyses, have yet to be investigated. Nevertheless, economic approaches to evaluating environmental and non-priced resources benefits do point the way towards establishing rigorously derived and valid estimates which will give a more balanced picture to enable both the appraisal of projects and decisions on resource use to be made more objectively and with less bias towards market-based activities.

The third methodological area of relevance to sustainable tourism is the means (methods) for dealing with externalities which are both detrimental and beneficial, and the associated instruments (techniques) for implementing environmental policies. Traditionally, economists favour the use of market based methods for they are seen as being more efficient in achieving given objectives. Thus, to mitigate externalities, instruments such as taxes equivalent to the social cost generated should be levied, whereas activities which confer social benefits should be subsidized. In economic terms, the market price of a good or service should equal the total cost of production to reflect the costs of all inputs, including the environment, even if it carries no price in the market. Other market based instruments, such as user or entrance charges, licences/permits and tradeable quotes, have been advocated.

However, where a tax or charge is combined with a standard, for example, allowing the discharge of a specific quantity of effluent into a water course by purchasing a licence to do so, then the impact on market price is more indirect. This is because the supplier certainly incurs increased production costs which, depending on the elasticity of demand, may be passed on to the consumer in the form of a higher selling price. This is the principle of the 'polluter must pay', whether the polluter is the supplier who internalizes the externality generated and absorbs all or a proportion of the additional cost, or the consumer who demands the product. Nevertheless, the economic view is that the unfettered market should be controlled because of the loss of consumer welfare and the possible damage to the environment, which is likely to increase production costs irrespective of the instruments designed to reflect the true costs of provision of the good or service. In effect, the economic instruments overcome the failure of the market and, in theory, should secure an optimum level of supply and demand where the marginal social cost equals the marginal social benefit.

An alternative stance in economics to the market-based means of reaching an economic optimum is the notion that if property rights can be defined the problem of market failure does not arise. This supposes (Coase, 1960) that, regardless of in whom property rights are vested, the perpetrator and sufferer of an externality can determine an optimal position by bargaining, but in practice there are a number of reasons, which cannot be pursued here, as to why such a solution may be difficult to achieve. Suffice it to say, however, that given the open-access resource problem, which is a fundamental issue in tourism, it is a device worth pursuing, particularly with resources of global importance which attract tourists. Through international agreements with the force of law, it may be possible to restrict access to fragile environments.

The command and control method and its instruments, on the other hand, do not necessarily take account of economic factors. They may ban the use of certain materials and/or discharges or emissions of specific waste products in order to meet an environmental standard or target. Such a method is effectively regulation in a physical sense for the control is exercised in terms of a standard, for example, the maximum parts per million of, say, nitrate discharge into water or kilograms of emission of sulphur into the atmosphere. The problem of regulation, in economic eyes at least, is that it is inefficient in that the state or an appointed agency has to set up systems for gathering information and monitoring the activities of suppliers to ensure compliance. Also, it is an arbitrary approach because the costs of control vary in different firms in the same industry as well as spatially. In any event, if the cost of control is greater than the benefits achieved then resources used to mitigate the problem are wasted.

As has been argued with respect to the concept of implementing sustainability, the application of policy instruments gives rise to practical problems. The attributes of a good instrument, apart from economic efficiency, are that it should be acceptable, not be highly regressive (fall more severely on lower income group or poor nations), be flexible enough to adapt to the dynamics of the real world, and be effective in meeting the objectives for which it was devised. These attributes are not fully found in either marked-based or regulatory instruments and, it must be remembered, were originally intended for pollution abatement so that they are not necessarily appropriate if applied in other fields. Undoubted, however, is the uncertainty surrounding not only the identification and evaluation of the environmental damage but the ramifications of the impact of the instrument. This is especially marked where instruments are employed differentially in firms or countries. Those to which the instruments are applied stringently are likely to be disadvantaged *vis à vis* others operating in milieus in which regimes are more lax or measures are not introduced. While intuitively this argument is appealing, in practice this may not be so. Although the evidence is scanty (Institute of Business Ethics, 1994) it is possible to show that in fact environmental protection actions may not only reduce costs but enhance sales.

It is possible to envisage the application of a number of the instruments to tourism, for the reasons already stated that, as a major industry with respect to the consumption of productive and energy resources, generation of wastes and use of unpriced natural and man-made resources, it has a substantial environmental impact. This aspect of tourism has been examined by a number of researchers (Middleton and Hawkins, 1993; Beioley, 1995; Dingle, 1995). The imposition of charges on the use of inputs and production of waste will make both more expensive and encourage businesses to economize, for example, taxes on packaged products, deposit-refund schemes for recyclable materials and charges for the use of waste services. With respect to natural and man-made environments, the application of taxes and/or user charges, perhaps, in the form of tradeable permits/quotas, or simply regulation by physical quotas, can be applied to restrict access to the tourism resource base. Entry into a country or certain areas within it can be controlled in this way by charging for entry or by setting an upper limit on the numbers allowed. Instruments such as airport or harbour taxes (landing fees) are examples used to raise revenue to be devoted to environmental protection. However, the danger of charges or taxes, as opposed to regulation, is that the revenue raised is not applied to the purpose for which they are implemented, i.e. they may be unhypothecated. In practice, whether the command and control or market based methods are effective, used singly or in combination, is not always apparent. Tourism businesses, in common with many other forms of economic activity are primarily concerned with their self-interest. Response to penalties or incentives will only be positive if there is a net gain. For example, given a choice between a charge, or investment to meet a specific environmental target, which would effectively eliminate the problem, the charge would be chosen if it proves to be the cheaper option. In certain countries, from a business viewpoint, it may even be appropriate, as evidenced in the contributions to this book, to make payments to individuals or organizations to circumvent environmental protection mechanisms.

This outline of three key areas of economic analysis of environmental issues serves to demonstrate the contribution the discipline can make to the concept of sustainable tourism (further reference is made to an economic stance below). The value of the concepts and principles of economics is that the subject can provide both a framework and perspective by which to assess environmental initiatives and indicate their impact, particularly where it is imperative to draw comparisons and to make decisions along a monetary, or at least quantitative, dimension. Indeed, in most instances such a measuring device is the only feasible way.

The interpretation of the economics element by contributors to this text ranges from a conventional theoretical and analytical one by trained, largely academic, economists to a business one by practitioners in the industry, while those from other spheres, such as anthropology, sociology or public sector bodies, offer little or no explicit indication of how economic factors relate to sustainability.

The economists' line of argument as to the role of their subject reflects that outlined in this section with the emphasis being on its methods of analysis and the assessment of the impact of tourism. Practitioners think in terms of economics as meaning the market environment and the performance and management of tourism businesses subject to constraints appertaining to environmental protection and ways of mitigating them through offering new but acceptable forms of tourism.

The Environment

As with the conference themes concerning the ethics and economics of sustainable tourism, it is desirable to delineate what is meant by the environment. To most people it immediately conjures up natural resources, such as oceans, lakes, rivers, forests, beaches, mountains, the rural landscape, and wildlife habitats. However, in relating it to sustainable tourism, a much more all-embracing understanding needs to be grasped, even when considering the environment in a physical sense.

Pointers as to its many dimensions emerge in identifying its functions as: the source of productive inputs, both in terms of its provision of materials and energy and as a resource itself; a life-support system through its assimilative capacity made possible by its bio-chemical processes; a sink for waste products and its aesthetic/psychic appeal giving a sense of wellbeing to human beings. This view underlines the essential symbiotic relationship between economic activities and the natural environment which has engendered the sustainable development issue.

Some examination of the tourism resource base has already been undertaken to suggest that it includes the man-made, often built, environment as well as the natural one. Both can be perceived as the primary base where the former attracts tourists, for example, historic monuments and buildings, art galleries and museums, but may also constitute the secondary base because it consists of such facilities as hotels, restaurants, bars, banks, car hire establishments and infrastructural elements, for instance, railways, roads, and telecommunications. These are tangible resources but, once one starts to include man-made ones, which include historic and cultural artefacts, then the intangible aspects, for instance, music, literature, language, customs and social mores, become part of the cultural environment and consequently components of the tourism resource base. The last example, however, could be considered as part of the social environment, which sociologists certainly would argue can be seen as a separate environment, but this also perhaps may be nested within or subsumed under the political environment, an important influence on tourism. Lastly, the economic or business environment is an important factor likely to have an impact on the tourism industry.

These different dimensions to the environment are not always acknowl-edged so that the impact on the pursuit of sustainability is seen rather oversimplistically. This is true in both a domestic and an international context where the problems of translating what are readily accepted principles of sustainable tourism are formidable. It is not just a question of considering the economic, cultural, social and political dimensions outlined here, but also an organizational one, underpinned by scientific and technical environments.

In the contributions to this book, the interpretations as to what the environment signifies do range over the dimensions identified. Under-standably, emphasis tends to be placed on the natural environment and concerns over the impact of tourism on it. However, the twist often imparted is the business self-interest one that, unless tourists and the industry seek to protect it, the enjoyment and livelihood gained from it will be degraded if not destroyed. On the other hand, some reference is made to the potential for tourism to enhance the protection and quality of the environment. It is, however, an indication of the extent to which the trend is towards despoliation that little evidence can be found for the beneficial effects of tourism. Overall the general impression gained from the current tourism literature is how the environment can best be utilized as a resource.

Sustainable Tourism

It is not the intention to rehearse at length the definitions of sustainable development on which the tenets of sustainable tourism should be based, nor is it proposed to consider in detail the debate as to the nature and the direction of initiatives to implement it, as a number of contributions in this book do this admirably in different tourism contexts. However, it is necessary to outline the principal issues and to suggest how tourism is performing in terms of the broad, perhaps largely from an ecological and economic viewpoint, acceptance of what sustainability means.

Most commentators on the meaning of sustainable development, from whatever perspective they are considering it, agree that definitions of it are rather vague and open to different interpretations. Pearce *et al.* (1989) indicate that the number of definitions runs into hundreds. This vagueness, or perhaps, more sympathetically, generality, is both a strength and a weakness. Generality makes the concept all-encompassing thus covering most eventualities and facilitating adaptability and flexibility. On the other hand, it allows the principle to be hi-jacked and applied to whatever purpose is thought fit. This is certainly what has happened in its adoption for achieving sustainable tourism and a number of interpretations have been offered and examined in the literature; these have been well reviewed in Hunter and Green (1995), Inskeep (1991) and McKercher (1993). The argument raging in tourism circles appears to be more about the extent to which eco- or green tourism is sustainable and the management of the resource base to conserve its attractions and allow continued development of the industry, with passing reference to the implications of its growth on local communities' socio-cultural and economic structures. The rather pessimistic inference to be drawn is that market driven tourism is incompatible with sustainability. This view is reinforced, if not justified, by a supposition, alluded to earlier, that effectively sustainable tourism is a notion from the developed world which is being imposed on developing countries and that destination communities are likely to be deprived of improved living standards if they are denied the opportunity to develop their tourism. This is in contrast with the circumstances in developed countries where often conflict occurs between the interests of the tourism industry and those of the local community. Moreover, the tourism industry itself, on the whole, equates sustainability with viability, interpreting the term in two ways. First, it wishes to ensure the long-term survival of tourism businesses and so considers the sustainability of the market and how best to maintain conditions conducive to the profitable operation of firms. Second, as part of the maintenance of an appropriate market environment, tourism firms acknowledge that the resource base should be sustained so that it will continue to be attractive to tourists. Thus the industry should have a vested interest in protecting destination environments if only so that it can achieve its own objectives.

Accordingly there is an increasing tendency for both thinking academics and practitioners to concede that a trade-off between environmental and market goals is inevitable. This is certainly not how deep green ecologists, and indeed many economists, consider that sustainability initiatives should be viewed. It is instructive to go back to precisely, perhaps with an economic bias but bolstered by an ecological stance, what sustainable development, which underpins sustainable tourism, means. It does not suggest, on the one hand, minimal environmental protection consistent with continued economic development as measured by increases in gross national product (GNP) per head, nor, on the other hand, does it signify economic stagnation with the implementation of draconian protective measures. Two lines of argument are propounded by economists who largely accept that some form of market intervention is necessary. The first is that the stock of capital, consisting of both natural and man-made elements, should not be depleted or degraded. The second is that the environment, together with that capital, should be in such a state as to sustain a flow of income to meet the needs of the present and future generations. There is no fundamental disagreement among economists on the second as it is more a question of how this is achieved in terms of the nature of the capital stock. It is this issue which has given rise to sustainability stances, ranging from the very weak to the very strong, which are determined by the view taken as to the role of the market in securing environmental protection and sustainable development. Several environmental economists (Pearce *et al.*, 1989; Daly and Cobb, 1990; Constanza and Daly, 1992; Turner *et al.*, 1994) suggest that the very strong, ecocentric position presupposes that there is no substitution of natural capital (environments) for man-made capital. Indeed, it has been argued that there is a requirement to reinstate already degraded environments. At the other extreme, the very weak anthropocentric stance, the environment is treated no differently from any other form of capital so its substitution for man-made capital is allowable. In between these extremes, various stances are possible which can be identified by considering their operational implications, but they will not be pursued here.

Taking the extreme ecocentric position first, this involves quite severe constraints to conserve the environment using standards and regulation and possibly even control of birth rates, perhaps leading to a no growth outcome which is not necessarily explicit but more a consequence of the implementation of ecocentric policies. The very foundation of this very strong position is that ecosystems, including biodiversity, are essential to support life and welfare and if degraded will lead to the collapse of human existence if not all life forms. The implications of this for economic and tourism development are not necessarily over-gloomy. The advocates of ecocentrism perceive economic and human activity as being small scale, labour intensive focusing on the quality of life rather than growth. With respect to the use of resources and waste generation, as far as possible economies should attempt to use renewable resources, minimizing polluting discharges and emissions. The watchwords of the ecocentric stance are the 'precautionary principle' and 'safe minimum standards' signifying respectively that should, for example, the impact of the introduction of new technologies and processes be uncertain then they should be restricted and/or phased in slowly and pollution should not exceed the capacity of the environment to assimilate it. At present rates of increase in the global population, achieving such a goal seems very unlikely, the sheer numbers of humans being seen as the fundamental impediment to achieving the necessary balance between development and sustainability. While the advocates of a kind of Schumacherian (Schumacher, 1973) Utopia may appear to be naive, smacking of the noble savage in harmony with the environment (there is, however, much evidence to show that former civilizations were not so; see for example, Diamond, 1992), it is glaringly obvious that there is an ultimate limit to the earth's resources and continued population growth. Perhaps, therefore, a more realistic and workable interpretation of the ecocentric stance is that present trends should move in a different direction or some, at least, be slowed down if not reversed, in order that economies can work to institute true sustainability, which can be achieved through new non-polluting technologies which effectively render all resources renewable.

The anthropocentric stance, while not in complete contrast with the ecocentric one, allows for substitution between natural and man-made capital and places greater emphasis on this kind of trade-off, because it considers human needs as paramount. At the global level this stance takes a line that economies, as in business practice, should aim to maintain constant assets, which means ensuring that investment is undertaken to offset capital consumption; in effect, to put economies into a position where they can be sustained. The weak stance looks more strongly to market mechanisms and the mitigation of market failure through applying price based instruments and defining property rights more clearly. However, this tends to lead to the degradation of the environment, as demonstrated earlier.

It is not surprising that the stances on sustainable development have confused those concerned with issues surrounding sustainable tourism, the ecocentric one appearing to lead to the curtailment and eventual demise of tourism while the anthropocentric suggests carrying on operations in much the same way as at present. However, in considering how tourism stands in relation to just one aspect of the two stances outlined here regarding the balance between natural and man-made capital, its reliance on natural environments as its primary resource base must compel it to move in the direction of ecocentrism. If the resource base is degraded then tourism will inevitably decline. There is essentially no dispute over the need to reduce the consumption of exhaustible resources and the generation of waste and pollution. Therefore, the industry, whether it does so itself or is compelled to by regulation and monetary means, both nationally and internationally, must move on from acceptance of a set of principles and codes of practice to implementation in order to minimize the impact of its activity. To do this, in common with the more general view for minimizing resource use and waste, which includes natural environments, it needs to consider increasing its reliance on renewable resource use. This would invoke the bioeconomic principle of maximum sustainable yield, a long-held concept in agriculture, forestry and fishing, for maintaining the productivity and conservation of resources (Conrad, 1995). To facilitate this, the economic methods outlined in the previous section are relevant in that it is necessary to consider any sustainability initiative within an appropriate analytical appraisal framework so that the benefits and costs be identified and evaluated to establish the net effect before the introduction of the requisite instruments. Clearly there is no point in taking action if it results in there being net costs, i.e. the cost of mitigation outweighs the benefits.

What has not been widely examined in the literature is the process by which sustainable tourism might be pursued. Some indication is given by contributors to the literature on environmental impact assessment (Hunter and Green, 1995) and environmental auditing (Goodall, 1992), which emphasizes the need for state of the environment appraisals as a prelude to action. Both impact assessment and environmental auditing constitute the prerequisites for introducing management systems and procedures in what should be an ongoing process. There is a tendency in the industry to assume that, once the capital investment has been made, the goal of sustainability is achieved and the process will take care of itself (Wight, 1994). Clearly, however, it is necessary to measure the impact of the systems and procedures adopted and to assess their effectiveness. This involves continuous monitoring and periodic review, especially in the light of possible new and improved technologies for achieving environmental protection. Currently, the moves to do this have been somewhat hesitant and fragmented, the hospitality sector being an example where larger, international businesses have begun to take a lead (International Hotels Environment Initiative, 1993). This then is an aspect of the implementation of sustainable tourism that warrants further investigation since it is an essential element of translating principle into practice.

The more economically and what can be termed 'philosophically' orientated chapters in this book take up the wider implications of pursuing sustainable tourism and indeed consider some of its contradictions. This tends to mark off the academics from the practitioners for the latter tend, understandably, to be more pragmatic by relating sustainability to the needs of the industry and what it can achieve in the short run. Accordingly, the contributors suggest ways in which environmental and cultural protection can secure tourism's resource base and facilitate the long-term viability of the industry, including its potential to engender economic development. Emphasis is therefore placed on responsible behaviour by both tourists and businesses and a number of means of attaining the industry's objectives are proposed, such as a better quality product, lower density, small-scale and dispersed tourism development, infrastructural improvement, reinvestment to mitigate the effects of the destination cycle, education and training, and economic, social and cultural programmes for local communities, especially to involve them in the operation of the industry and protection of their cultural and built and natural environment.

Content and Format of the Book

The chapters underline the importance of tourism as an economic activity and that as a consequence it is at the forefront of the development or growth versus sustainability debate. The challenge facing the industry, and the perspective taken by many of the contributors, is how to translate principles into practice and to attain the seemingly irreconcilable objectives of tourism development and the long run conservation of physical, ecological and socio-cultural environments. Concentrating on destination areas, the authors, who include both academics and practitioners, consider how far the issues and problems can be resolved by the tourism industry itself. The book is in four parts, the order of which, to an extent, follows the transition from principles of sustainability to practice and policy.

Inevitably the division of the book into parts involves a degree of arbitrariness for many chapters transcend one or more of the boundaries which have been delineated to encompass the related topic areas identified. Far from being a weakness, it can be argued that an overlap between both the chapters and the four parts is a strength for its signifies the commonality of perception by those concerned with the interrelated nature of tourism development, the industry, the community and natural environments. Each part is preceded by an editorial introduction which indicates the key features of the chapters of which it is comprised and attempts to reinforce the supposition of the interrelatedness of the issues arising from tourism activity and sustainability. The first part contains contributions concerned with general issues regarding concepts, theories and methodologies, which includes chapters which critically examine and comment on the themes of the conference, particularly the ethical questions arising from the sustainability concept which have been touched on above.

In the second part, the chapters, a significant number of which have been contributed by academics with close links with, and practitioners in, the industry, reflect the responses which are or could be made by the industry to the need to pursue sustainable tourism. Naturally there is a tendency for sustainability to be interpreted in terms of the survival of firms and the industry or the ability to secure development. The objectives of businesses and the extent to which they collectively or individually can achieve such objectives are discussed. Approaches involving marketing, community involvement, environmental management strategies and education and training are analysed.

Again with contributions from academics and practitioners, there is a strong link between the second and third parts where attention is turned in the latter to the types of tourism likely to be compatible with minimizing environmental impacts, for example eco-, quality and cultural tourism. An interesting aspect of this is the role that these types might play on the one hand in changing the pattern and effect of destination life-cycles but on the other hand in exacerbating them. In a sense, in this part, the chapters represent aspects of sustainability in which the industry response needs to take account of wider issues by considering not so much ameliorative actions, in terms of a reactive response to problems engendered by what might be called industry-centrism, but a proactive one. By this is meant that the chapters focus on more preventive measures to avoid detrimental environmental effects. The coverage is by no means comprehensive but rather illustrative, especially as a number of authors give case studies which highlight problems or, more positively, suggest the direction in which tourism development could and should move. To an extent the perspective is from a standpoint a little more removed from an industry one than Part 2 and this gives a sharper critical edge to the contributions.

The final part contains chapters which consider the policy implications, in both the business and public sectors, of sustainable tourism principles leading to an examination of the effectiveness of public policies in balancing the inherent conflicts of development and conservation of the tourism resource base.

Throughout the collection, many case studies are given which, to varying degrees, illustrate the economic, ethical and sustainability issues of tourism. A number of examples are drawn from the UK with, naturally, some centring on Cumbria where the conference was held. Others consider quite well established destinations, for instance, Crete and Spain. However, most studies are concerned with problems in developing countries, examining circumstances in destinations such as Belize, The Gambia, Goa, Indonesia, Madagascar, Malaysia and the Seychelles in which tourism is developing rapidly and also where ecological and socio-cultural environments are not only very vulnerable but also, because of inadequate legal and political systems, most at risk from being degraded.

References

Allison, G., Ball, S., Cheshire, P.C., Evans, A.W. and Stabler, M.J. (1996) *The Value of Conservation: A Literature Review of the Economic and Social Value of the Cultural Built Heritage*. English Heritage, London.

Bateman, I., Willis, K.G. and Garrod, G. (1994) Consistency between contingent valuation estimates: a comparison of two studies of UK National Parks. *Regional Studies* 28, 457–474.

Beioley, S. (1995) Green tourism – soft or sustainable? *Insights* May, B75–89.

Brown, G.Jr. and Henry, W. (1989) *The Economic Value of Elephants*. London Environmental Economics Centre Paper 89–12, University College, London.

Clawson, M. and Knetsch, J.L. (1966) *Economics of Outdoor Recreation*. Johns Hopkins University Press, Baltimore.

Clewer, A., Pack, A. and Sinclair (1992) Price competitiveness and inclusive tour holidays in European cities. In: Johnson, P. and Thomas, B. (eds) *Choice and Demand in Tourism*. Mansell, London.

Coase, R. (1960) The problem of social cost. *Journal of Law and Economics* 3, 1–44.

Combs, J.P., Kirkpatrick, R.C., Shogren, J.F. and Herriges, J.A. (1993) Matching grants and public goods: a closed-ended contingent valuation experiment. *Public Finance Quarterly* 21(2), 178–195.

Commission of the European Communities (1993) Towards sustainability: a European Community programme of policy and action in relation to the environment and sustainable development. *Official Journal* No. C138: 5–98, 17 May. CEC, Luxembourg.

Conrad, J.M. (1995) Bioeconomic models of the fishery. In: Bromley, D.W. (ed.) *Handbook of Environmental Economics*. Blackwell, Oxford.

Constanza, R. and Daly, H. (1992) Natural capital and sustainable development. *Conservation Biology* 6, 37–46.

Daly, H. and Cobb, J. (1990) *For the Common Good*. Greenprint Press, London.

Diamond, J. (1992) *The Rise and Fall of the Third Chimpanzee*. Random House, Vintage edition, London.

Dingle, P.A.J.M. (1995) Practical green business. *Insights* March, C35–45.

Garrod, G. and Willis, K. (1991) The environmental economic impact of woodland: a two-stage hedonic price model of the amenity value of forestry in Britain. *Applied Economics* 24, 715–728.

Goodall, B. (1992) Environmental auditing for tourism. In: Cooper, C.P. and Lockwood, A. (eds) *Progress in Tourism, Recreation and Hospitality Management*, Vol. 4. Belhaven, London.

Hanley, N. and Ruffell, R. (1992) The valuation of forest characteristics. *Queen's Discussion Paper*, 849.

Hanley, N. and Spash, C.L. (1993) *Cost-Benefit Analysis and the Environment*. Edward Elgar, Aldershot.

Hunter, C. and Green, H. (1995) *Tourism and the Environment: A Sustainable Relationship?* Routledge, London.

Inskeep, E. (1991) *Tourism Planning: An Integrated and Sustainable Approach*. Van Nostrand Reinhold, The Hague.

Institute of Business Ethics (1994) *Benefiting Business and the Environment*. IBE, London.

International Hotels Environment Initiative (1993) *Environmental Management for Hotels: the Industry Guide to Best Practice*. Butterworth Heinemann, Oxford.

Lancaster, K.J. (1966) A new approach to consumer theory. *Journal of Political Economy* 84, 132–157.

Lichfield, N., Hendon, W., Nijkamp, P., Ost, C., Realfonzo, A. and Rostirolla, P. (1993) *Conservation Economics*. International Council on Monuments and Sites (ICOMOS), Sri Lanka.

Lockwood, M., Loomis, J. and DeLacy, T. (1993) A contingent valuation survey and benefit-cost analysis of forest preservation in East Gippsland, Australia. *Journal of Environmental Management* 38, 233–243.

Loomis, J.B., Creel, M. and Park, T. (1991) Comparing benefit estimates from travel cost and contingent valuation using confidence intervals from Hicksian welfare measures. *Applied Economics* 23, 1725–1731.

McKercher, B. (1993) The unrecognised threat to tourism: can tourism survive sustainability? *Tourism Management* 14, 131–136.

Middleton, V.T.C. and Hawkins, R. (1993) Practical environmental policies in travel and tourism. *Travel and Tourism Analyst* No. 6, 63–76. London, Economist Intelligence Unit.

Pearce, D.G. (1989) *Tourism Development*, 2nd edn. Longman, Harlow.

Pearce, D.W., Markandya, A. and Barbier, E.B. (1989) *Blueprint for a Green Economy*. Earthscan Publications, London.

Schumacher, E.F. (1973) *Small is Beautiful: a Study of Economics as if People Mattered*. Blond and Briggs, London.

Smith, V.K., Palmquist, R.B. and Jalkus, P. (1991) Combining travel frontier and hedonic travel cost models for valuing estuarine quality. *Review of Economics and Statistics* 63, 694–699.

Stabler, M.J. (1995) Research in progress on the economic and social value of conservation. In: Burman, P., Pickard, R. and Taylor, S. (eds) *The Economics of Architectural Conservation*. Institute of Advanced Architectural Studies, University of York, York.

Turner, R.K., Pearce, D.W. and Bateman, I. (1994) *Environmental Economics: An Elementary Introduction*. Harvester Wheatsheaf, London.

United Nations Conference on Environment and Development (1992) *Agenda 21: A Guide to the United Nations Conference on Environment and Development*. UN Publications Service, Geneva.

Wight, P. (1994) The greening of the hospitality industry: economic and environmental good sense. In: Seaton, A.V. *et al.* (eds) *Tourism: the State of the Art*. Wiley, Chichester.

Willis, K. and Garrod, G. (1993) The value of waterside properties: estimating the impact of waterways and canals on property values through hedonic price models and contingent valuation methods. *Countryside Change Unit Working Paper 44*. University of Newcastle, Newcastle.

Willis, K., Garrod, G., Saunders, C. and Whitby, M. (1993) Assessing methodologies to value the benefits of environmentally sensitive areas. *Countryside Change Unit Working Paper 39*. University of Newcastle, Newcastle.

World Commission on Environment and Development (1987) *Our Common Future*. Oxford University Press, Oxford.

A Critical Appraisal of the Sustainability Concept: Some Theoretical and Methodological Issues

The five chapters in this part, although primarily concerned with general issues, include appropriate cases which support the contentions made in what are largely adversely critical evaluations of the current perspectives on sustainable tourism. They consider the ethical implications of both tourists' and the tourism industry's attitudes to and behaviour with respect to sustainability and examine some of the difficulties of operationalizing its principles. Butcher (Chapter 2) and Wheeller (Chapter 3) take a broad view while Fyall and Garrod (Chapter 4) and Slee, Farr and Snowdon (Chapter 5) focus more on methodological problems from an economic standpoint, whereas House (Chapter 6) examines feasible means by which both tourists and providers can achieve the common goal of sustainability.

The chapter by Butcher offers a hard-hitting critique of the assumptions which underpin sustainability in the face of increasing demand for tourism by the inhabitants of developed countries. He focuses on the cultural and environmental degradation and economic dislocation of tourism development but, above all, argues that destinations, especially in developing countries both *desire* and *need* economic development. Western nations are accused of double standards in wanting all the advantages of material goods and services for themselves but denying them to developing countries on the grounds that their own life-styles will destroy traditional cultures which they wish to experience and enjoy. There are, however, commentators in developed countries who go further and, while they perceive their own culture as being degraded and therefore advocate the preservation of authentic cultures, overlook what indigenous societies want, which is not what 'Western crusaders' consider is good for them.

Butcher takes up a number of issues, and gives a view which challenges the strong sustainability stance. He puts such questions as: is it not natural for human beings to place themselves above all other forms of life (countering the deep green tenet of sustainability); should not economic progress be acceptable; is materialism necessarily bad for the many suffering poverty; tourists may be boorish/loutish but surely are merely expressing fundamental human traits which have always existed? Thus, his review is disturbing in that he implies to an extent that matters are already out of hand in attempting to achieve both sustainable development and sustainable tourism.

With refreshing candour in a direct and racy language, Wheeller, who is not averse to the occasional personal anecdote to demonstrate his ambivalence as a traveller, punctures the balloon of what he perceives as the inflated hypocrisy of the self-professed ecotourist and provider alike in their often selfish and short-sighted stance on sustainable development and sustainable tourism. In the light of the entrenched positions of market orientated operators on the one hand and interventionists (greens) on the other, he is pessimistic as to whether such opposed tenets can be reconciled to resolve the acute problems of attaining sustainable tourism. He perceives that green/ecotourism is a cloak for mass tourism and that the concept of sustainable tourism is superficial and shallow. While there is a need for a holistic approach, currently tourism is nowhere near achieving this. International and national regulation at a government level is at an impasse with self-regulation by the industry. A fundamental rethinking of policy is required.

After a brief review of the sustainable development/sustainable tourism concepts, Fyall and Garrod consider the debate in economics and, equally relevant to sustainable tourism, how the subject seeks to translate principles into practice. The chapter takes the specific methodological issues of how best to account for (measure) natural and man-made capital stocks and flows and what the unit (monetary) of measurement should be. Various techniques are outlined, which have been applied elsewhere in economics, to show how intangible benefits and costs can be estimated to indicate the non-market value of the environment and the impact of tourism on it, particularly whether that impact is detrimental or beneficial. The chapter essentially sets out an agenda and initial programme for the measurement process.

In their chapter Slee, Farr and Snowdon consider first the difficulties of defining both the spatial and industrial sector boundaries of tourism, particularly the difficulties of delineating rural tourism development, as a prelude to analysing the nature of provision in two areas in the UK from a strongly economic perspective. It is rightly asserted that, to assess the likelihood of green tourism also being sustainable, it is necessary to ascertain the economic impact of tourism development to establish whether it is indeed viable. Considerable attention is paid to the input–output methods for measuring the impact of tourism in estimating the benefits of two styles of development, namely 'soft' and 'hard' tourism. Alternative or green or soft or responsible tourism is seen as being more conducive to sustainability, the accent being on small scale. Hard tourism is equated with mass tourism, which has a much greater impact on the economies, cultures, social structures and environments of communities.

The results of the studies are somewhat inconclusive for in one area the income and employment effects of hard and soft tourism are similar, while in the other soft tourism has a more marked effect, particularly in terms of employment and its capacity for it to be local. The chapter is one which defines sustainability in a broad sense, covering social, cultural and environmental as would be expected but concentrating on the economic, which some would be more likely to refer to as viability.

House examines the interpretation and implementation of sustainable tourism, in which she evaluates applications which seek to 'reform' (reduce social, cultural and environmental impacts) rather than 'structurally' execute fundamental change. She investigates the ethos and rationale of case studies of alternative tourism which aim at the latter. A very interesting discussion is engendered on the operation of selected locations providing a form of tourism which seeks to inculcate alternative environmental values into the holiday behaviour of guests, as well as influence their lifestyle, philosophy and ethical stance in a socially responsible way. She questions the efficacy of 'eco-'/'green' tourism and its 'viability'. The approaches of the locations studied are traced back to the 1960s 'Environmental Movement' and 'Communes'.

Sustainable Development or Development?

2

J. Butcher

Faculty of Hospitality Management, Birmingham College of Food, Tourism and Creative Studies, Summer Row, Birmingham B3 1JB, UK

A Spectre Haunting Our Planet?

'A spectre is haunting our planet: the spectre of tourism.' The opening lines from a recent book on sustainable tourism (Croall, 1995, p. 1) sum up the negative view of tourism emanating from this school of thought. Since Swiss academic Jost Krippendorf (1987) wrote about tourism's role as a potential burden on cultures, economies and environments in his seminal work, sustainable tourism has become the orthodoxy for those studying travel and tourism. 'Sun, sea and sand', or 'mass' tourism, is out of fashion. Sustainable, eco- and green tourism are the new buzz words. Even Magaluf is going 'green', recently blowing up a number of mass market hotels in a quest to rediscover its past beauty.

Global bodies such as the United Nations, governments, local councils and tourism non-governmental organizations, such as Alpaction and Tourism Concern, have adopted sustainable tourism as their outlook. The Federation of Nature and National Parks in Europe (1993, p. 5) recently defined sustainable tourism as activity which '... maintains the environmental, social and economic integrity and well being of natural, built and cultural resources *in perpetuity*' (author's italics). In the face of the perceived tourist threat, advocates of sustainable tourism – and there are many, from Britannia Airways to Baroness Chalker – point out three areas of concern: cultural degradation, environmental destruction and economic dislocation.

The new school problematizes the act of tourism, an activity traditionally seen in a positive light. It can '... ruin landscapes, destroy communities, pollute the air and water, trivialise cultures, bring about uniformity and

generally contribute to the continuing degradation of life on our planet'
(Croall, 1995, p. 1). If catastrophe is to be avoided it will be necessary to step
back and balance the desire to travel with a recognition of the damage done.
To quote the title of one influential report on the issue, people are 'loving to
death' areas of natural beauty (Federation of Nature and National Parks in
Europe, 1993). The discussion of the problems caused by tourism leads on
to a parallel debate on tourism ethics – interpersonal conduct is deemed to
have a vital role to play in sustainable tourism.

The primary purpose of this chapter is to question the assumptions
implicit in the sustainability paradigm. It will be argued that the emphasis on
sustainable development has ruled out of order a discussion of what many of
the less developed countries to which it is applied need most: thoroughgoing
development. It will also be argued that the parallel discussion of ethics
denigrates the modern mass tourist.

Untenable Assumptions

Undoubtedly the development of the tourism industry has some destructive
side effects, but it is worth looking at the assumptions implicit in sustainability to judge the merit of being environmentally correct on holiday.

The sustainability school emphasizes the problems thrown up by
tourism and sees limits to its development, limits that society has over-
stepped. The rapid growth in tourism numbers, especially in the post-war
period, is seen as having imposed a heavy burden on cultures, environments
and economies. Travel has increased throughout history. There are now over
500 million tourists annually, compared to 25 million in 1950. Whilst this is
a massive growth, it still represents only about one in ten of the world's
population. If it is true that there are such limits to travel, then this is a bleak
prospect for the 90% who are not tourists. Neither is it an exceptional
growth. Other consumer goods such as cars and televisions have also
expanded at least as rapidly in this period.

But the problem is claimed to go far beyond numbers. It is commonplace
for the notion of respect for local or indigenous culture to be raised as an
argument against too many tourists. Croall (1995) believes that tourism has
had an '. . . adverse effect on traditional ways of life, and on the distinctiveness
of local cultures' (p. 1). Opposition to the displacement of the nomadic
Masai, to make way for Kenya's national parks, and to the erosion of the
traditional Masai way of life has become something of a *cause célèbre* in
sustainability circles. The 'needs' of the Masai are seen by proponents of
sustainable tourism as being best met by preserving the status quo. The sort
of changes that Western societies have undergone, leading to vastly better
living standards, are considered neither realistic nor desirable in Kenya.

The championing of 'authentic' culture in the face of commercialism reflects a lack of an expectation of growth, and implicitly accepts that these societies are going nowhere. For example, Gurung and De Coursey (1994) writing on a sustainable tourism project in Annapurna, Nepal, comment that '... village youths are easy prey to the seductiveness of Western consumer culture as tourists are laden with expensive trappings such as high tech hiking gear, flashy clothes, cameras and a variety of electronic gadgetry' (p. 179). But what is really so wrong in aspiring to own a camera and wear fashionable clothes? Clearly the indigenous people sometimes have less 'respect' for their own culture than tourism academics would wish.

The assumption here is that Nepalese culture is to be respected and sustained. This presumably means that Nepal will remain outside of the 20 countries from which 80% of tourism is generated. In other words, it will remain poor. The defence of indigenous culture sounds radical, but, by elevating cultural difference to a determinant of development, an acceptance of underdevelopment, 'in perpetuity', is reinforced. Such a static conception of culture would seem out of place if applied to Western societies, but has become commonplace in relation to analyses of less developed countries (LDCs). Effectively, the concept of development has become culture specific. In this context, concepts such as appropriate development and sustainable development only reinforce an acceptance of the divide between developed and less developed countries.

There is some truth in the view that the arrival of tourism into an area can upset subsistence economies. In Goa, India, according to Tourism Concern, 'five-star tourism' has denied local fishermen access to the coastline, whilst rice paddies, cashew plantations and pasture land are under threat from six planned golf courses (Donovan, 1995). But, despite the column inches devoted to issues such as this, it is the *lack* of development that characterizes the Indian economy, rather than ill effects arising from tourism. Emphasizing the need to sustain local economies in the face of unplanned development rules out of order a discussion on the sort of thoroughgoing economic development that is all too often needed to reduce grinding poverty.

Pressure on infrastructure is another area of concern. Often noted, from the Spanish Costas to the Caribbean, is the strain put on local sanitation systems by tourists. Less often considered is the notion of rebuilding these facilities to cope with the needs of locals *and* tourists. Sustainable tourism, hence, seems to fit well with today's low expectations for economic advance.

False Critiques

Some writers have challenged the 'growth is bad' arguments of the cruder environmentalists. However, they also share the underlying assumption that the problem to be dealt with is the conservation of what exists, rather than the transformation of economies and societies. Resources are deemed to be finite, and therefore natural limits exist, limits that have now been reached. The critics of the sustainability school are often at pains to accept the underlying premise of sustainability – that the problem is tourist numbers. Brian Wheeller, for example, an outspoken critic of ecotourism, does so on the basis that it is used as a marketing ploy providing a cosy, environmentally friendly feel-good factor for Western tourists. Meanwhile, global capitalist growth, and the subsequent growth in tourist numbers, continues unchecked (Wheeller, 1992). The emphasis of many critics is on 'appropriate' technology or growth, rather than none. This sounds a useful idea for LDCs. However, it also involves the denial of society's scientific advances (many of which are taken for granted in the West) to people in LDCs. As such, it serves to recreate a divide between the developed and less developed world, not in the language of colonialism, but in the politically correct language of sustainability. Why should the technology enjoyed in developed countries be deemed 'inappropriate' to people in other parts of the world? This can only be the case if it is accepted that culture is a determinant of the most appropriate level of economic activity. Whilst biological theories of difference have been unacceptable to most since the experience of the Second World War and the Holocaust, it seems that cultural difference has been elevated as a new way of rationalizing poverty and underdevelopment in the post-colonial era (Malik, 1996).

To take this argument further, some sustainability writers imply that less developed economies are in fact in some way positive; that somehow people in the West are sullied by consumer society and have a lot to learn from a simpler, less complex, existence. To be precise, they would deny that such economies are less developed, but suggest that they are *differently* developed. The criteria by which development might be measured in the UK, such as gross national product per capita, telecommunications capacity, levels of education, healthcare and the ability to travel itself, are deemed to be inappropriate when looking at societies with different histories and cultures. It could be argued that it is possible to look at development in terms of objective standards, rather than as a subjective, relative phenomenon. The dissagregation of humanity into defining cultural groups involves the loss of common universal standards. Ultimately, a sense of humanity is lost.

Tickell (1994) disagrees. He believes that humanity should '... glory in our differences rather than subordinate ourselves to some grey middle standard' (p. ix). With reference to the Himalayas, the Amazon basin and Egypt, he argues for the '... preservation of such environments and cultures ...' (p. ix). All three areas have a range of problems including severe poverty and a lack of modern medical provision. None of the three enjoy levels of wealth, educational provision, telecommunications or electricity provision comparable to developed countries. None of the three enjoy levels of material wealth that enable their citizens to travel widely. Few Amazonian

Indians enjoy the benefit of Tickell's high office. Such cultural relativism involves a blatant rejection of the notion of human progress, and says more about the nervous, backward looking 1990s than it does about any real limits being reached with regard to travel, development or anything else. Modern society seems preoccupied with limits and suspicious of science and progress. Human agency is denigrated – it 'does more harm than good'. 'Respect' and caution have replaced experimentation and progress as the key terms for our time. The notion that mankind is 'simply a part of nature' – so popular amongst environmentalists and sustainability writers – downplays what is specific to human beings, that which places them above animals. The history of human development has involved mastering, harnessing and utilizing nature. This is what makes men and women distinctly human, and has led to the dramatic developments in science and technology that enhance the lives of many. It is easy to sit back in the comfort of an armchair, and type the words 'humans are simply a part of nature', and transmit them almost instantaneously around the globe via the internet. The very fact humans can conceptualize 'nature', never mind engage in academic debate about the future of tourism in this way, proves that they are far more than this. The denigration of human progress embodied in the sustainability paradigm is likely to hold back humanity from facing up to and solving the problems of poverty and underdevelopment. It is hence a far bigger problem than some of the troublesome by-products of unplanned tourism development.

Sustainability in Vogue

It is notable that the emphasis on sustainable development, and the broader emphasis on environmental factors, coincides with a period of *lack* of systematic development in the economy as a whole. Ironically, the less developed the region, the more sustainability writers tend to problematize development. For example, at a time when large parts of the continent of Africa have become delinked from the world economy, with dire con-sequences for its inhabitants, it is *growth* that is castigated. The notion of developing LDCs in an all round and thoroughgoing fashion has been removed from the agenda. Sustainable tourism, in emphasizing the natural or cultural limits to development, seems to fit in well with contemporary low expectations. Sustainable development points the finger at unplanned eco-nomic growth, but provides no prospect of liberating people from their poverty. So why has sustainability become such a key concept in contempo-rary thought? The answer has little to do with any given limits – natural, cultural, economic or environmental. It has a lot to do with the broader trends in society.

The combined, yet uneven, development of the world economy has consistently reproduced division throughout contemporary history. Division has been legitimized and rationalized in the past by racial thinking. In the last century Africa was 'the White Man's Burden', considered incapable of developing itself in any sense at all. In the post war world, divisions have persisted, and in many cases intensified. Dependency theorists, neo-Marxists and Marxists, amongst others, put forward views that embodied a recognition of a problem at the level of society; be that in the sphere of distribution (dependency theorists, neo-Marxists), or in the sphere of production (Marxists). Society and, more specifically, the way society is organized (unequal terms of trade, capitalist profit) were cited as the barriers to third world development. Solutions offered were also solutions at the level of society, be they aimed at reforming the market relations that reproduced inequality, or even overthrowing capitalism completely. Limits were a product of the organization of society and could be tackled through an active critique of that society. Such a notion was evident in radical movements, liberation struggles and political parties.

Today things look quite different. There is no popular critique of society. The collapse of all things connected with the project of any form of social change has pulled the rug from under such debate. In Britain, for example, the decline of the labour movement has reinforced the idea that 'there is no alternative' to the market. And, of course, all this is underwritten by the victory of the market over 'communism' in the Cold War.

Alongside these developments, Western societies are experiencing unparalleled insecurities bred from recession, the end of a worldview characterized by the East–West division, and the failure of a New World Order to emerge. Emblematic of such times are Fukuyama's *End of History* thesis (1992) and the notion of the end of ideology.

It is this twin crisis, at the level of any alternatives to economic and political orthodoxy, and the crisis of contemporary society itself, that breeds a deep sense of limits. Society has stopped looking forward to new possibilities. People are more likely to look back to a mythical 'Golden Age' in the past. This is embodied in the contemporary preoccupation with the environment. Whilst modernity and human greed are blamed for war, unemployment and poverty, environmentalism suggests society takes a step back from the modern abyss. After all, humans are 'only a part of nature'.

A century ago, Thomas Cook said it was only a matter of time before there would be tourists to the moon. Today, the United Nations discusses limits to travel (Cowe, 1995) along with limits to population ... in the third world of course. It is the contemporary, and hopefully temporary, decline of the idea of progress and human agency that has brought many to a position where sustaining what exists seems a better option than transforming it into something better. Unless society is really willing to accept that a world where millions do not eat properly and the majority are denied the opportunity to travel is the best humanity can achieve, the notion of natural limits should be rejected.

Local Needs

'Think globally – act locally.' The 'local' is elevated to a point of principle in much of the literature. What is local is particular to a distinct area: the culture, the economy, the environment. Yet the most pressing needs are not local at all. They are universal needs, such as ample food, employment, good healthcare, good housing, clean running water, etc. Problems described as 'global' can best be dealt with globally. The restriction of human agency to what is local marks a low opinion of the human potential. Is it really only possible to act upon what is local, or immediate? Human engagement in society can go far beyond the local. Human comprehension and action have the capacity to transcend direct experience, and address goals far wider than 'local needs'.

The 'local' is inhabited by 'the community'. Community involvement is normally part and parcel of sustainable projects. It is held that input from the local population means that local needs will form part of the planning process, and will not be overlooked. The community is often a myth. Communities are often divided between groups with divergent interests, such as employers and workers, the old and the young. The exposure of Nepal to Western consumer goods may be exciting to the youth, but problematic to older Nepalese, who see their authority undermined. The interests of one may not coincide with, or may be completely opposed to, the interests of another. Community representatives are often salaried for their troubles, giving them a direct stake in a sustainable tourism project. And besides this, the notion of building democracy into sustainability projects through consultation with local people implies that people are being given a choice. If this choice is between unplanned development and small scale environmentally sensitive tourism, it may amount to a choice between the devil and the deep blue sea. The option of structured, planned development is seldom on offer in a slump riven economy.

Double Standard

The involvement of global institutions in sustainable tourism projects exhibits a staggering hypocrisy with regard to LDCs. The United Nations and the European Union have committed themselves to the cause of sustainable development. In the case of the United Nations, sustainability has become a central theme, formalized through the 1992 Earth Summit in Rio. Projects to sustain local cultures in the face of mass tourism are supported. Meanwhile, bodies such as the World Bank and the International Monetary Fund milk LDCs dry, and the United Nations starves children through enforcing sanctions on Iraq. This is a double standard second to none. It represents an old phenomenon – the denigration of the third world by the 'civilized' West – but in a new politically correct and morally coded guise. Cultural difference is now explicitly linked to economic potential through the sustainability discourse. Hence it provides a rationale for the persistence of underdevelopment. The view of Western colonialism and exploitation being to blame for underdevelopment in LDCs is replaced by a focus on the culture in LDCs as the primary determinant.

The Tourist in the Frame

The assumption that limits are being exceeded leads to a subsidiary question – who is doing the damage? Both at home and abroad, the tourists themselves fall neatly into the frame, especially those engaged in 'mass tourism'. Annual visitors to the Lake District in the UK, home of that upholder of the rural idyll, William Wordsworth, now approach 100,000 annually (Croall, 1995). No doubt Wordsworth is turning in his grave. In response to such perceived 'invasions' ethical tourism has become the vogue. Ethical tourism is summed up in the tourist motto 'Take nothing but pictures, leave nothing but footprints, kill nothing but time.' Tourist behaviour is seen as crucial to sustainable tourism. Consequently, the annual opportunity to do exactly what one wants has now become an area of life where one is presented with a surfeit of advice and guidance. 'Good tourist' codes include the following advice: cycle or walk; do not take the car; use local resources; learn about the local culture, and respect it. In addition, there are the customary warnings about sex (diseases), sun (skin cancer), crime (foreigners generally) and, inevitably for the British 'lagerlout', alcohol consumption.

On a recent edition of the 'Central Live' television programme in the UK, during a debate on Club 18–30 holidays, two male Club 18–30 tourists were referred to as 'Orang Utans'. The term seemed to stick amongst those participants who clearly felt the two were crude, unsavoury characters, only interested in sex, sex, beer, sex … and more sex. But what is the problem with sex? What is the problem with hedonism? Why has interpersonal contact between host and tourist come to be seen as a cultural minefield? To prevent tourists from damaging indigenous culture, the notion of 'respect' was wheeled out once again. And the 'Orang Utan' has simply no respect for anything.

The preoccupation with young people on their holidays mirrors the preoccupation with young people in just about every aspect of life. They are either 'at risk', or putting others at risk. They are either victims, or perpetrators in the 'Risk Society' of the 1990s (Beck, 1992). Youth are the perpetrators of anti-social acts whilst under the influence of drink … what's new? This is a fairly universal phenomenon. It used to be part of growing up but now it qualifies one as part of the animal kingdom. The labelling of working class youth on holiday as animals, tribal, and so on, is reminiscent of the openly racist language applied to Britain's colonial subjects in the past. It reduces a section of humanity to a race apart from respectable society. It is different from the past in that the masses themselves replace the British elites as colonizers, whilst the elites in the West struggle to deal with the crude habits of their common people. Perhaps the real problem with 'Orang Utans', though, is that they are social animals – they drink with, play beach volleyball with and shag with anyone – with absolutely no sensitivity or restraint. They ignore moral codes. They stand in sharp contrast to the wary, cautious approach of the new tourism. They live life to 'the max', and have little time for stakeholding in a society that offers them little. To the elites, they are out of control, and therefore need controlling.

If one believes codes of conduct are needed, though, there are certainly plenty of them about. The environmental group Ark published a magazine titled *The Ark Guide To Sun Sea Sand and Saving the World* – a pretty big burden to bear when one is trying either to get laid, pissed, a tan or to find the bus time back to the hotel. Perhaps as a concession to youthful exuberance, they add that 'we can still have fun' … Thanks! Tourism Concern's code for backpackers in the Himalayas even suggests that tourists can help the locals respect their *own* environment, by advising them to help guides and porters follow conservation measures. Alison Stancliffe of Tourism Concern admits that holidays are about escaping the stresses and strains of working life – but this involves '… closing your eyes to the things you normally care about' (Croall, 1995, p. 56). Hedonism, once a virtue of tourism, becomes a threat. Caution and wariness are characteristic of the new tourism.

The Great British Tourist

The problems caused by tourists are apparently especially severe when one considers the British tourist. In 1870, the Reverend Francis Kilvert commented in his diary that, 'of all noxious animals, the most noxious is a tourist; and of all tourists the most vulgar, ill bred, offensive and loathsome is the British tourist' (Kilvert, 1870). So what is new? According to Martin Graham, Chairman of the Federation of Tour Operators, very little. Sustainable tourism '… hasn't really made any difference to the great unwashed. A lot of people muck up their own back yard, and do just the same on holiday. There's a much greater awareness of the issues in Germany, Austria and Switzerland' (Croall, 1995, p. 56). One difference is that at least in the last century Thomas Cook, pioneer of the package holiday, was prepared to defend his tours against the allegation that they devalued tourism by making it more widely available. He referred to his tours as 'agencies for the advancement of human progress' (Boorstin, 1992, p. 88). After all, he said, '… railways and steamboats are the result of the common light of science, and are for the people ...' (Boorstin, 1992, p. 88). Today, this advancement is held to have proceeded too far. The liberating, positive aspect of travel is played down – 'The effects of our freedom ... threaten to engulf us', according to Krippendorf (1987, p. xiv).

The Problem with Culture

Cultural differences exist – how one eats, manners, etc. – both between countries and within them. But cultural differences are not the defining characteristic of humanity. People have a mutual interest in communicating and cooperating. It is not lively tourists that hold this back, but material factors – the severe lack of resources, especially in less developed countries. The celebration of difference and cultural diversity effectively dissagregates humanity into distinct segments, and makes communication and cooperation less easy. It does not tackle the denial of material resources that underlies North–South divisions. It implicitly reinforces an acceptance of these divisions by rationalizing them in cultural terms. After all, *'we are all different'*.

Anyway, the elevation of culture is a Western liberal concern. It is of little concern to young Nepalese who want Nike, Levis and Coca Cola. The fact they aspire to a greater level of material wealth, even if in a degraded fashion, is something positive. They do not recline in their uncomplicated, unsullied existence, at one with nature, but strive to free themselves from poverty and drudgery. The fact that McDonalds and Levis cannot deliver is not an argument for rejecting materialism. This is akin to throwing the baby out with the bathwater.

The Proyecto ... or On the Pull

Typical of the many sustainable tourism projects is the Proyecto Ambiental Tenerife, a European Union funded rural development charity based in the mountains of Tenerife. Amongst its aims are whale and dolphin conservation. Tourists are encouraged to only use local boats that adhere to the project's code of conduct. Amongst other activities, volunteers to the project can help preserve traditional farming techniques by getting involved in compost making, and can promote the survival of local culture by researching the mythologies of the goat herders. To volunteers and tourists it is stressed that '... the taking of knowledge by outsiders without any form of compensation becomes tantamount to theft or rape' (Proyecto Ambiental Tenerife, 1995). The project's mission statement pledges them to '... help sustain the rich diversity of human life on earth by providing support to traditional communities under threat'. The mass tourism that we associate with Tenerife is claimed to have '... devastated the rural communities and forced ... an age old culture to the edge of extinction' (Proyecto Ambiental Tenerife, 1995).

Alternatively, if one is part of the 'guilty masses', one might prefer a Thomson's package to the Hotel Las Vegas in Puerto de la Cruz on Tenerife's north coast, with its minigolf, sun terrace, aerobics and, according to the 1996 Thomson Summer Sun brochure, 'plentiful nightlife'. It may not be nirvana, but at £309 halfboard for 7 nights in May, it may prove a value for money alternative to sanctimonious advice from well meaning killjoys.

In conclusion, then, sustainable tourism is a concept with little to offer the tourist. Cultural difference has been elevated above commonality within the debate to the extent where both the tourist and the host are denied their humanity – they are seen as almost inhabiting two different worlds with a cultural divide in between. Problems associated with underdevelopment are rarely addressed from a perspective embodying a commonality of interest – what is desirable for one is deemed undesirable for another. Moreover, the implications of the sustainability debate for less developed regions are depressing. It would appear that, in the light of problematic and badly planned development, the very desirability of development itself is often denigrated. The championing of culture in LDCs has become an excuse for failing to address the pressing development needs of these countries. At worst, the emphasis on preserving what exists over investigating the potential for development succeeds in sustaining and excusing poverty in perpetuity.

References

Beck, U. (1992) *Risk Society: Towards a New Modernity.* Sage, London.

Boorstin, D.J. (1992) *The Image: A Guide to Pseudo Events in America*, 1st Vintage Books edn. Vintage Books, New York.

Cowe, R. (1995) Tourism concedes restraint may be needed to help environment. *Guardian*, 6th September, p. 19.

Croall, J. (1995) *Preserve or Destroy: Tourism and the Environment.* Calouste Gulbenkian Foundation, London.

Donovan, P. (1995) Tourism on someone else's land. *Guardian*, 11th July, p. 16.

Federation of Nature and National Parks in Europe (1993) *Loving Them to Death?* Grafenan (Germany).

Fukuyama, F. (1992) *The End of History and the Last Man.* Hamish Hamilton, London.

Gurung, C.P. and De Coursey, M. (1994) The Annapurna Conservation Area Project: a pioneering example of sustainable tourism. In: Cater, E. and Lowman, G. (eds) *Ecotourism: A Sustainable Option.* Wiley, Chichester, pp. 177–194.

Kilvert, Rev. F. (1870) *Diary.* Quoted in Croall, J. (1995) *Preserve or Destroy: Tourism and the Environment.* Calouste Gulbenkian Foundation, London, p. 72.

Krippendorf, J. (1987) *The Holiday Makers.* Introduction and Part 1. Heinemann Professional Publishing, London.

Malik, K. (1996) *The Meaning of Race.* Macmillan, London, pp. 128–144.

Proyecto Ambiental Tenerife (1995) Information pack.

Tickell, C. (1994) Foreword. In: Cater, E. and Lowman, G. (eds) *Ecotourism: A Sustainable Option*. Wiley, Chichester, pp. ix–x.

Wheeller, B (1992) Alternative tourism – a deceptive ploy. In: Cooper, C.P. and Lockwood, A. (eds) *Progress in Tourism, Recreation and Hospitality Management*. Belhaven, London.

Here We Go, Here We Go, Here We Go Eco

3

Brian Wheeller

Centre for Urban and Regional Studies, University of Birmingham, Edgbaston, Birmingham B15 2TT, UK

Perhaps, first, the title should be explained. Originally it had been Here We Go, Here We Go Eco, Here We Go Echo. This was on the basis that the day the abstract was being written, the author received a letter from a Turkish student asking to be informed of all that was known about echotourism, which caused confusion, until it was noticed that it was addressed to Brain Wheeller. However, it was then thought that maybe echo might be inappropriate, not because of any cuddly animal connotations of Echo and the Bunnymen, but because it would be echoing the same points made five years ago. This, it was considered, might put a lot of people off, which led to the dropping of the echo. However, since the paper was submitted the author is not so sure. Simply, and the word is used deliberately, because issues raised by critics of ecotourism over five years ago have yet to be actually addressed. Some claim that since then the debate has now moved on. Well, maybe it has but in so doing it has avoided, rather than answered, the basic problems and dilemmas of ecotourism, in particular the conundrum of numbers and the ego trap (Wheeller, 1993a). Echoes of the past should come back to haunt, but do not appear to.

These diversionary tactics were brought home at a recent sustainable tourism weekend forum held for Voluntary Service Overseas delegates, where the author was asked to speak, ably partnered by Melvyn Pryer from the Birmingham College of Food Tourism and Creative Studies, against sustainable tourism and against Bernard Lane, well respected proponent of sustainability and sustainable tourism. Bernard, in an uncharacteristically dismissive phrase, described critical views of sustainable tourism as being 'simplistic' and, by inference, worthless. As a critic, this very simplicity, far

© CAB INTERNATIONAL 1997. *Tourism and Sustainability*
(edited by M.J. Stabler)

from being a damning indictment, can be seen as a virtue. Simple common-sense surely screams that sustainability is a completely futile exercise.

The possibility of achieving sustainability in general, and sustainable tourism in particular, is as close, or likely, as the UK is to becoming John Major's classless society. There are no sensible moves (the People's Charter laughingly springs to mind here) towards the latter, though maybe it is, in fact, being approached but only in the sense of a society lacking class or style. Attempts to achieve the former are little more than platitudes.

During a paper on biodiversity at the 1995 World Congress on Adventure Travel and Ecotourism (Bahamas), it was, as usual, graphically pointed out how many species were being eradicated, man being the culprit. Then, as the clincher, the speaker cited the example where (paraphrasing) 'only last week we had a sad case when a turtle choked to death on a plastic bag that had carelessly been thrown into the sea. The poor turtle had mistaken it for a jellyfish'. Among the delegates there were audible sighs of grief and, maybe, even a few tears – an appalling response. Do not be misled, the author is not insensitive to the plight of these creatures, their depletion, nor the frightening speed of events; far from it. It is just that one abhors and rails against the age old fact that if something is nice and attractive it is nurtured, while those of an ugly appearance never get equal attention. Beauty is only skin deep. There is much to be gleaned here in the eco-/mass tourism debate and the superficial conclusions often drawn. It always depends on how one looks at the situation/problem, from what perspective and to what depth of analysis. Had the delegates themselves been jellyfish out for an afternoon sojourn round the bay, they would be pretty pleased that the turtle had, in fact, eaten the bag rather than themselves. They would have gratefully thanked their new, man-made camouflage, and also noted that, in future, only to go out for a cruise round when there were plastic bags about. The incident could be seen as symbiotic relationship: man and jellyfish together outsmarting the turtle. This is not just a jocular irrelevance, it is a serious consideration – as was the absurdity of the 'dolphin débâcle', also witnessed at the conference (see Wheeller, 1996a).

Two incidents on recent trips to the Amazon make one question further just what 'being at one with nature' actually means. The author can draw here on an unpublished paper, 'Eco/egotourism. A muddled model' prepared for the Tourism Down Under Conference in New Zealand, 1994. Whilst in Brazil, the author had the privilege of being taken out by caboclos, night fishing – the 'noble savage' poised, spear in hand hanging on the bow, with the alien being (the author) ensconced in (it must be said in a rather fetching red) life-jacket, apprehensive, to the rear. With deft, graceful expertise fish were speared and hoisted, flapping and wheezing (the fish not the fisherman) into the canoe. In the midst of this wonderful, humbling experience the mood was somewhat broken when one of the numerous flying fish, flitting harmlessly over the surface around the canoe, apparently chose to change direction and, in the complete blackness, hit the alien squarely between the eyes. He was, to say the least, alarmed at this unexpected juncture. Having recovered his composure, his first reaction was to find the unfortunate fish, which was still floundering where it had fallen between his feet, and to return it alive to the water. As he was attempting this rescue mission, the torch revealed the other speared, dead fish to the front of the boat. The irony of his futile attempt of being 'kind to animals' again hit him straight between the eyes.

The second event occurred on returning to the hut one night. There was an intruder present, a huge, white spider on the wall. It was ignored until the first sound of it scurrying along the wall proved too much: the guide's assistance was sought. Being on an eco-friendly holiday one was reassured when the guide casually picked up a brush and returned to the hut. On spying the said dinner plate on the wall he proceeded to clout and flatten the unfortunate insect with one blow. It was then realized that the spider's white appearance had in fact been the result of a sack which, ripped open by the blow, disgorged a hoard of very much alive baby spiders that cascaded around the room. It was a seminal moment: doubtless a sign, but one is still not at all sure quite what of.

An important point made recently was one raised by Buhalis and Fletcher who stated:

> The environmental damage so often discussed in articles relates only to the direct environmental damage. The full environmental cost of tourism development like the full economic activity can only be truly estimated if the direct and indirect impacts are assessed. If tourism development is 'sold' to destinations on the basis of its strong backward linkage with other sectors of the economy, then the environmental damage emanating from their supporting industries, as a result of tourism activity must also be brought into the equation.
>
> (Buhalis and Fletcher, 1995, p. 3)

The authors are referring to physical perspectives of 'environment'. If environmental is taken to be all embracing, incorporating social, cultural, physical, etc., and if, where appropriate, one adds 'induced' into their statement, then surely Buhalis and Fletcher highlight an absolutely crucial point in the sustainability debate. It just is not tenable to argue that the benefits (usually economic) of tourism are derived from direct, indirect and induced linkages without acknowledging the corollary that so too are the costs. They go on to recognize that the principles (of sustainability) 'should also be applied to those enterprises indirectly related to tourism which support and supply the tourism sector' (Buhalis and Fletcher, 1995, p. 17).

This is an obvious line of reasoning that is often, conveniently, overlooked by both hard line tourism industry practitioners and indeed by many advocates of sustainable tourism. Why is this so? The mainline tourism industry might argue that it would be impossible to expect all aspects of the tourism industry to demonstrate exemplary green sustainable credentials and therefore, in a sense, irrelevant to what they can be expected to achieve as they attempt to put their own house in order. Fair enough until this is looked at in a little more depth: doing so quickly exposes the flaws of sustainability.

British Airways (BA) continues to be heralded as a worthy example of a company endeavouring to achieve green respectability. To their credit, they have appointed a scientist of the calibre of Hugh Somerville rather than a public relations person to lead their environmental unit – a point which was overlooked in the clamour to criticize BA at the Commonwealth Institute Conference on Sustainability in November 1995. Under Dr Somerville positive action has been attempted. To debate the pros and cons of this here is not appropriate. British Airways is taken as a convenient example in a wider context to try to make a point.

According to recent press reports, British Airways has joined up with Disney as the number one airline for EuroDisney (*Times*, 15th February 1996). Nothing wrong with this, making, it would appear, economic sense. No doubt some of the profits generated can be ploughed back into the environmental unit. But just a minute. Are not Disney being criticized (rightly or wrongly) for culturally undermining France's heritage – first with EuroDisney itself and now more recently with their sugary version of Hugo's *Hunchback of Notre Dame*? (Also, there is the question of course of *Pocahontas*.) Is not cultural integrity part of the sustainability gambit, one of the codes of practice often cited? Therefore, how can British Airways, now in cahoots with Disney, presumably for economic reasons, be deemed sustainable? Obtuse reasoning it may be, but reasoning which is important if the sustainable tourism argument has any credence – which, of course, it has not. If sustainability does not cover all aspects of a tourism operation it cannot be sustainable in the true sense of the word and, as seems obvious, it cannot possibly cover all aspects. Or is one only considering the physical environment and, therefore, excluding cultural heritage from the tourism product?

Some of the proponents of sustainability, it is suspected, recognize that this all embracing approach would be expecting too much. Choosing, by default, not to enter into the fray, they ignore it. But is it expecting too much? Not if sustainable tourism is the supposed goal. If this notional dream is what is desired then all tourism, inclusive of its wide web of linkage, must fully adhere to the nebulous principle of sustainable tourism. If it does not then, significantly, it cannot be aiming for sustainable tourism. This same selective, rather than comprehensive, interpretation of sustainability is also evident if the subject is approached from a spatial perspective.

Exhortations are heard to consider the world as a global village, to think global, act local. An issue of concern with eco/egotourism then is which particular environment is being considered – the global, the local or both.

> The ecotourist (and the supposedly eco-friendly firm), so concerned to ostentatiously behave sensitively in the vulnerable destination environment, is not generally so concerned about the danger to the overall environment they cause in actually reaching that destination. Here convenience takes precedence over conscience — a car to the airport and a jumbo jet are hardly paradigms of virtue in the environmental stakes.
>
> (Wheeller 1993a, p. 125)

Even at the loosely defined 'destination' this dichotomy over when, and where, the tour and the tourist can be deemed eco-friendly remains vague, a vexed issue. Take a trip to Cuba – 'The Green Lizard' for example. On arriving in Havana, from the pampered luxury of the Hotel Nationale or the Ingletarre, the concerned ecotourist can book a trip to the eco-resort of Guama, then, as part of the package, travel on to the culturally appealing Trinidad, a World Heritage Site recently restored to its former glory with UNESCO funding, followed by 'R & R' on the sands of Varadero, before returning refreshed, to the capital. Back in Havana, one can then reflect on just which part of the completed internal trip through the island one actually was the mythical ecotourist. Maybe ecotourists can fool themselves that it was while strolling through the splendid streets of Trinidad, taking in the cultural wonders, careful to be sensitive 'travellers', at one with the 'natives', while their coach was parked literally inches from someone's open front window, the noisy engine (left on to ensure air-conditioned comfort on return to their cocooned transport) incessantly belching out toxic fumes. Incidentally at Varadero, in their high-rise beach hotel (which, of course, is 'naturally' out of bounds to Cubans – employees apart), they can switch on their TV, and be a little confused by the promotion video on the tourism channel, Sol, extolling the virtues of Cuba as a haven, a sanctuary for wildlife, while simultaneously and somewhat alarmingly also actively promoting it as a prolific venue for game fishing and hunters.

The example of Cuba is, certainly, not an isolated one. The problem of just which part of the holiday can be deemed eco-friendly is, if anybody cares to think about it, a common one. Perhaps the difficulties are most apparent and transparent with the growing trend towards two centre holidays.

> A number of the supposedly eco-friendly holidays seem to be two-centre destinations with, in the case of photo-safaris one week in the bush being supposedly eco-friendly followed by one week recovering afterwards in pampered luxury on the beach — a sort of 'let us spoil you in unspoilt Africa'. No doubt, for image purposes, the package as a whole would be deemed eco-friendly and statistically categorised under nature tourism (always a safe bet, politically).
>
> (Wheeller, 1994, p. 651)

Take another example. A recent advertisement for Iberian Airways highlights the unspoilt nature of the venue, the Amazon, while stressing the luxury the traveller can expect getting to and from the destination. The caption reads 'Totally unspoilt in Latin America, totally spoilt on Iberia'. The advert then includes 'The outstanding natural beauty of the towering Andes ... the magnificence of the mighty Amazon ... the enigmatic Aztec carvings', phrases which are juxtaposed with the 'unashamed luxury of First or Business Class'. In Amazonia itself there is a proliferation, a mushrooming of ecolodges, ecovillages, etc. Visitors from Europe often recover from 'roughing it' with a week in the rapidly developing resort of the North East Coast, or stretched out on Copacabana for a few 'daze'. Even those tourists flying direct to the Amazon from the States will usually have several days acclimatizing in a luxury hotel in Manaus, the Tropicana, for example. Here they can be picked up and dropped off, at their convenience, before or after their 'jungle jaunt'.

Cohen's 'environmental bubble' has been well-developed elsewhere (Cohen, 1972). However, an interesting slant on it was given during discussions with an extremely helpful and interesting Ms Mameud of the Anavilhanas Creek Lodge, Amazonia, on the Brazilian stand at the World Travel Market 1994. She firmly believed that for travellers really to appreciate the beauty of the Amazon they had to be in 'good health and safety'. To this end, the argument was that they had to be cosseted in accommodation where they were comfortable and 'at one with their own nature' (i.e. feeling at home) before they could venture out into the jungle, to be at one with nature there. Initially, these observations were dismissed as the usual smokescreen to disguise the 'big sell' but, on reflection, perhaps the arguments were genuinely valid and there is some truth in this after all. Certainly they were given further credibility when this same reasoning was evident, in a similar eco-context, in a recent television programme highlighting the filming of both brown and white bears in the Canadian wilderness. The film crew 'had hot showers, a cooking tent and lived in relative luxury in remote spots. We needed the comfort so we could work well' (Turner, 1994). And we know the significance of travail to tourism (Boorstin, 1962).

Returning to the exhortations to 'think global, act local'. Is there any real substance in this as a solution to tourism's problems? The phrase is somewhat easier on the ear than 'think macro, act micro' but is nevertheless just as clichéd, futile and flawed. The problem with this 'look after your pennies and the pounds will look after themselves' mentality, leaving aside the suspicion that the actual currency is being continually devalued, is that in tourism planning it is just not feasible automatically to transfer policies applicable at a local level to the wider arena. This is simply because successful ecotourism policies seem inevitably to be based on the restriction of numbers at the local level. At the macro level the number of tourists is rapidly escalating and doubtless will continue to do so. So how can acting local, in essence applying restrictions, be the solution to the burgeoning macro problem?

Also at a micro level, examples of supposed sustainable tourism are being highlighted. This is taken as being a plus factor, even though this is not the case on a global scale. For every 'good' sustainable project and practice (if such a thing exists) there may be as many as 30 examples of 'bad' sustainable tourism, created under the auspices and patronage of sustainable tourism. Therefore, the overall effect of sustainable tourism is negative not positive. The response to this, of course, is exhortation to eliminate any rogue practices. As if this were possible.

The crux of the issue lies in selfishness and incessant demand for 'more', aptly summed up in an understatement from 'Mr Personality', Graham Kelly, Chief Executive of the Football Association – 'There is a problem of impatience in our culture at all levels' (Kelly, 1995, p. 25). It is this desire for more – now, immediately, if not sooner – that completely dominates contemporary society, and, specifically here, tourism demand. Two examples illustrate this assertion. According to a recent article on game fishing, there is

> a new dawn that is not driven by superpower avarice and politics. Now, all anglers, whether American, British, German or Japanese are coming together. There is nothing now to stop progress towards a new and infinitely more exciting horizon. Bhutan, Ladakh, Nepal and Tibet are now open for the game fisher, bursting with the spirit of adventure
>
> (Bailey, 1994)

There are obvious strong parallels here with the spread of ecotourism. It is this overwhelming desire for more, for new horizons and the equal determination on the part of the tourist industry to satisfy and further fuel this demand (as long there is enough money to be made from it), which means ecotourism is never going to succeed in being anything other than a niche market based on the standard business motive of short term profits. Bailey's idea of 'all' is, irrespective of his caveat, narrowly limited to a superpower perspective and their anglers' immediate interests. His view that there is nothing to stop their progress is so heavily laden with a value judgement. So, too, it can be contended, are the ecotourists, who, when it really comes down to it, adopt a similar 'What is in it for me now' attitude.

Under the appropriate banner 'Sign of the Times' a biting, dry article highlights the superficiality of the 'do gooder' factor in aspects of contemporary youth travel. After outlining some of the realities of the hedonistic new 'Grand Tour', Llewellyn Smith (1996) succinctly exposes the 'bogus' claims that 'it is to help those less fortunate than oneself and in the process to become a better and wiser person'. She continues 'strangely, one seldom hears of 18 year olds, finding themselves by helping out in a home for battered wives in the suburbs of Leicester' and concludes by quoting 'most of my friends could tell you everything about Guatemala City ... but none of them has a clue what's going on in a council estate in Hull'. The sentiments of her article are readily transferable to the superficiality of eco-/sustainable tourism, where philanthropic aspirations mask hard-nosed, immediate self-interest (Wheeller, 1993b).

This brings to mind an excellent quotation that captures the true spirit of the eco-/sustainability sham, 'Anyone can sign up for sustainable development so long as it requires no specific commitment to do anything that will threaten their material interests' (Blowers, 1994, p. 2). One might also wish to question just what precisely sustainability actually is. There is obviously room for considerable variation in its interpretation (and hence manoeuvre) here. Another quotation, this time from a US Military spokesperson, sums it all up. In introducing a new computerized, digital helmet, he said 'This one is more lethal, more sustainable' (CNN, 1994).

The ostentatious need for travellers to go ethnic is apparent in their desires, when abroad, to patronize local restaurants and use local transport. Is it not strange that at home they eschew public transport and have probably never ventured into a transport cafe? They are not travellers: they, like most visitors, are tourists. Similarly the advertisement showing the less adventurous, but equally self-deluding, intrepid explorer, setting out on a railcard trip of Europe, being given a mobile phone by her father, reinforces Boorstin's assertion that when the traveller's risks are insurable he or she has become a tourist (Boorstin, 1962).

Despite glaring counterforces, there are continual exhortations on the need to adopt a holistic approach to the subject of tourism development, planning and sustainability. The spell, and spelling, of this notional ideal has always been intriguing. Would not 'wholeistic', all incorporating, be more apposite? The holistic argument championed in support of sustainability always, as the spelling may suggest, seems to have a gaping (black?) hole in it, down which that not so insignificant dimension, reality, invariably seems to disappear. The dream of sustainability and sustainable tourism will remain just that so long as, to all intents and purposes, it continues effectively to ignore the rather powerful pressures generated by short term materialistic desires of a large proportion of the world's ever increasing population. Some advocates of sustainable tourism recognize this as the root cause of the problem. Yet they choose to ignore it in their search for the Holy Grail. Holy? A truly holistic approach would be one that embraces realism. Sustainable tourism unfortunately fails, at the practical level, even to acknowledge it.

Brief mention has been made here of just how Utopian tourism sustainability is to be reached. As pointed out in a recent short paper (Wheeller, 1996b) business interests appear to be advocating the message 'No outside regulation, we can regulate ourselves'. Paraphrasing the immortal Mandy Rice Davies 'they would say that wouldn't they'.

Those who study it must now all be familiar with the long running 'long run' arguments, emanating from the industry that: the tourism product is the environment, tourism depends on the environment for its continuing success and it is in tourism's interest to preserve and enhance the environment. Easy, the industry means to ensure its long run self interest. Most seem to accept this line of reasoning, the debate increasingly focusing on the optimum means of preserving/enhancing the venerated Golden Goose – the product, the environment. As usual, fundamental questions remain unanswered. What is successful? What is the long long run? Is the environment all embracing? It is not convincing that the tourism product is synonymous with environment, certainly not just the physical environment. One cannot even be so sure of the apparently straightforward logic of tourism's self interest in preserving its product in the long run. Is not London still the UK's top tourism attraction, with increasing numbers of visitors, despite recent rises in crime figures, suggesting that aspects of the environment there might be deteriorating, and is not Blackpool, hardly the most salubrious or sustainable of resorts in some green-eyed observers' opinions, sustaining its position as one of Britain's top resorts? In fact as Andy Lyon, a former student of the author's, once so rightly said: 'the Goose is alive and well and living in Blackpool'. On the Golden Mile?

The spring 1996 television advertising campaign for France had a raucous background chorus of 'here we don't go, here we don't go.' As visions of beautiful French landscapes unfold before the eyes, the 'boys' are faded out to be replaced by the little sparrow, Edith Piaf. Then there is the comforting voice-over saying something along the lines of its nice to go where others do not, or words to that effect. But where does this tie in with the 'fact', if statistics are to be believed, that France is the world's number one tourist destination (Elliott, 1996)?

Incidentally, on the subject of Blackpool, and in the context of evolving destinations, a 'must' for all those interested in sustainable tourism planning is a brilliant, succinct piece 'There'll Always Be A Blackpool' by I. Brown. Originally written for the *Observer*, 13th May 1945, it was reprinted in *Tourism*, Summer 1995. Fifty years ahead of its time the original article contains the classic line 'The sublimity of such Beauty Spots may eventually become their ruin … all Spot and no Beauty.' Is this a new dimension for those planners who enjoy squeezing spots?

Meanwhile the means to an end, the regulation or self-regulation debate, has reached something of an impasse. Entrenched in the philosophy of relatively free-market forces, those in favour of self-regulation rely on old arguments. The retort from advocates of a more interventionist approach, equally entrenched in their own views, is also in danger of taking on a rather jaded hue. Old chestnuts such as 'in the long run we are all dead', 'the proof of the pudding' and 'so far, so bad' are all employed. It is unlikely, despite dialogue, that these fundamentally opposed tenets can be brought together; a Herculean task, indeed. The basic difference comes down to one of philosophy. The author does not happen to share the tourism industry's aversion to increased government involvement and legislation as regards business ethics and enforced environmental control. As global tourism continues to grow, the impasse clearly favours the non-interventionist approach.

Ecotourism must surely be seen as nothing more than astute short term business practice, part of the conventional tourism industry which utilizes the same infrastructure, is driven by the same motivation, namely profit, and that 'everybody's doin' it, doin' it, doin' it'. Predictably, the white elephant of ecotourism has metamorphosed into the equally deceptive oxymoron of mass ecotourism, a mythical beast of mammoth proportions. Displaying, beneath a woolly coat, all the familiar, inherent characteristics lurks the same old monster – mass tourism. It has been driven into the circus of the eco-arena. On its back, dressed appropriately for the occasion, is the Emperor, in his new clothes.

The author's nihilistic views on sustainable tourism have been recorded elsewhere. At best, what can perhaps be aimed for is merely something better than currently exists. This has little to do with the concept of sustainability, except in terms of its vagueness. Rather surprisingly there is, however, hope. A recent *Times* article, headed reassuringly: 'Miserable people make more sense', stated 'research by a team of eminent psychologists has shown that happy people cannot think straight. The glum do much better' (Leake and Lawrence, 1995). Couple this with the old maxim: 'Cheer up, chum: Don't be glum: We all know there's worse to come,' then maybe, if everyone is a clear-thinking, gloomy pessimist, something positive will materialize. Now there's real (eco) logic.

References

Bailey, J. (1994) New horizons. *Salmon and Trout* Sept/Oct, 76–78.

Blowers, A. (1994) *Inlogov Informs On Sustainability.* Issue 1, Vol. 4, Institute of Local Government Studies, University of Birmingham.

Boorstin, D. (1962) *The Image. What Happened to the American Dream?* Weidenfeld and Nicolson, London.

Brown, I. (1945) There'll always be a Blackpool. *Observer,* 13th May.

Buhalis, D. and Fletcher, J. (1995) Environmental impacts on tourist destinations: an economic analysis. In: Coccossis, H. and Nijkamp, P. (eds) *Sustainable Tourism Development.* Avebury, Aldershot, pp. 3–24.

CNN (1994) CNN Television News, 14th December.

Cohen, E. (1972) Towards a sociology of international tourism. *Social Research* 39(1), 164–182.

Elliott, H. (1996) France and Spain head world tourist league. *Times,* 15th February.

Kelly, E. (1995) Coal industry continues without a strike. *Times,* 18th November.

Leake, J. and Lawrence, L. (1995) Miserable people make more sense. *Sunday Times,* 10th December.

Llewellyn Smith, J. (1996) Sign of the times. *Times,* 23rd February.

Turner, J. (1994) Natural neighbour. *BBC TV,* 22nd October.

Wheeller, B. (1993a) Sustaining the ego. *Journal of Sustainable Tourism* 1(2), 121–129.

Wheeller, B. (1993b) Willing victims of the ego-trap. *In Focus, Tourism Concern* No. 9, 14.

Wheeller, B. (1994) Tourism and the environment. A symbiotic, symbolic or shambolic relationship? In: Seaton, A.V. *et al.* (eds) *Tourism. The State of The Art.* Wiley, Chichester, pp. 647–655.

Wheeller, B. (1996a) World congress on adventure travel and ecotourism, Conference Report. *Tourism Management* 17, 383–385.

Wheeller, B. (1996b) Regulation – rail on regale. *In Focus, Tourism Concern* No. 19, 14–15.

Sustainable Tourism: Towards a Methodology for Implementing the Concept

4

A. Fyall[1] and B. Garrod[2]

[1]Southampton Business School, Southampton Institute, East Park Terrace, Southampton SO14 0YN, UK; [2]Faculty of Economics and Social Sciences, University of the West of England, Frenchay Campus, Coldharbour Lane, Bristol BS16 1QY, UK

Introduction

Over the past decade or so, a widespread recognition of the sustainable development imperative has emerged. The common perception of the tourism sector as being one that is fundamentally 'soft' in terms of its relationship with the natural environment, having neither major impacts on the natural world nor a significant reliance on natural resources, has largely been rejected. Indeed, tourism has recently been described as 'a voracious consumer of resources' which 'represents an insidious form of consumptive activity' (McKercher, 1993a, p. 8). The view that is emerging from this re-evaluation of tourism's relationship with the environment is that sustainable development represents a considerable challenge to the tourism industry.

In sharp contrast with this gathering acceptance of the need for sustainable development, however, is considerable disagreement regarding exactly what the industry should be doing in order to ensure that sustainable tourism will ultimately be achieved. The impression that is given by many of those involved in the management of the tourism industry, both academics and practitioners, is that the major challenge in implementing sustainable tourism is to establish sustainable development as the primary strategic objective for the tourism industry. To the extent that a large and growing part of the tourism industry has already abandoned the environmentally-naive 'growth paradigm' and adopted instead the principles of sustainable development, the argument seems to be that the critical step in achieving sustainable tourism has in fact already been taken. Often the assumption is that, once this new paradigm has been accepted, any activity that is undertaken within it must, by definition, be sustainable tourism. Much of the literature of sustainable tourism is of this genre, often concentrating on how the tourism industry, or parts of it, can succeed in establishing the concept of sustainability. Discussion of the practice of sustainable tourism is very often relegated to a cursory list of 'guidelines' or 'codes of good practice' which, in the opinion of the authors concerned, should be adopted by would-be sustainable tourism organizations.

Applying the concept of sustainable development is, however, essentially a four-stage process:

- defining and establishing the concept of 'sustainable tourism'.
- determining the conditions for sustainable tourism to be achieved.
- developing a framework for measuring progress towards sustainable tourism.
- developing a set of techniques to make sustainable tourism operational.

It can be argued that only in the case of the first of these stages, defining and establishing the concept of sustainable tourism, has the tourism industry really begun to grasp the nettle. The growing acceptance of the principles of sustainable development among those actively involved in the tourism industry is no small achievement and it is not the intention of this chapter to belittle the considerable progress that has been made thus far, but it should be recognized that defining the concept of sustainable development and establishing it as an objective of the tourism industry are really only the first step in achieving genuinely sustainable tourism. Further steps must be taken before tourism can truly be said to have become sustainable tourism.

Defining and Establishing the Concept of Sustainable Tourism

While the conceptual basis for sustainability has been established in the academic literature for a considerable period of time, the past two decades have witnessed a vast proliferation of both academic debate and industry interest in the concept of sustainable tourism (Department of the Environment, 1991; Inskeep, 1991; Hawkins, 1994).

Despite its unchallenged position on the tourism agenda of the 1990s, however, sustainable tourism is a concept which still defies uniform understanding and a common, and perhaps more importantly, *working* definition. Indeed, a large number of working definitions have been proposed in the literature. Among these, perhaps the most widely accepted are those of Bramwell and Lane (1993), Department of the Environment (1991) and the World Commission on Environment and Development (1987). Nevertheless, each new contribution to the literature of sustainable tourism seems to start with a new and slightly different definition of the concept.

Mutability of the sustainable tourism concept

The purpose of this section is to warn against this continual redefinition of the concept; it is time to move on to the much more important issue of how to *implement* sustainable tourism. There is a danger that, while lingering at the definition stage, the concept remains 'mutable', that is, capable of being translated into action in a wide variety of ways, not all of which will necessarily be compatible with the true sustainability of the tourism sector. Nevertheless, in order to appreciate fully the preconditions for implementation, it is instructive to elaborate on the outline of the mutability of the concept given in the first chapter.

Indeed, a number of authors on the subject of sustainable tourism make this very point. Hunter and Green (1995), for example, argue that, although sustainable development does in fact have a fairly specific meaning in the case of the tourism industry, the concept often remains 'oversimplified and vague' (p. 57) in the literature. The problem is that while the concept remains 'oversimplified and vague' it can be moulded to fit widely differing approaches to environmental management. This has almost certainly contributed to the widespread misunderstanding of the concept by the tourism industry and helps to explain the rather limited extent to which sustainable tourism is currently being implemented.

An illustration of the different implementation strategies that can evolve from a single definition of sustainable tourism is provided by Inskeep (1994), who asks whether tourism should be 'market led' or 'product led'. The former requires that tourism attracts a broad market regardless of the impact of the development, while the latter involves developing only those forms of tourism that are the most compatible with the host environment and culture. Both seem possible within the broad definitions of sustainable tourism that are commonly forwarded.

If anything, this apparent mutability of the concept of sustainable tourism lends weight to the view that it is merely becoming a mediating term which bridges the gap between the developer and the environmentalist. The beguiling simplicity and apparently self-evident meaning of the term have tended to obscure its inherent ambiguity (Hunter and Green, 1995). On the one side are the resource exploitative, growth orientated tourism developers, while on the other are the resource preservationist, zero growth orientated conservationists. It can be argued that, at this moment in time, the concept of sustainability is merely keeping these two polarized and fractious constituencies at bay, both believing that sustainable tourism will serve their cause well. McKercher summarizes the situation well when he writes that

> The inherent vagueness of 'sustainability' is its great weakness. At present it is being used by both industry and the conservation movement to legitimize and justify their existing activities and policies although, in many instances, they are mutually exclusive. Rather than acting as a catalyst for change, sustainability may serve to entrench and legitimize extant policies and actions, thus exacerbating rather than resolving development/conservation conflicts.
>
> (McKercher, 1993b, p. 131)

This suggests that the idea of marketing merely serving to specify the benefits tourists are seeking and the strategies that can be applied to serve both the tourist and the host community represent only an elusive ideal at this stage in the sustainable development debate. From the industry perspective, the enormous economic benefits that quite clearly do still accrue from successful tourism development are too big an incentive for them to lose interest. This undoubtedly inhibits the full adoption of sustainable tourism practices. There is also the additional problem that it is not only tourism developers who religiously adopt the principles of the marketplace. Although the much quoted principles of sustainable tourism purport to protect the interests of the host community, and insist on their involvement in all stages of the planning process, for many in the less developed nations tourism, of almost any kind, is regarded as an economic saviour and one that should be nurtured at all costs in order to maintain their economic growth potential.

Green influences

Developing the argument put in Chapter 1, several writers have noted that the 'green' issue has largely failed to make an impact on tourism. It is argued that tourism is, in its widest sense, simply the marketing of the environment. If this is so, then it is clearly in the interests of the industry to be a leading and genuine force in environmental protection. But at the same time tourism is a notoriously fragmented and often marginally profitable sector, characterized by small businesses. As such, perhaps it is unrealistic to expect the tourism industry to put the interests of the environment too high up on its strategic agenda. This view is supported by Müller (1994), who, in addition to questioning the industry's ability to regulate itself with regard to sustainable development, argues that not only is the hedonistic philosophy still apparent in large numbers of tourists but also that the combined pressures of demand and the fact that 'there are too few agents with too few resources and too little time to act' (p. 132) serve as significant barriers to the eventual implementation of 'green' sustainable tourism practices.

Walley and Whitehead (1994), meanwhile, follow the view of Porter (1991), who argues that the adoption of green principles should be regarded as a key selling point and as a means of product differentiation rather than as a cost or 'tourist tax' on the environment. However, this is only of use if the tourist concerned is fully aware of the benefits that this may bring him/her. Porter's attempt to divert interest away from the cost element of sustainability is nevertheless significant in the sense that the paymaster of sustainable tourism policies has yet to be identified. It has been argued that in the case of tourism

> not only should the tourists have to pay for the use of environmental resources but the developers of accommodation units, etc. must be made to pay the full costs of measures for environmental maintenance ... As countries charge foreign mining companies for extracting their oil and other minerals, so foreign tourists should be charged for extracting the benefits of an ecotourist site.
>
> (Burns and Holden, 1995, p. 224)

But this view may be somewhat naive in the specific case of the tourism industry; the elastic demand for many tourism products will ensure that the price factor will always be high on the tourism agenda. This raises the question of whether those operating within the industry are really willing to make short-term economic sacrifices (through lost business) to ensure the long-term achievement of sustainable development.

Determining the Conditions for Sustainable Tourism to be Achieved

The previous section has emphasized a number of difficulties with the concept of sustainable tourism. Yet, to some extent, this may be the easiest part of achieving sustainable tourism. As Pearce and Turner (1993) point out, 'defining sustainable development is not the same thing as searching for the necessary and sufficient conditions for achieving it' (p. 8). The next stage in actively implementing sustainable tourism, then, must be to determine the appropriate conditions for achieving it. In seeking to do so, the tourism industry might do well to look to the academic discipline of environmental economics for some guidance.

Indeed, considerable efforts have been made by environmental economists to determine the necessary and sufficient conditions for sustainable development. Pearce (1992), for example, provides a useful translation of the definition of sustainable development in arguing that, succinctly put, sustainability means that 'we should ensure that the average wellbeing of the global population rises over time' (p. 3). He goes on to argue that, in order to achieve this condition, present economic activities must not be allowed to result in costs which must be borne by future generations and for which no sufficient compensation has been made. One of the major ways in which these 'intergenerational costs' can be imposed on future generations is through economic activities that deplete or degrade present stocks of natural resources (or 'natural capital'). Any such depletion or degradation of natural capital stocks will deprive future generations of the natural capital they will inevitably require as the basis for their own economic activities. In order, therefore, for sustainable development to be achieved it must be ensured that full and effective compensation is made to future generations for any loss of wellbeing they suffer through the activities of today's generation.

If this view is accepted, inevitably the next issue that arises is what form should this compensation properly take. In response, Pearce (1992) argues that 'the way to compensate the future is to ensure that we pass on to the next generation a stock of ... capital assets no less than the stock we have now' (p. 4). The difficulty, of course, as identified in Chapter 1, is in determining the composition of that capital stock. For example, is it legitimate to pass on a smaller stock of natural capital provided that the shortfall is compensated for by a greater stock of man-made or human capital being inherited? Moreover, *how much* additional man-made or human capital would be required to effect this compensation fully?

This is where much of the current debate on the economics of sustainable development lies. Broadly speaking, academics are divided into two main camps: those advocating 'weak sustainable development' and those advocating 'strong sustainable development'. Figure 4.1 illustrates the two contrasting options.

According to the weak sustainable development view, the form in which the capital stock is in when it is passed on to future generations is relatively unimportant. Advocates of the weak sustainability viewpoint suggest that there are many situations in which one form of capital can be substituted for by another form without any loss of human wellbeing. An example might be

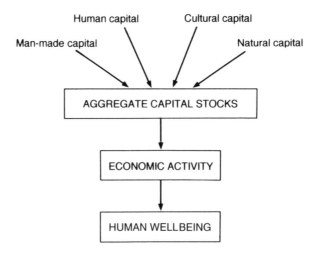

Fig. 4.1. Capital stocks and human wellbeing.

the loss of countryside associated with the construction of a trunk road bypassing a town. Provided that the lower wellbeing brought about by the loss of natural capital (the use of that part of the countryside) can be compensated for by the additional wellbeing that will be associated with the trunk road (faster commercial journey times, a quieter town, fewer accidents, etc.), then the aggregate capital stock can be said to have remained *at least constant*, hence meeting the 'constant capital rule' suggested by Pearce and others.

According to the strong sustainable development approach, meanwhile, it is argued that the form in which capital stocks are passed between generations is of vital importance. In short, advocates of strong sustainability argue that natural capital cannot be substituted by any other form of capital, only by other forms of natural capital. Hence the sustainability rule for supporters of strong sustainability must be adapted to specify that not only must the aggregate capital stock remain at least constant as it is passed from one generation to the other but also that the natural capital stock should remain at least constant.

This is not to suggest that there are no significant differences of opinion within these two broad perspectives on sustainable development, or shades of 'greenness' in between (Turner *et al.*, 1994). For example, Turner (1992) also identifies 'very weak' and 'very strong' versions of sustainable development. This serves to widen the spectrum, so that the two extremes become even more polar. Nevertheless, it should be noted that, whichever form of sustainability one chooses to support, in order to make the principle of sustainable development workable it will be necessary to measure the relevant capital stocks, both natural and man-made, in some way. Unless it is possible to measure changes in the size of the various capital stocks that are involved, then sustainable development (whichever form is preferred) is destined to remain simply a concept to which economics in general, and the tourism industry in particular, may or may not choose to subscribe.

The point here is that it is actually irrelevant whether or not the tourism industry has embraced the theoretical concept of sustainable development when there is at present no means of:

- determining whether the industry's efforts in achieving sustainable tourism are paying off.
- judging whether the industry's efforts are concentrated on the most pressing environmental concerns, or on their most environmentally-damaging activities.
- assessing the industry's progress towards, or potentially away from, the goal of achieving sustainable development.

This suggests that in order to achieve the effective measurement of stocks and flows of capital two important conditions must first be met:

1. there must be an appropriate *accounting framework* within which the stocks and flows of capital employed by the tourism industry in supporting its economic activities can be measured.

2. there must therefore be a set of *measurement techniques* available in order to evaluate the magnitude of these capital flows. Clearly if capital flows are to be compared with one another, which is the fundamental prerequisite for determining whether or not full compensation has taken place, then a *common unit of account* will be required.

These preconditions correspond to the third and fourth of the stages identified in determining how genuinely sustainable tourism might be accomplished in practice and are considered in turn below.

Developing a Framework for Measuring Progress Towards Sustainable Tourism

It has been argued that, in order for the concept of sustainable tourism to be implemented in any practical sense, some form of accounting framework must be established in which progress towards the attainment of sustainability, whether in strong or weak format, can be evaluated in a meaningful way.

What might not be so apparent from the basic definitions of weak and strong sustainable development given in the section above is that *both* definitions require the measurement of both man-made and natural capital. In the case of weak sustainability, the objective is to maintain at least constant aggregate capital stocks. In other words, weak sustainable development requires that the size of the combined stock of natural and man-made capital remains at least constant between generations. The strong sustainability rule, meanwhile, requires not only that the aggregate capital stock is held at least constant but also that the overall size of the natural capital stock remains at least constant over time. Clearly in both cases it is essential that changes to not only the man-made capital stock but also the natural capital stock are in some way measured.

This, of course, is where the major sticking point in establishing an accounting framework for the measurement of progress towards sustainable development appears. While man-made capital lends itself readily to measurement, being easily quantifiable in both physical and monetary terms, measuring natural capital in a meaningful way is far more problematic. Natural capital does not lend itself readily to being quantified or valued in monetary terms. This is not to suggest, however, that quantifying or attaching monetary values to natural capital is either impossible or undesirable.

It is possible to identify at least two different methodological approaches that might be considered appropriate in establishing a framework for natural resource accounting:

- maintaining full stock and flow accounts for various forms of natural capital.
- augmenting existing balance sheet approaches with information on natural capital flows.

An example of the first of these approaches is provided by Repetto *et al.* (1989), who provide a hypothetical example relating to the forestry sector.

In the example illustrated in Table 4.1, the opening stock of 100 units (presumably of biomass) has experienced an overall reduction of 15 units in physical terms over the time period concerned. However, over that same time period the unit value (i.e. price less extraction cost) has risen from £1 per unit to £3. Assuming an average unit value of £1.60 over the time period, the value of the stock has increased from £100 to £255, signifying an increase in the value of the capital stock over the period concerned.

Note that in the above example both physical and monetary values relating to the size of the natural capital stock are used. Regarding the appropriateness of each of these two general approaches, Pearce *et al.* (1989) suggest that, while physical approaches have the advantage that they are easier to implement (it is easier simply to measure physical flows without having to attach monetary values to them), physical approaches are 'limited because they lack a common unit of measurement and it is not possible to

Table 4.1. Sample forestry accounts (after Repetto *et al.*, 1989).

	Physical units	Unit value	Value (£)	Basis of calculation
Opening stock	100	1.00	100	
Additions				
Discoveries	20	1.60	32	
Revisions	(30)	1.60	(48)	
Extensions	15	1.60	24	
Growth	0	1.60	0	
Reproduction	0	1.60	0	
Reductions				
Production	(20)	1.60	(32)	
Deforestation	0	1.60	0	
Degradation		1.60	0	
Net change	(15)	1.60	(24)	
Revaluations				
Opening stock	–	–	200	100 × (£3.00 − £1.00)
Transactions	–	–	(21)	15 × (£3.00 − £1.60)
Closing stock	85	3.00	255	

gauge their importance relative to each other and to the non-environmental goods and services' (p. 115). Clearly this would be a critical requirement of any accounting system for the measurement of progress towards sustainable development. This is most easily seen in terms of the attainment of weak sustainable development, although the notion holds good for strong sustainable development also. Weak sustainability requires that the aggregate capital stock does not diminish over time. In order to verify whether or not this condition holds it is of critical importance that both the natural and manmade components of the capital stock can be expressed in a form that permits comparison and hence aggregation. The common unit of account need not, of course, be monetary. However, in view of the fact that man-made capital is *already* measured in monetary values and, moreover, that it is possible *at least in principle* to express natural values in monetary terms, monetary values would seem the most sensible ones to adopt.

An example of the second major methodological approach in establishing an accounting framework for the measurement of progress towards sustainable tourism is illustrated in Table 4.2 (Adger and Whitby, 1993). The table shows how the conventional economic measure of a sector's contribution to the national income can be modified to account for changes in natural capital stocks. In the case depicted, the value of the environmental damage attributed to the UK agriculture and forestry sector in 1988 (£69 million) was less than the environmental services either provided by or supported by the sector (£1034 million) in 1988, thus 'Green Net National Product' (gNNP) is larger than conventional Net National Product (NNP). The implication is that the UK agriculture and forestry sector is a *net contributor* to natural capital stocks and, hence, that the sector is a 'sustainable' one (using either the weak or strong version of the concept). However, as Atkinson and Pearce (1993) point out, the adjustments that have been effected are strictly partial in their coverage. For example, the damage costs associated with slurry incidents and the biodiversity losses associated with some modern farming practices are excluded from the analysis. It can also be argued that another major omission of the analysis is the impact of the sector on energy reserves. This is not to suggest, however, that account of such losses could not be incorporated given the application of sufficient time and expertise.

The major advantage with this second methodological approach is that it avoids the requirement to measure directly the various forms of natural capital stock associated with the economic activity in question. Largely for this reason, many academics prefer this latter approach, which has been applied both at the sectoral and national levels for a number of years. Uno (1988), for example, presents an instructive series of adjusted Gross National Product (GNP) data for Japan. Atkinson and Pearce (1993), meanwhile, develop and apply a more formal 'sustainability test' for the UK based on natural capital accounting principles. Their results, which relate most closely to the notion of 'weak' sustainable development, suggest that the UK as a whole was not on a 'sustainable growth path' for much of the early 1980s. The authors attribute this to an exceptionally high rate of natural capital depreciation associated with the exploitation of North Sea oil during the early part of the 1980s, coupled with the tendency for the UK to consume rather than reinvest the associated 'rents'. While this test is applied at the macroeconomic level in this case, there is no reason why the test, or one similar to it, could not be applied at the sectoral level.

In terms of the achievement of sustainable tourism, it would be fair to argue that the importance of establishing and maintaining natural resource accounts or 'environmental balance sheets' has not been widely recognized. Where the concept is recognized, its central importance as a precondition for achieving a sustainable development strategy is not generally highlighted. Inskeep (1991), for example, provides a series of lists making recommendations regarding how sustainable tourism development should be implemented. The suggestion that governments should 'apply sectoral and/or regional environmental accounting systems for the tourism industry' is certainly made but it is relegated to being one point in a list of over 40 other recommendations, including the suggestion that tourists themselves should travel 'refraining from inappropriate behavior which negatively affects the host community' (p. 466), with which it is apparently accorded equal weighting. While Cooper et al. (1993) refer to a number of attempts to establish an environmental balance sheet for tourism, it appears that nothing practical has yet come of these attempts.

Table 4.2. UK agricultural and forestry accounts (after Adger and Whitby, 1993).

Sustainable net product	£ 88 million
Gross National Product (GNP)	5498
minus Depreciation on man-made capital	1470
equals Net National Product (NNP)	4028
minus Degradation of natural capital (water pollution)	-11
minus Defensive expenditures (government expenditures to maintain landscape and conserved areas and to clean up pollution)	-58
equals ENVIRONMENTAL DAMAGE	-69
plus Net carbon fixing	146
plus Biodiversity	94
plus Amenity (of which: green belt +642, national parks +152)	794
equals ENVIRONMENTAL SERVICES	1034
equals Green Net National Product (gNNP)	4993

Measurement Techniques in Achieving Sustainable Tourism

The previous section sets out a case for establishing some form of 'environmental balance sheet' for the tourism sector. It was argued that, in order for sound accounting principles to be applied, a *common unit of account* will be required. Although it would at first appear 'more ethical' not to place monetary values on natural capital, in practical terms the use of physical values is not at all helpful. There are two related reasons why this is so.

1. Different natural resources lend themselves to being measured using different physical units; fish stocks in tonnes of biomass or fresh-water reserves in cubic metres. This makes it impossible to compare aggregate changes to the natural capital stock, since it is not possible to add values measured in different units together. But this is clearly a precondition for measuring progress towards either weak or strong sustainable development.
2. By the same token, if natural resources are to be measured only in physical terms, it will not be possible to determine the size or direction of changes in the aggregate capital stock, since man-made capital tends to be measured in monetary terms rather than physical terms. But again this is a precondition of measuring progress towards achieving either form of sustainable development; both require that the overall capital stock is monitored.

In fact, using monetary values as a common unit of account does not imply, as many opponents of attaching monetary values to natural capital suggest, that economists believe that anything may legitimately be bought and sold, or indeed that 'everyone has his/her price'. Economists are merely pointing out that people do have preferences towards the environment being in various 'states' (for example a pristine state versus a degraded one) and that it is perfectly legitimate to measure the magnitude of such *preferences* in monetary terms. That people generally have a 'willingness to pay' in order to demonstrate their preference towards a certain state of the environment is indisputable; the contributions (of money) made by individuals to conservation funds alone serves to demonstrate this point.

In any case, it can be argued that using money as a common unit of measurement is more a matter of pragmatism than ethics. Pearce (1995), for example, argues that 'the use of money as the measuring rod is a convenience: it happens to be one of the limited ways in which people express preferences' (p. 40). The implication is that if there were another unit of measurement sufficiently flexible and universal to measure people's preferences towards both natural and man-made capital, then that unit of measurement would be equally suitable in the present context. Suffice to say no such measure currently exists.

If it is accepted that attaching monetary values to changes to the natural capital stock is an inevitable requirement for achieving sustainable development, then the next question is how these values are to be measured. Environmental economists have proposed three major methods. In addition to that given in Chapter 1, a helpful overview of these is provided by Green *et al.* (1990). Further detail is provided in, for example, Turner and Bateman (1990). Meanwhile Bull (1991) provides an interesting initial review of how these methods might be applied in the tourism context. See also Tribe (1995).

It is not proposed that the ground that has already been covered in Chapter 1, in which the three main valuation techniques were reviewed, should be inspected again. However, it is useful to summarize their principal features and this is done in Table 4.3. It is also of value to add illustrative examples of their application in tourism which reinforce the relevance of their role in the measurement of sustainability.

The hedonic pricing method

The hedonic pricing method (HPM), as has already been indicated, has largely been applied to explaining house prices, although other real-world markets can and have been used as surrogate markets in the HPM. The basic premise is that there will be a measurable price differential in the housing market associated with changes in the state of the environment. Take the example of aircraft noise: it is reasonable to expect on a priori grounds that houses badly affected by the noise will attract a lower market value than those located in areas less affected by aircraft noise. The task assigned to the hedonic pricing method is simply to isolate this price differential. The example of aircraft noise is particularly apt in the case of the tourism industry. The demand for foreign holidays has increased at an unprecedented rate since the development of the package holiday in the 1970s and with this has come a greatly increased derived demand for air travel. A useful example of hedonic pricing is a study by Pennington *et al.* (1990), which uses the HPM to estimate the impact of aircraft noise on residential property values adjacent to Manchester International Airport.

The travel cost method

The exposition of the travel cost method (TCM) as another 'surrogate market method' of attaching monetary values to people's preferences towards the continued existence or state of an environmental asset, given in Chapter 1, indicated that people tend to react in a similar way to increased travel costs as they would to the resource or activity being 'priced' (e.g. an entry fee being charged to the heritage site or the payment of a licence fee to fish a particular stretch of water).

Table 4.3. Economic monetary valuation techniques.

	Hedonic pricing method (HPM)	Travel cost method (TCM)	Contingent valuation method (CVM)
Basis of measurement	Income-compensated demand curve	Income-compensated demand curve	Uncompensated demand curve
Type of preference	Revealed preference	Revealed preference	Expressed preference
Method	Surrogate market	Surrogate market	Experimental market
Example	Pennington et al. (1990)	Willis (1991)	Penning-Rowsell et al. (1992)
Subject	Aircraft noise	Forest recreation	Beach recreation

The TCM has a substantial track record of successful application and is widely used in evaluating the environmental benefits associated with recreation of various kinds and, given the emphasis accorded to recreational tourism by tourism companies of late, its application in the context of sustainable tourism is clear. An interesting aspect of this is aptly illustrated in one of the many studies conducted by Willis (1991), using the travel cost method. His evaluation of the recreational benefits associated with the Forestry Commission's forestry estate suggests that the values associated with the recreational use of forest resources are generally significant, in many cases outweighing the conventional development (timber) value of the forest. This is an important point for it indicates that a hitherto undervalued amenity use of a resource, because it is not traded in the market, is greater than its commercial use. This confirms that the valuation methods can be employed in the decision-making process where there are potential or actual competing uses for resources.

The contingent valuation method

The contingent valuation method (CVM) as an expressed preference valuation method has been used in the UK since the early 1970s and, of the three valuation methods discussed in this section, is probably the one that is most widely favoured by economists on both conceptual and methodological grounds, the major reasons being that:

- as a 'direct' method of valuing environmental goods it generates a 'true' measure of willingness to pay.
- it does not require the existence of a suitable surrogate market.
- the considerable data-intensiveness of both the travel cost and hedonic pricing methods usually renders contingent valuation the most cost-effective valuation method.
- it can be used to derive not only use values but also non-use values such as 'intrinsic value', depending on how it is conducted.

An instructive example of CVM in practice is provided by Penning-Rowsell *et al.* (1992), who use the CVM to establish values for changes in user benefits resulting from coastal erosion and coastal protection measures at Hastings. As a popular seaside resort, Hastings attracted some three times its resident population in day visitors during in the early 1980s, a major attraction being its well-developed seafront. However, over the entire period between the 1830s, when the present seafront was constructed, and the early 1990s, when the study was carried out, erosion had gradually been affecting the quality of both the beach and the promenade, with the result that some visitors had already switched their allegiance to other beaches in the area. Without measures to improve the existing coastal protection scheme, it was anticipated that continued erosion would result in decreasing visitor rates to the town.

In order to determine whether the massive expenditure on coastal protection was indeed warranted, clearly some estimate of the associated benefits of protecting the seafront would be required. The results of the contingent valuation study suggested a mean willingness to pay of £5.58 per annum, with 77% of participants being (at least in principle) willing to pay through the proposed payment vehicle of 'increased rates and taxes'.

Issues for Further Consideration

The foregoing discussion raises a number of issues which warrant further serious consideration:

1. at what level should an environmental balance sheet for the tourism sector be established? Clearly there are advantages to conducting the balance sheet approach at the regional level; yet this approach does little to ascertain the position of the national tourism industry as a whole regarding environmental sustainability.

2. should, indeed, the balance sheet be applied at the industry level or is a wider, macroeconomic approach required?

3. in conducting the balance sheet approach, the major emphasis is generally placed on the impact of new developments. How, if at all, should the impacts of the existing activities of the tourism sector be accounted for? Is a retrospective life cycle analysis of existing tourism activities required?

4. how might appropriate analytical boundaries be established? For example is the travel industry to be included within the scope of an environmental balance sheet for the tourism industry? Given the enormous environmental impacts associated with private road travel and air travel this could make a substantial difference to whether we believe the tourism industry to be on the path to sustainable development.

5. an overarching consideration is which form of sustainable development should be considered suitable as a goal for the tourism industry to adopt? Will 'weak' sustainable development suffice or should the 'strong' perspective be taken?

Concluding Remarks

This chapter argues that the idea of 'sustainable tourism' has yet to be opened out fully by the academic tourism literature. For the most part all that has as yet been done is to start to apply the sustainable development concept to the tourism context. On some occasions this has been done with more rigour than others. If the tourism industry is to treat sustainability seriously then there is much that could be done in terms of providing a framework for measuring progress and in establishing and improving non-market valuation methods. Unfortunately, what seems most at doubt at present is the tourism industry's commitment to carry these proposals through.

References

Adger, W.N. and Whitby, M. (1993) Natural resources accounting in the land-use sector: theory and practice. *European Review of Agricultural Economics* 20, 77–97.

Atkinson, G. and Pearce, D.W. (1993) Measures of economic progress. In: Pearce, D.W. (ed.) *Blueprint 3: Measuring Sustainable Development*. Earthscan, London, pp. 28–47.

Bramwell, B. and Lane, B. (1993) Sustainable tourism: an evolving global approach. *Journal of Sustainable Tourism* 1(1), 1–5.

Bull, A. (1991) *The Economics of Travel and Tourism*. Longman, Melbourne.

Burns, P.M. and Holden, A. (1995) *Tourism: A New Perspective*. Prentice Hall, London.

Cooper, C., Fletcher, J., Gilbert, D. and Wanhill, S. (1993) *Tourism: Principles and Practice*. Pitman, London.

Department of the Environment (1991) *Maintaining the Balance*. Government Task Force Report, HMSO, London.

Green, C.H., Tunstall, S.M, N'Jai, A. and Rogers, A. (1990) Economic evaluation of environmental goods. *Project Appraisal* 5(2), 70–82.

Hawkins, R. (1994) Towards sustainability in the travel and tourism industry. *European Environment* 4(5), 3–7.

Hunter, C. and Green, H. (1995) *Tourism and the Environment: A Sustainable Approach*. Routledge, London.

Inskeep, E. (1991) *Tourism Planning: An Integrated and Sustainable Development Approach*. Van Nostrand Reinhold, New York.

Inskeep, E. (1994) *National and Regional Tourism Planning*. Routledge, London.

McKercher, B. (1993a) Some fundamental truths about tourism: understanding tourism's social and environmental impacts. *Journal of Sustainable Tourism* 1(1), 6–16.

McKercher, B. (1993b) The unrecognized threat to tourism: can tourism survive 'sustainability'? *Tourism Management* April, 131–136.

Müller, H. (1994) The thorny path to sustainable tourism development. *Journal of Sustainable Tourism* 2(3), 131–136.

Pearce, D.W. (1992) *Towards Sustainable Development Through Environmental Assessment*. CSERGE Working Paper PA 92–11, CSERGE, Norwich and London.

Pearce, D.W. (1995) *Blueprint 4: Capturing Global Economic Value*. Earthscan, London.

Pearce, D.W. and Turner, R.K. (1993) Defining sustainable development. In: Pearce, D.W. (ed.) *Blueprint 3: Measuring Sustainable Development*. Earthscan, London, pp. 3–14.

Pearce, D.W., Markandya, A. and Barbier, E.B. (1989) *Blueprint for a Green Economy*. Earthscan, London.

Penning-Rowsell, E.C., Green, C.H., Thompson, P.M., Coker, A.M., Tunstall, S.M., Richards, C. and Parker, D.J. (1992) *The Economics of Coastal Management: A Manual of Benefit Assessment Techniques*. Belhaven, London.

Pennington, G., Topham, N. and Ward, R. (1990) Aircraft noise and residential property values adjacent to Manchester International Airport. *Journal of Transport Economics and Policy* 25(1), 49–59.

Porter, M.E. (1991) America's green strategy. *Scientific American* 264, 168.

Repetto, R., Magrath, W., Wells, M., Beer, C. and Rossini, F. (1989) *Wasting Assets: Natural Resource in the National Income Accounts*. World Resources Institute, Washington DC.

Tribe, J. (1995) *Economics of Leisure and Tourism*. Butterworth-Heinemann, Oxford.

Turner, R.K. (1992) *Speculations of Weak and Strong Sustainability*. CSERGE Working Paper GEC 92-26, CSERGE, Norwich and London.

Turner, R.K. and Bateman, I.J. (1990) *A Critical Review of Monetary Assessment Methods and Techniques*. Environmental Appraisal Group Report, University of East Anglia, Norwich.

Turner, R.K., Pearce, D.W. and Bateman, I. (1994) *Environmental Economics: An Elementary Introduction*. Harvester Wheatsheaf, Hemel Hempstead.

Uno, K. (1988) *Economic Growth and Environmental Change in Japan: Net National Welfare and Beyond*. Institute of Socio-Economic Planning, University of Tsukuba, Japan.

Walley, N. and Whitehead, B. (1994) It's not easy being green. *Harvard Business Review* May–June, 46–52.

Willis, K.G. (1991) The recreational value of the Forestry Commission Estate in Great Britain: a Clawson–Knetch travel cost analysis. *Scottish Journal of Political Economy* 38(1), 58–75.

World Commission on Environment and Development (1987) *Our Common Future*. Oxford University Press, Oxford.

Sustainable Tourism and the Local Economy[1]

5

W. Slee, H. Farr and P. Snowdon

Department of Agriculture, University of Aberdeen, 581 King Street, Aberdeen AB24 5UA, UK

Introduction

Tourism is a major industry in many developed and developing economies. At a global level it is growing in size and in its relative importance *vis-à-vis* other sectors of the global economy. As such, it is of great interest to development planners, for it offers substantial potential for economic betterment in a variety of settings ranging from decaying industrial areas (e.g. heritage tourism), to rural areas undergoing economic restructuring (e.g. green/soft tourism), to almost pristine undeveloped environments in such places as New Guinea (e.g. ecotourism). The rapid growth of tourism, the multiplicity of forms that it takes, and its wide-ranging ramifications on the economy, society and environment make the search for more sustainable forms an urgent task.

This chapter focuses on the economic impacts of different styles of tourism on the local economies of rural regions of the UK. Whereas most studies of sustainability and tourism development are framed within the utopian abstractions of environmentalists, this study takes as its starting point the desirability of creating and sustaining an economically viable tourist sector in rural Britain. It is recognized that different styles of tourism have the capacity to generate very different effects on the local host economy, society and environment. Given the wide range of rural settings in which

[1] The work on which this project was based was funded by the European Commission under the AIR programme, Project No. AIR1.CT 00447.

rural tourism is being promoted by local, national and transnational institutions, especially the European Union (EU), there is a strong case for comparing and contrasting the effects of different styles of tourism on the sustainable development of the rural socio-economic system.

First, some introductory comments on the challenge of creating more sustainable tourism are offered. Second, the problems of defining the tourism sector are reviewed, with particular reference to rural regions. Third, the economic literature on sustainability is explored, with a view to providing a conceptual framework for considering the relative sustainability of different styles of tourism. Fourth, the sustainable tourism debate is linked to the debate about alternative (endogenous) models of development. Finally, some results are presented from a recent wide-ranging economic study of rural tourism which indicate the very different impacts of different styles of tourism on the local economies of Speyside (specifically, the district of Badenoch and Strathspey) in northern Scotland and Exmoor (the Exmoor National Park) in south west England.

The Challenge

It is important to be aware of the nature of the tourism industry. It includes a huge variety of businesses ranging from transnational corporations to part-time enterprises on farms. It has a marked tendency to be self destructive, in that over-exploitation or misuse often threatens the well-being of the physical, cultural and biological resource base upon which tourism so frequently depends. This problem arises, *inter alia*, because the ownership of the accommodation sector and the resource base is often in different hands and the accommodation providers are effectively free-riders on the resource base. High levels of competition in highly volatile and often oversupplied tourism markets mean that the industry is characterized by high levels of business failure. It is an industry with low average rates of profit and low wage rates and with a high proportion of unskilled labour. Like most products, tourist destinations tend to experience a life cycle and the decline phase in this cycle will be characterized by many tourist firms chasing few visitors, a decline in profits and a run-down of the associated tourist infrastructure. An additional problem is that certain practices in the tourism industry, such as casual work and prostitution, fall outside the formal economy, making it a very difficult industry to regulate. Tourists travel increasingly long distances, inevitably depleting fossil fuel reserves and creating pollution in so doing, and the most affluent tourists often place the greatest demands on extremely fragile physical, biological and socio-cultural environments. This does not look like fertile ground on which to sow the seeds of ideas about sustainability.

Tourism development is also influenced by a wide range of institutions.

Tourism is a vast, highly complex and heterogeneous industry that, because of its potential for growth, inevitably attracts the interest of private, public and voluntary sector institutions. The majority of these institutions comprise profit-seeking firms. Other institutions include tourism support bodies funded by state or international bodies and other public sector agencies which influence the business environment within which firms operate, for example planners, regional development agencies and conservation agencies, and the bio-physical surroundings within which tourists recreate, for instance, national park agencies. In addition, it is important to recognize the role of voluntary bodies like the Worldwide Fund for Nature on a global scale, or the National Trust or other national conservation organizations at a national or local level, in shaping the environment of tourist opportunity through their activities as providers of attractions or as lobbyists influencing government interventions. The pursuit of more sustainable tourism practices demands a recognition of the multiplicity of institutions related to tourism and of the way in which these institutions operate individually or interact to shape more or less sustainable outcomes.

Definitions of Tourism and their Implications

Any social sciences text on tourism (see, for example, Smith, 1989; Ryan, 1991) recognizes the need to define the object of enquiry. The object of enquiry can be the person (the tourist), or the activity (tourism), or the visit. Tourists are normally thought of as those visiting somewhere other than their normal place of residence for more than 24 hours and less than one year. Certain categories of people who are routinely away from their normal places of residence, such as students, soldiers, migrant workers and diplomats, are excluded from the definition. Tourists may be international or domestic; and they may be principally involved in leisure (holiday tourism) or business (business tourism), although the line between business and holiday tourism is not always clear. A distinction is normally made between tourists and excursionists (or day visitors), the latter comprising those who are on a trip of less than 24 hours away from home. Tourist destinations can be considered both as the places visited by the tourists or excursionists and the centres of accommodation provision.

A tourism trip occurs each time the tourist is away from the normal place of residence for more than 24 hours. The nature of the trip can be classified in many ways, for example by reference to transport used, duration, accommodation sector used, purpose of trip, season of trip.

Smith (1989, p. 31) asserts that tourism 'does not have a real, objective, precise and independent existence' although he admits to the need for a generally accepted definition for planning and policy purposes. The United Nations Conference on Trade and Development (1971, quoted in Smith,

1989) defines tourism as 'the sum of those industrial and commercial activities producing goods and services wholly or mainly consumed by foreign visitors or domestic tourists'. Powell (1978, p. 1) suggested that tourism should include 'all the elements that combine to form the tourism consumer's experiences and (that exist) to service his needs and expectations.' Definitions of tourism as an industry are inevitably based on the sum of activities providing directly for tourist needs, but many tourist needs are not satisfied by specifically tourist businesses. For example, the local resident and the tourist will both use restaurants, food shops and entertainment industries. Although it will almost always be necessary to draw a line around a bundle of economic activities to describe the tourism sector of an economy, such a definition will routinely underestimate the economic, socio-cultural and environmental impacts of tourism.

It is preferable to see tourism as an industry with a core comprising 100% tourism and a series of surrounding zones which are progressively less shaped by tourist spending. Thus, the core will include the accommodation sector and day visitor attractions, whilst gift shops and the transport system occupy the next zone. Beyond this can be found a wide array of other retail establishments, such as newsagents and tobacconists and petrol stations. In addition, the tourist industry will create a set of knock-on demands on a wide array of suppliers of goods and services.

Any definition that denies the contribution of tourist spending over a (uniquely?) wide range of service industries and fails to explore tourists' impacts on the wider economic, socio-cultural and bio-physical environments will be inadequate. Equally, any definition that takes on board the breadth of economic activities that satisfy tourist needs will tend to fall partly outside government statistical service definitions of tourism. These definitional questions are of more than esoteric significance, for any attempt to move the tourism sector (the suppliers of tourist products) or the tourists (the demand side of the equation) towards more sustainable practices requires some definitional boundaries.

Rural tourism has the same characteristics as tourism in general in that it is highly heterogeneous and consists of core components which are exclusively tourist-related, with outer zones on which tourism has a progressively weaker economic influence. There are, however, a number of differentiating features which merit attention.

1. Rural tourism is associated closely with the quality of the bio-physical environment, compared to the built environment associated with urban tourism.
2. There is probably a higher degree of pluriactivity in rural tourism, with tourist enterprises forming part of wider businesses which are often land based.
3. Local culture is both vulnerable to external influence from tourist activity

and an important part of the tourism product.

4. By its location, rural tourism is often part of fragile rural economies in which decline is a common characteristic.

Sustainability and Economic Development

General issues concerning sustainability, particularly with respect to tourism, have been outlined by the editor in Chapter 1; here it is examined as a major component in discussions relating to economic development. Sustainability is a highly ambiguous and much debated concept and those who have tried to turn the widely quoted Brundtland definition of sustainable development (World Conference on Environment and Development, 1987) into an operational set of principles have found the definition malleable. The emerging economic literature on sustainability (Pearce *et al.*, 1989; Pearce and Turner, 1990; Jacobs, 1991; Turner *et al.*, 1994) recognizes the breadth of ideas encapsulated in the term 'sustainable development' and the challenge of operationalizing them into guiding principles. Jacobs (1991) identifies three core ideas relating to sustainability. First, he argues that institutions should incorporate environmental considerations into policy and practice in a logical and consistent way. Second, he reaffirms the principle that no depletion of the natural resource stock is justifiable: that there should be intergenerational equity with respect to the natural capital of the planet. Finally, it is argued that sustainable development is not the same as growth and that it is essential to consider community well-being, cultural values and the intrinsic value of the environment in any comprehensive examination of development. Many aspects of these guiding principles are open to variable interpretation.

With regard to the first principle, governments can be seen to be developing environmental policies and creating the preconditions in which other economic institutions are obliged to consider environmental implications of their actions. There is now a burgeoning literature on environmental audit, for example Welford and Gouldson (1993), and government guidelines on how to proceed with such tasks in the UK (British Standards Institute, 1992). Some regard environmental auditing as a tool to apply to policy instruments as well as to business strategy.

With regard to the second principle, there is often a dilemma facing those seeking to promote development in that more development, at least as conventionally measured in terms of Gross Domestic Product (GDP), can most easily be achieved in the short run by eating into reserves of natural capital, as has happened so frequently in developing countries. The dilemma of development is encapsulated in the Trade Off Model (Pearce and Turner, 1990), which suggests that development gains are necessarily at the expense of the capital stock. However, it is essential to have a clear idea of the constituent components of the capital stock and the extent to which there might be substitutability between these different components. Furthermore, it is not inconceivable that, at different stages in economic development, an economy may move from a more sustainable mode of operation to a less sustainable mode, or vice versa.

The second principle is thus subject to negotiation. At one extreme, it is possible to find a sustainability principle that demands no more than future generations being compensated for the loss of natural capital by some gain in other forms of capital such as man-made capital. As economies develop and become less obviously dependent on natural capital, it may be relatively easy to substitute man-made capital gains for natural capital losses. However, others would argue for a 'stronger' sustainability principle which identifies some elements of natural capital as 'critical natural capital' and allows no diminution in the quantity of these. Critical natural capital is usually considered to be that which is essential to support human life and within this category air, water, warmth and shelter and the maintenance of a (genetically diverse) food-producing system comprise some basic elements. A still stronger version of sustainability would allow no depletion at all of the stock of natural capital.

There is widespread recognition that the biophysical resource base is endangered by development. This can arise through the depletion of natural resource stocks, through the excessive use of renewable resources (i.e. where harvest exceeds sustainable yield), and through pollution effects. The threat to resource stocks may be localized, national or global, (for example, local: Caithness (north east Scotland) peatlands; natural: UK water supplies; global: whaling). The resource base can be threatened in various ways, including the wilful exploitation of a resource for short term gain, ignorance of the consequences of particular actions, or market failure enabling one economic agent to reduce the resource stock or quality of other economic agents where, for example, a factory pollutes a watercourse. A normal minimum position is that certain types of natural capital are seen as critical, and no depletion of those elements of natural capital which are essential to support life is regarded as acceptable.

The third principle, that sustainable development is not the same as growth, and that self determination and cultural well-being are core elements of sustainability, is contentious and reflects the concern of many that behind more sustainable economies there should be a recognition that development requires more than just an increase in national income as measured conventionally. Such an assertion, which moves the debate about sustainability firmly into the realms of social policy, has received widespread endorsement from the social science community (for example, Bryden, 1995).

In summary, it might be asserted that, for there to be sustainable development from an economic perspective, gains in GDP are required, where the increase is not threatened by negative feedback from biophysical systems (pollution, etc.) or adverse social and cultural impacts and, dependent upon whether it is a strong or weak sustainability criterion being used, where any depletion of the natural resource stock of non-renewables is compensated by increases in the stock of renewables and where no critical natural capital is depleted. There is abundant evidence that in practice these conditions are rarely satisfied (Cairncross, 1991; Turner *et al.*, 1994). The challenge is thus to identify less unsustainable forms of tourism, accepting that, by any rigorous definition, sustainable tourism, or indeed any other form of economic activity, is unachievable.

Sustainable Tourism and Rural Development

As in the wider arena of development, so in the field of tourism there has been a parallel debate about desirable forms of tourist development. Alternative/green/soft /responsible tourism has been identified as a potentially more important component of the tourist industry of the future. A number of official publications in the UK (Scottish Office, 1991; Tourism and the Environment Taskforce, 1991) and at a global level (McIntyre *et al.*, 1993) have alluded to the need for more sustainable approaches to tourism, but consist of little more than statements of platitudes, backed by glossy images. More recently there has been a trend towards providing indications of good practice (Countryside Commission *et al.*, 1995). The launch of a new journal dealing exclusively with sustainable tourism (*Journal of Sustainable Tourism*) and the presence of a large number of papers on sustainable tourism at a recent 'state of the art' conference on tourism (Seaton *et al.,* 1994), the conference from which this book was compiled, and the numerous and sometimes acrimonious articles in other tourism publications, for example, Wheeler (1992), are evidence of the considerable contemporary interest in the subject.

There is a growing literature on sustainability and tourism, for example, Hunter and Green (1995), which brings into discussion many of the central difficulties both of defining sustainability and of applying its concepts to the tourist sector of the economy. However, at the heart of almost all tourist activity there are usually commitments of resources which make it impossible for tourist activity to satisfy any semi-rigorous definition of sustainability. This does not of course imply that sustainability gains cannot be made; merely that, except in exceptional circumstances, it is impossible for tourism to satisfy many environmentalists' notions of sustainability.

These difficulties of satisfying any rigorous definition of sustainability have not stopped a variety of agencies from endeavouring to produce sets of principles for sustainable tourism. Some of these sets of principles are the product of state bodies or 'quangos', whilst others are a product of environmental pressure groups and tourism institutions. An example is given in the Appendix. Most of them use rather flexible definitions of sustainability that, at best, would satisfy a weak definition of sustainability and, at worst, are a combination of pious platitudes and seriously contradictory statements.

Over the last decade, traditional models of rural development have been subjected to a wide-ranging critique. For example, Buller and Wright (1990) conceptualize development as gains in well-being, gains in the access to the means to sustain that well-being, and gains in self-determination, rather than simple gains in GDP. The concern that opportunities for self determination be increased and that sustainable development be promoted are evident in the work of Chambers (1983). There is a prominent thrust in development thinking in both developed and less developed countries which questions dominant development theories and argues that that top-down development strategies have frequently been responsible for unsustainable development practices. A significant body of research has been carried out exploring endogenous development strategies (van der Ploeg and Saccomandi, 1994), which argues that such strategies are more environmentally benign (and thus more sustainable) and that they protect social and cultural capital whilst creating opportunities for gains in GDP.

There are very close connections between the underlying objectives of alternative development and alternative tourism as articulated, amongst others, by Lane (1988, 1989, 1994), who has identified a number of characteristics of soft tourism, including:

- embedding of tourism in a wider functioning economy;
- operation at a level which places no unacceptable demands on the environment;
- employment of local people in work that enhances their self esteem;
- empowerment of local people to enable them to control their own destinies;
- use of local products in ways that enhance a sense of place;
- respect for local cultural traditions.

However, it is important not to conflate soft rural tourism with sustainable tourism, as many authors seem to be in constant danger of doing, for it is highly dependent on a definition of sustainability based on community involvement and a sense of place, rather than other elements of the concept of sustainability. Soft tourism is thus not possible where there are concentrations of large scale tourist specific enterprises in resorts. However, it is also possible to enhance sustainability in large-scale tourist enterprises and hard tourist regions by, for example, using environmental audits constructively and introducing waste minimizing and pollution reducing measures.

Soft tourism, as identified by Lane, has many of the characteristics sought by those seeking the development of more sustainable forms of tourism. However, these forms of tourism represent a relatively small subsegment of the tourist industry. The bulk of the tourism industry consists of, and will continue to consist of into the foreseeable future, mass tourism provision. The mass tourism market is, of course, highly segmented, but it has a number of characteristics which distinguish it from soft tourism. It is different in scale; it tends to be geographically concentrated; it tends to be economically dominant in the regions in which it is located; it is often characterized by a relatively high degree of penetration by external capital; it is often associated with an itinerant workforce; it tends to be linked strongly to the wider international economy, rather than being strongly embedded in local economies; and, lastly, it tends to be more environmentally destructive and less sustainable than soft tourism.

Nonetheless, soft/sustainable tourism has been identified as a major component of development strategies for rural regions of Europe and other parts of the world. The extent to which soft tourism provides a model solution in Europe is evidenced by the extent to which the EC-funded LEADER projects have established tourism-related projects. If a broad definition of sustainability is adopted, following Jacobs (1991) and Bryden (1995), it is easy to see the attraction of soft tourism, although the current interest in rural tourism seems to be based more on political expediency in seeking a possible solution to endemic decline in certain rural regions, than on any clear evidence as to the sustainability of soft rural tourism.

Soft tourism is also seen as a means of maintaining cultural landscapes and maintaining semi-natural ecosystems associated with extensive farming, a means of valorizing a way of life which is threatened by a combination of socially, politically and economically induced changes. In an agricultural context, this might be interpreted as a need to retain the richness of understanding in local knowledge systems relating, *inter alia*, to traditional agricultural practices and food processing. In relatively remote regions, there may still be significant levels of locally based, artisanal activity which can potentially be revitalized by tourist demand. In developing countries, soft tourism can enable local populations to engage in tourism, as in the Selous area of Tanzania, where top-down tourism planning strategies have been replaced by policies to enable local populations to manage and harvest the game on which tourism in the area largely depends. In other developing countries, there is significant evidence of the vitality of traditional artisanal production arising from tourist demand.

A concern for cultural capital is a major component of the debate about alternative models of rural development (Hettne, 1990). In a tourism context, such concerns about cultural capital might suggest a need to ensure that tourism development respects and allows for the retention of local cultural attributes. This raises questions about whether there can be forms of tourist development which have neutral or beneficial effects on local culture. Whilst there are likely to be different cultural effects of different types of tourist development, a degree of depletion of cultural capital, or at least the modification of cultural capital, seems to be an almost inevitable concomitant of tourist development.

The model of soft rural tourism identified by Lane and others appears to offer a potential solution to some of the challenges facing certain areas today. However, it is necessary to consider whether there is an economic rationale for this apparently utopian solution. It is, at first sight, difficult to see how rural tourism with the following diagnostic characteristics (Lane, 1994): small settlements; weak infrastructure; small establishments; local ownership; few guests; and amateur management, can compete effectively in the face of competition from an urban or resort based, relatively concentrated, professionally managed tourist industry, operating increasingly in international markets.

Several possible explanations can be offered. First, for a variety of reasons, capital investment in full-time, single-purpose tourism ventures is likely to be unprofitable when other forms of economic organization such as simple commodity production can succeed. This could arise where smaller part-time tourist operators can provide the fixed capital needs of the enterprise at low cost, for example, spare rooms in a farmhouse, or where the tourist enterprise is able to be based on cheap family labour. This may be true in some instances, but, should the enterprises develop successfully, it is likely that external capital will be attracted to attempt to extract profit.

Second, there may be potential for environmentally friendly, culturally sensitive, rural development initiatives based on an endogenous model of development. Van der Ploeg and Saccomandi (1994) argue that small-scale, endogenously based enterprises may be capable of operating competitively in a mature capitalist economy. For example, family firms may be able to offer niche tourist products that cannot be imitated successfully by larger businesses.

Third, it is possible that demand-led changes have created new opportunities which cannot readily be captured by larger scale tourism ventures. If there is a demand for 'authentic' farm or forest holidays it may be easier for the farmer or forester to develop tourist enterprises than for the tourist enterprise to develop a pseudoauthentic farm. If these niche markets are relatively small and dependent on the provision of a complex product bundle including cultural attributes, this is likely to offer advantageous opportunities to small-scale operators.

Finally, traditional tourist resorts may be burdened by semi-redundant or redundant fixed capital, creating an ambience of senescence and decay. In effect, the tourism product in these areas is in the decline phase of the product life cycle. It may be easier to develop new tourism ventures in relatively undeveloped tourist destinations than to revitalize the old tourist infrastructure in traditional destinations. It is also possible that Butler's (1980) six-stage model may not apply in areas that never become dominated by tourism, thus avoiding the potential problem of a customized but redundant tourist infrastructure in the decay phase.

There are thus sound reasons for believing that there are possibilities for the development of economically viable, sustainable tourism initiatives in rural areas. There remains, however, a residual doubt that this style of tourism is being promoted for non-economic reasons, which have more to do with the tourism preferences of soft tourism advocates than the development needs of rural regions. These doubts can be eased by an economic analysis of soft tourism which explores whether there are beneficial economic ramifications of this type of development on the host economies.

The Economic Analysis of Rural Tourism: a Study of Two Rural Areas

Economic viability is an essential and necessary precondition of sustainability, but is not in itself a sufficient condition. Economic viability can be considered at the level of the firm, the region or at other levels. It is not inconceivable that economically viable firms will operate in ways that are predatory on other competing firms (thus creating displacement effects) and on the public good resources of the rural environment and in ways that undermine existing linkages in rural economies. Intuitively, this predatory behaviour might be expected to be associated with tourist enclaves that parasitize the neighbouring economic, bio-physical and socio-cultural environments, because of the numerous external benefits that these environments provide for the enclave. Consequently, it is important to consider economic sustainability within the context of regional or subregional economies and their associated sets of social, cultural and environmental resources.

The principal aim of the research reported on here was to compare the economic effects arising from different styles of rural tourism development. Therefore, it was necessary to disaggregate the tourism sector. In each study area, sectors of the tourism industry were identified as being 'hard' or 'soft'. The hard sector was defined as all accommodation with ten or more bedrooms or units, i.e. hotels and holiday villages. The soft sector was taken to denote all accommodation on agricultural and other land-based businesses, namely farm bed and breakfast, farm self-catering and farm and forest camping and caravan sites. Expenditure data for the previous 24 hours were collected from visitors staying in hard and soft accommodation in the core areas.

The overall economic impact of tourism can be explored through regional economic modelling approaches. Regional input–output analysis has long been recognized as an appropriate method of economic analysis which can identify the local economic impacts on income and employment of an industry. However, the diverse nature of the tourism industry makes it almost impossible to use conventional input–output tables produced at a national level as the basis for the analysis. Therefore, survey-based methods are necessary. A full input–output analysis requires data on all sectors of the economy. The costs of such analysis usually make this an impractical solution. Alternative methodologies have therefore been developed which are based on partial input–output analysis. The Proportional Multiplier Method was initially developed by Archer (1973) and subsequently modified by Henderson and Cousins (1975), Walker and Vaughan (1992) and Vaughan (1994). It differs from a full input–output analysis in that it analyses only those aspects of the economy related to the industrial sector under consideration, in this case, tourism.

This methodology adopts an incremental approach, combining partial input–output analysis with traditional Keynesian multiplier analysis. It therefore measures the whole impact of economic change in the sector of the economy under scrutiny, not just the initial first-round effects. The income and employment generated is the result of income circulating within the local economy. The first-round effect, the direct impact, is the impact of tourist spending on incomes and jobs at the businesses where tourists spend their money. The second-round effect, the indirect impact, results from transactions between local businesses arising from the tourist spending. The induced impact is the effect on incomes and jobs of the spending of income earned from tourist expenditure.

The multipliers which are produced are a measure of the additional benefit generated in the economy from an injection of capital. Another principal difference of the Proportional Multiplier method from traditional multiplier analysis is in the expression of the multiplier. The traditional Keynesian multiplier is a ratio of the total (direct + indirect + induced) impact divided by direct impact, whereas the Proportional Multiplier is the total impact divided by per unit of visitor spending. Thus an income multiplier of 0.22 indicates that, for every £1 spent in the local economy by a tourist on accommodation, attractions or leisure activities, 22 pence accrues as direct, indirect and induced income to the local economy.

The data requirements of the Proportional Multiplier Method are:

- the amount of visitor spending and the distribution of that spending between different business types,
- the pattern of expenditure in direct and indirect businesses,
- the expenditure patterns of local residents.

For the study reported in this chapter, the first two data sets were obtained from extensive surveys of visitors and tourism related businesses in two study areas during the summer of 1994. Expenditure patterns of local residents were based on published data relating to household expenditure patterns. In the UK, this is the Family Expenditure Survey (Central

Statistical Office, 1995), an annual government sponsored survey.

The project also investigated the relationship between the local and regional economy. Studies using multiplier analysis have shown that small local economies have greater leakages than larger more diversified economies and therefore have lower multipliers (Coppock *et al.*, 1981; Mathieson and Wall, 1982). In order to examine this, the study areas were divided into the central 'core' area and the 'extended' area, i.e. the area within 25 km of the core area boundary. Data were collected from direct and indirect businesses in the core area and from indirect businesses in the extended area. Thus, it was possible to estimate a local multiplier and a subregional multiplier for each area.

Results

The results given in this section compare hard and soft tourism in the two UK study areas. Table 5.1 shows that there are distinct differences between the two areas. In Exmoor, the hard and soft multipliers are relatively similar, particularly when the extended area is considered. However, in Badenoch and Strathspey, soft tourism has a multiplier which is one third greater than that produced by hard tourism.

This pattern is repeated for the employment multipliers (Table 5.2). Hard and soft tourism produces very similar levels of employment in Exmoor, with soft tourism being marginally greater. Soft tourism in Badenoch and Strathspey generates significantly more employment per £100,000 of visitor spending.

Table 5.3 shows the three different levels of impact which are combined to produce the total economic impact in the core + extended area. In all cases, the greatest impact is the direct effect. As would be expected from the results in Table 5.2, there are similarities between hard and soft tourism for Exmoor. In Badenoch and Strathspey, the difference between the soft and hard tourism is due to both the direct and the indirect impacts being higher for soft tourism. The impact on employment shows the same pattern (Table 5.4).

Tables 5.5 and 5.6 give the total economic impact in each study area

Table 5.1. Income (£) generated per unit of visitor spending for two UK study areas.

	Hard		Soft	
	Strathspey	Exmoor	Strathspey	Exmoor
Core area	0.22	0.23	0.30	0.21
Core + extended area	0.29	0.34	0.42	0.34

Source: own survey, June–October 1994.

Table 5.2. Employment (FTEs*) per £100,000 of visitor spending for two UK study areas.

	Hard		Soft	
	Strathspey	Exmoor	Strathspey	Exmoor
Core area	2.8	2.6	3.6	2.7
Core + extended area	3.5	3.4	4.8	3.6

Source: own survey, June–October 1994.
* All employment figures are given as standardized jobs, i.e. full time equivalents (FTE) per £100,000 of tourist spending.

Table 5.3. Direct, indirect and induced income multipliers per unit of visitor spending in the core + extended area of the UK study areas.

	Hard		Soft	
	Strathspey	Exmoor	Strathspey	Exmoor
Income (£)				
Direct	0.20	0.23	0.27	0.22
Indirect	0.09	0.10	0.15	0.11
Induced	0.003	0.005	0.004	0.004

Source: own survey, June–October 1994.

Table 5.4. Direct, indirect and induced employment multipliers for the core + extended area generated by £100,000 of visitor spending in the UK study areas.

	Hard		Soft	
	Strathspey	Exmoor	Strathspey	Exmoor
Employment (FTEs)				
Direct	2.2	2.2	2.9	2.3
Indirect	0.8	0.7	1.3	0.9
Induced	0.5	0.5	0.6	0.4

Source: own survey, June–October 1994.

resulting from spending by hard and soft tourists in the summer of 1994. It reveals striking differences between the two study areas. Exmoor is predominantly a soft tourism destination, with approximately three times the number of soft visitors than hard. This therefore generates higher levels of income and employment in the core and core plus extended areas. In Badenoch and Strathspey, the reverse is true. Hard tourism has over five times the number of visitors than soft and accounts for a much greater level of income and employment in the area despite having lower multipliers (Tables 5.1 and 5.2).

Table 5.5. Total income resulting from hard and soft tourist expenditure in the UK study areas, June–October 1994.

	Hard		Soft	
	Strathspey	Exmoor	Strathspey	Exmoor
Visitor numbers	91,618	16,989	15,746	51,433
Core area	3,155,261	823,222	664,758	1,118,832
Core + extended area	4,251,751	1,200,554	937,648	1,754,148

Source: own survey, June–October 1994.

Table 5.6. Total employment resulting from hard and soft tourist expenditure in the UK study areas, June–October 1994.

	Hard		Soft	
	Strathspey	Exmoor	Strathspey	Exmoor
Visitor numbers	91,618	16,989	15,746	51,433
Core area	515	113	99	176
Core + extended area	630	149	133	235

Source: own survey, June–October 1994.

The results of this analysis can be used to predict the effects of increasing the numbers of visitors to an area, or increasing the length of stay. Tables 5.7 and 5.8 present the income and employment that could be expected from 100,000 visitor days, all other things remaining equal. In both cases, hard tourism generates more local income and employment, though there is less difference in Badenoch and Strathspey than in Exmoor.

Discussion

The analysis reveals some significant contrasts between the local and subregional economic effects generated by different styles of tourism. In Badenoch and Strathspey, where there is a marked contrast in the size of the hard and soft tourist businesses, it is apparent that the soft sector generates more jobs and higher incomes per unit of visitor spending. This marked difference is not apparent in Exmoor, where the average size of hard tourist businesses is smaller and the hard sector lacks the enclave characteristics of this sector in Badenoch and Strathspey.

Soft tourism is associated with rural land management units. In consequence, it is in the interests of the land mangers to ensure that the environmental services delivered by land management for the tourist

Table 5.7. Income (£) generated from 100,000 days of visitor spending in hard and soft tourism in the two study areas.

	Hard		Soft	
	Strathspey	Exmoor	Strathspey	Exmoor
Core area	776,002	998,235	480,404	339,493
Core + extended area	1,045,671	1,455,786	667,615	532,271

Source: own survey, June–October 1994.

Table 5.8. Employment (FTEs) generated from 100,000 days/nights of visitor spending in hard and soft tourism in the two study areas.

	Hard		Soft	
	Strathspey	Exmoor	Strathspey	Exmoor
Core area	127	137	72	53
Core + extended area	155	180	96	71

Source: own survey, June–October 1994.

experience are maintained. Whereas the hard tourist business tends to parasitize the public goods of the environment, the public good externality of landscape is effectively internalized in the case of the soft tourist provider, thus ensuring the more sustainable management of the resource.

Soft tourism employs more people per unit of visitor spending in both study areas. This suggests that rural land management units have a greater capacity per unit tourist spending to provide local jobs. Furthermore, the soft sector is more likely to use local labour, whereas, in contrast, larger scale tourist enterprises are more likely to import labour, which will reduce the local respending because of remittances sent out of the area. Such imported labour may have little understanding of or affinity for local culture. Soft tourism is also more likely to be associated with self-employed individuals and there is likely to be a greater capacity for self-determination than in the case of hired labour, which is more prevalent in the hard sector. Given that self determination is seen as an essential characteristic of socio-economic sustainability, the higher levels of self determination should be regarded as a positive feature of this sector.

The definitions used in this analysis are definitions of convenience and the a priori categories 'hard' and 'soft' were not always appropriate for the enterprises examined. For example, some of the smaller hotels studied, which were categorized as 'hard' businesses, had characteristics that would normally be considered as more typical of soft enterprises. For example, a number of hard businesses consciously purchased inputs from local suppliers and employed local labour. A number of the land-based enterprises, especially camping and caravan sites, were quite large and exhibited some of the characteristics of hard businesses. The categories hard and soft are thus more a feature of the attitudes and business strategy of the owner/manager, rather than related to whether the business was located on a rural land management unit. Nonetheless, there was still a clear difference in the behaviour of the hard and soft tourist businesses in Speyside, which created substantially higher local economic benefits from soft tourism in the local and sub-regional economies.

The soft tourism businesses surveyed normally charge less per visitor night than hard tourist businesses and it is thus difficult to label this type of tourism as being elitist as some have endeavoured to do (Wheeller, 1992). Simply because a particular type of tourism does not accord with majority tastes does not make it elitist. However, there may well be some types of green/sustainable/soft tourism that can legitimately be described as elitist as the very high prices charged for such activities as deer stalking mean that it satisfies the demand of a small but affluent clientele.

Most socio-economic analysts examining sustainability in rural development identify at least three major facets of sustainability that merit scrutiny: economic sustainability; socio-cultural sustainability; and environmental (bio-physical) sustainability. This study of small-scale soft rural tourism has shown that not only is soft tourism preferred from a local economic development perspective, but it also has socio-cultural features and bio-physical effects that may give it an advantage in sustainability terms over many other types of tourism. However, these non-economic benefits have been inferred rather than measured and there remains a strong case for further analysis of the relative sustainability of different styles of tourism.

The promotion of small-scale soft rural tourism may thus constitute a legitimate element of agency actions to support the more integrated development of rural economies and to provide an alternative source of well-being for households that are likely to experience diminishing returns from their land-based activities in the future. Tourism is certainly not a panacea for the current problems of the rural economy, but appropriate tourism developments have an important part to play in securing more sustainable sources of well-being for a significant number of rural people.

Appendix: Principles for Sustainable Tourism

1. The environment has an intrinsic value which outweighs its value as a tourism asset. Its enjoyment by future generations and its long term survival must not be prejudiced by short term considerations.
2. Tourism should be recognized as a positive activity with the potential to benefit the community and the place as well as the visitor.
3. The relationship between tourism and the environment must be managed

so that the environment is sustainable in the long term. Tourism must not be allowed to damage the resource, prejudice its future enjoyment or bring unacceptable impacts.

4. Tourism activities and development should respect the scale, nature and character of the place in which they are sited.

5. In any location, harmony must be sought between the needs of the visitor, the place and the host community.

6. In a dynamic world some change is inevitable and change can often be beneficial. Adaptation to change, however, should not be at the expense of any of these principles.

7. The tourism industry, local authorities and environmental agencies all have a duty to respect the above principles and to work together to achieve their practical realization.

Source: Tourism and the Environment Taskforce (1991).

References

Archer, B.H. (1973) *The Impact of Domestic Tourism.* University of Wales Press, Cardiff.

British Standards Institute (1992) *Specification for Environmental Management Systems.* British Standards Institute, London.

Bryden, J. (ed.) (1995) *Towards Sustainable Rural Communities.* University of Guelph and the Arkleton Trust, Guelph, Canada.

Buller, H. and Wright, S. (eds) (1990) *Rural Development: Problems and Practices.* Avebury, Aldershot.

Butler, R. (1980) The concept of a tourism area cycle of evolution. *Canadian Geographer* 24, 5–12.

Cairncross, F. (1991) *Costing the Earth.* Business Books and The Economist Books, London.

Central Statistical Office (1995) *Family Spending. A Report on the 1994–95 Family Expenditure Survey.* HMSO, London.

Chambers, R. (1983) *Rural Development: Putting the Last First.* Longman, London.

Coppock, J., Duffield, B. and Vaughan, D.R. (1981) *The Economy of Rural Communities in the National Parks of England and Wales.* Tourism and Recreation Research Unit, Edinburgh.

Countryside Commission, Rural Development Commission, English Tourist Board, and Department of National Heritage (1995) *Sustainable Rural Tourism. Opportunities for Local Action.* Countryside Commission, Cheltenham.

Henderson, D. and Cousins, L. (1975) *The Economic Impact of Tourism. A Case Study in Greater Tayside.* Tourism Recreation Research Unit, Edinburgh.

Hettne, B. (1990) *Development Theory and the Three Worlds.* Longman, London.

Hunter, C. and Green, H. (1995) *Tourism and the Environment. A Sustainable Relationship?* Routledge, London.

Jacobs, M. (1991) *The Green Economy.* Pluto Press, London.

Lane, B. (1988) Small scale rural tourism initiatives. The role of a British University.

Tourism and Leisure in Rural Areas. Council of Europe, Strasbourg.

Lane, B. (1989) The future of rural tourism. Unpublished manuscript.

Lane, B. (1994) What is rural tourism? *Journal of Sustainable Tourism*. 2 (1 and 2), 7–21.

McIntyre, G., Hetherington, A. and Inskeep, E. (1993) *Sustainable Tourism Development: Guide for Local Planners*. World Tourism Organization, Madrid.

Mathieson, A. and Wall, G. (1982) *Tourism: Economic, Physical and Social Impacts*. Longman, London.

Pearce, D., Markandya, A. and Barbier, E.B. (1989) *Blueprint for a Green Economy*. Earthscan, London.

Pearce, D.W. and Turner, R.K. (1990) *Economics of Natural Resources and the Environment*. Harvester Wheatsheaf, Hemel Hempstead.

Powell, J. (1978) *Report of the Tourism Section Consultative Task Force*. Department of Trade and Industry, Ottawa.

Ryan, C. (1991) *Recreational Tourism. A Social Science Perspective*. Routledge, London.

Scottish Office (1991) *Rural Framework*. Scottish Office, Edinburgh.

Seaton, A.V. Jenkins, C.L. Wood, R.C. *et al*. (1994) *Tourism. The State of the Art*. John Wiley & Son, Chichester.

Smith, S.L.J. (1989) *Tourism Analysis. A Handbook*. Longman, Harlow.

Tourism and the Environment Taskforce (1991) *Maintaining the Balance*. English Tourist Board, London.

Turner, R.K., Pearce, D. and Bateman, I. (1994) *Environmental Economics. An Elementary Introduction*. Harvester Wheatsheaf, Hemel Hempstead.

United Nations Conference on Trade and Development (1971) A note on the 'tourist sector'. *Guidelines for Tourism Statistics*. United Nations, New York.

van der Ploeg, J.D. and Saccomandi, V. (1994). On the impact of endogenous development in agriculture. In: van der Ploeg, J.D., Saccomandi, V., Ventura F. and van der Lande, A. *On the Impact of Endogenous Development in Rural Areas*. CERES, Wageningen, pp. 9–24.

Vaughan, D.R. (1994) The impact of visitor spending: a review of methodology. In: University of Aberdeen, Institut D'Etudes Politiques de Grenoble and Universidade de Tras-os-Montes e Alto Douro (eds) *Agrotourism and Synergistic Pluriactivity. First Progress Report*. University of Aberdeen, Scotland, pp. 31–90.

Walker, S. and Vaughan, D.R. (1992) *Pennine Way Survey 1990. Use and Economic Impact*. Countryside Commission, Manchester.

Welford, R. and Gouldson, A. (1993) *Environmental Management and Business Strategy*. Pitman Publishing, London.

Wheeller, B. (1992) Is progressive tourism appropriate? *Tourism Management* 13, 104–105.

World Conference on Environment and Development (1987) *Our Common Future*. Oxford University Press, Oxford.

Redefining Sustainability: A Structural Approach to Sustainable Tourism

6

J. House

Department of Social Science, University College of St Mark and St John, Derriford Road, Plymouth PL6 8BH, UK

Introduction

This chapter is concerned with exposing how interpretation of the concept of sustainability is stratified in its application to tourism development and behaviour. It is suggested that opinions, policies and actions regarding sustainability in tourism can be applied to a sliding scale. 'Reformists', whose ideology and actions are essentially in keeping with the status quo, can be placed at one end, while 'structuralists', who seek to challenge the methods and values of tourism development on a fundamental level, can be positioned at the other. This chapter uses such a framework to explore results derived from fieldwork conducted at three tourism destinations which appear to have adopted the structuralist model, both in ideology and *raison d'être* and in behaviour and policy. Results representative of what is described as the reformist model are derived from secondary source data. The overall aim of this chapter is to illustrate how, on the fringe of development, a new structuralist impetus exists, whereby reformist ideas of incorporating thrift and recognizable sensitivity into tourism activity appear superficial and short term, failing to address the critical agenda of the nature and future of tourism and whether the ideology that underpins current approaches can, in reality, be reconciled with sustainability.

Firstly, the research destinations and what constitutes an alternative approach will be briefly described to provide information on the nature of the tourism experience offered. The above framework is then discussed in more detail in the context of a brief exploration of the concept of

© CAB INTERNATIONAL 1997. *Tourism and Sustainability*
(edited by M.J. Stabler)

sustainability, what it has meant and to whom. It is not the remit of this chapter to discuss approaches, ideologies and case studies of sustainability at length and thus the issues around this important debate are only developed in terms of highlighting stratification within the concept. The discussion quickly moves on to review a selection of practices, policies and, hence, values of the three destinations, again as a means of exemplifying what has been termed a structuralist sustainable tourism approach. It is concluded that, although not necessarily well documented and largely on the periphery of tourism development, an alternative approach exists whereby assumptions about the nature of the tourist product and attitudes towards sustainability are being questioned and challenged.

Research Destinations and Research Practice

In general terms, the three tourism destinations on which this discussion is based can be described as alternative in form. That is, there appears to be:

> a shared concern on the part of ... participants to transform their own lives ... to create an alternative social order characterized by values counter to those that appear to dominate our present existence.
>
> (Rigby, 1974, p. 2).

In terms of tourism activity, this implies developments away from traditional mass tourism to a more organic, self sufficient approach emanating from within, rather than reflecting external values. Alternative tourism has been defined as those approaches which commonly adhere to an environmentally sensitive approach (Krippendorf, 1987; Frommer, 1988), are based on community values and positive 'host and guest' (Smith, 1989) interaction (Eadington and Smith, 1992), involve intellectual pursuit and are concerned with self-development and the actualization of participants' goals (Weiler and Hall, 1992). All such destinations do not claim to offer a mainstream product and most are unlikely to attempt to compete within what can be described as the corporate market. Each can be described as occupying a niche and thus is peripheral in terms of its market share in the wider industry. However, where the three destinations to be focused on here perhaps differ from many other destinations described as alternative is in the way tourism is envisaged more as a means to an end, rather than as an end in itself. That is, each was developed by individuals whose primary concern was to experiment with alternative existence approaches (see Rigby's 1974 definition above), hence tourism is seen more as a compatible vehicle through which the operation can be sustained and funded. This would seem contradictory to the majority of other tourism outlets, both special interest (Weiler and Hall, 1992) and mainstream, whose primary purpose is usually to offer a tourism product for economic gain. This structurally alternative value approach probably explains why such destinations offer a deeper and more fundamental approach in their application of sustainable tourism. Their founding model is one of sustainability, while, for many other operations, the adoption of sustainability means incongruously superimposing a new model on an already established and unsympathetic framework. Analogously, the destinations to be discussed can be depicted as an integral part, or resultant from the Environment Movement (see Dobson, 1990; Yearley, 1991). Their goals and approaches have been shaped by Environment Movement protagonists and philosophies, their approaches having evolved out of a concern for these issues; thus they, in a sense, have no compromise to make. Although it is not suggested that this type of tourism product can replace the more usual economically based tourism product, nevertheless these do raise interesting and fundamental questions from which lessons may be learnt.

Essentially each of the research destinations studied offers a similar type of experience to tourists: a combination of pupil to tutor skill and scholarly learning and experiential workshops in which individuals explore personal issues and practise newly acquired principles. The tutors, staff and a small number of volunteers at such destinations are often sympathetic to the ethos and *raison d'être* of the destinations and tend to consist of ex-participants of either the said destination or similar ones, who either stayed on or left to develop areas of expertise and have returned. The participants could be categorized mostly as members or offspring of the middle classes, or more specifically of the professional or new class (Inglehart, 1981; Eckersley, 1989), relatively extensively educated and tending to be liberal-socialist politically, although political interest was often relegated in favour of more personalized concerns.

It was recognized early in the research process that the experiences offered were alternative and thus possibly unorthodox; therefore ethnographic research was undertaken at each of the destinations, consisting of participant observation and interaction (Spradley, 1980). The decision to make the research covert was also taken, on the basis that participation in some of the experiences offered necessitated a sensitive and sympathetic approach, where the researcher's influence should be minimal, an 'insider's' perspective (Hammersley and Atkinson, 1983) being crucial if the experiences were to be fully understood. Obviously this raises certain ethical questions (Bulmer, 1982) not relevant to this discussion, but for this reason both destinations and individuals have been allotted fictitious names.

Owing to the time and length restraints inherent in a discussion of this sort, the first destination, Ecologic College, will be used to provide the majority of substantive evidence, the other two destinations being used mainly for purposes of elaboration and substantiation. Ecologic College, established in 1989, has a staff of approximately five salaried workers and a number of resident volunteers who exchange services for board and lodging. It is situated in a rural part of the south of England and can be described as alternative, experiential and experimental. It describes its ethos through its marketing literature as aiming to cultivate spiritual and ecological awareness, attempting to re-establish the importance of relationships and community within modern society. It is one of the attributes of a larger experimental project developed in the 1920s by two philanthropists aiming to regenerate and instigate sustainable industries for a specific rural community. Thus its heritage is one of innovation and rural environmental experimentation, but also one that recognizes the necessity for economic activity; the College was established through a consideration of both these attributes. Throughout the year it operates a succession of residential two to four week study courses, usually hosted by a well known scholar whose writings are linked to environmental or cultural insights and issues. Participants during that period engage in group seminars with the visiting scholar and attend a number of leisure and self-developmental activities such as nature walks, social evenings and dance and craft workshops.

The two supplementary destinations offer a similar product to the above and are also situated in the rural south of England. The first, Recton Farm, was originally established in the early 1970s and is a communally organized rented farm whose ten occupants have endeavoured to develop a self sufficient and environmentally sensitive lifestyle and production approach, their ultimate goal being stated as one of 'harmony with their environment'. In order to fulfil this ambition they have recently begun to host a variety of short weekly courses based on propagating sustainable farming, gardening and holistic lifestyle approaches (such as permaculture). The second, Alston Manor, perhaps the most commercial of the three operations, consists of a large jointly owned and managed country manor house and grounds. It is occupied by approximately six individuals who have invested in this venture in the hope of being able to exist within what is perceived as a more spiritually satisfying and meaningful value framework. The need to offer a commercial product in order to sustain this project underpinned the experiment from the outset. Thus, during its six years of existence, the operation has hosted both leisure and study courses internally and offered franchises externally to individual practitioners wishing to hire facilities and hostelry services sympathetic to their interests. All such courses are of a similar genre, that of focusing on self development, spiritual awareness and the raising of consciousness around personal issues, examples being meditation, holistic medicine and personal, psychological 'healing' methods. Both these two destinations also rely on contributions from volunteers, who are often ex-participants.

Stratification of Sustainability

As already suggested, the concept of sustainability is stratified, actions and attitudes varying from reformist to structuralist. In more detail, reformists can be categorized as reluctant to challenge the existing social, political and economic structures that underpin tourism development and behaviour. Their aim seems to be one of focusing on reducing physical environmental, cultural and social impacts but without necessarily questioning the underlying values that create them. In contrast, those advocating a structuralist approach are depicted as advocating not just the reforming of destructive, thriftless or irresponsible tourism behaviour, but as attempting to advance more challenging notions about the nature of tourism development and tourism itself. Thus their ideas could be described as fundamentalist, attempting to challenge the paradigms not only on which tourism has been based, but on which economic, social and political development itself has been based. Of course, just as with any scale, the majority of approaches are likely to fall somewhere in between these.

More generally, at its shallowest, sustainability can be described as a contemporary 'buzz word' which reflects concern, but not necessarily action to analyse and address the problem of human impacts on environments. At its most fundamental, it challenges the very foundations that underpin current paradigms of progress in the form of economic growth, and future expectation. A variety of approaches have been developed which reflect this awareness and most of these are concerned with the above aim of lessening the impact of people on natural environments and host populations. More specifically, the most commonly used terms are: 'eco-tourism' (Hanna, 1992) and 'green tourism' (Farrell and Runyan, 1991), whereby it is hoped nature and tourism can enter into a mutually beneficial partnership; 'soft' and 'real' tourism, which aims to provide authentic experience but in a sensitive environmentally aware framework (Weiler and Hall, 1992); 'active tourism', which supports pro-active pursuit but in relation to 'conservation ... and environmental awareness' (Heywood, 1990); and, finally, 'ethical tourism' (Frommer, 1988), which aims to incorporate an ethical concern for the likely social and cultural impacts of tourism activity. Superficially this paints an optimistic picture of the future of tourism and the environment. However as Valentine (1992) points out, such ideals can easily become confused: tourism that is merely 'nature based' being potentially damaging and not minimal in impact, the only ultimately truly sustainable model being perhaps one of *no* tourism. To sum up, to date arguments have tended to revolve around anthropocentric notions that tourism consumption is inevitable and that environment and host populations must somehow be reconciled to this, although some effort will be made to ensure the transition is not too painful or destructive. Similarly, the majority of definitions agree with the World Commission on Environment and Development Brundtland report (1987) that sustainability means attaining a balance between the immediate and the future. However, this definition also consists of an essentially anthropocentric discourse. It seems sustainability for many is defined either by preservation for future human generations or by attempts to find approaches that ensure that maximum human satisfaction can be obtained in the majority of situations. This neglects the concepts developed by Deep Ecologists (Devall and Sessions, 1985; Naess, 1989), influential in the Environment Movement, that all aspects of our planet are integral, interconnected and interdependent. The foundation on which assumptions about tourism development is largely based is still one of stewardship rather than integral partnership. Thus these approaches can be labelled reformist in nature, chiefly concerned with reforming behaviour, rather than focusing on being open to the possibility of structural, philosophical and cosmological change.

The logic of this argument is also reflected in the often fragmented way that sustainability is addressed. It could be argued that this results from interest in the area being relatively recent, or the recognition that lines have to be drawn somewhere on the basis of pragmatically approaching the problem. However, while these are plausible, there is still the issue that sustainability with regard to tourism is often discussed in relation to artificially isolated contexts which limit deep explorations of the issue as their boundaries constrain examinations of related, but perhaps essential factors. These areas tend to be manifested in the assessment of individual areas of countryside or particular rural communities, for example, The Lake District Sharpley, 1993), or in terms of reviewing business practice. This suggests areas of concern within sustainable tourism are polarized and are viewed in terms of landscape and nature resources (Sharpley, 1993) or in terms of human communities (Murphy, 1992) or in terms of resource management (Pigram, 1992). The integrity of the relationships between them, so central to the philosophy of the Environment Movement, the idea that 'the sum of the whole is more than the sum of the parts', although being touched upon is relatively neglected. Human and natural landscapes, undoubtedly important tourism resources, in particular are treated differently, which can only contribute in a limited way to the essential debate about making tourism sustainable.

Discussing sustainability in a broader context, Porter highlights an increasing reliance on mathematical and computer models for predicting future trends. He warns:

> There appear to be patterns in global uncertainties, but these defy capture on conventional statistical or mathematical grounds. Yet scrutiny of the practices of the global modellers, whereby boundaries are placed on systems simply in order to make conventional analysis possible, remain disturbingly absent from debate.
>
> (Porter, 1994, p. 60)

This indicates a blinkered approach and an inability to address the emergent disturbing and unpalatable trends and currents which seem to be implying a fundamental challenge to perceived stable social and economic systems. Although relating to mathematical models concerning global environmental issues, this would seem to mirror what can be seen as the inability of many sustainability approaches to engage effectively with what achieving sustainability implies. There is a sense of delusion here, a hiding behind rhetoric and the tweaking of models, in order that more acceptable and less threatening results can be obtained and people can be *seen* to be striking at the heart of crux issues. This is exemplified by the Davis (1991) definition of sustainable development, which is paraphrased by Vinten as:

> (a) renewable resources are used in preference to non-renewables; (b) technologies are environmentally harmonious, ecologically stable and skill enhancing; (c) complete systems are designed to minimise waste; (d) long-life products, easily reparable and recycled, reduce the consumption of scarce resources; and (e) there is maximisation of all the services that are energy- or material-intensive, but which contribute to the quality of life.
>
> (Vinten, 1994, p. 2)

This model can be described as reductionist in approach, aiming to 'clean up the act' of business and make methodical adjustments which fail to address the issue of production and consumption structurally. A comparative example within tourism can be seen in De Kadt's summary of documented definitions of alternative tourism;

> (it) is thought to consist of smaller developments, or attractions for tourists which are set in and organized by villages or communities. These are seen as having fewer negative effects, social or cultural, and a better chance of being acceptable to the local people than mass tourism.
>
> (De Kadt, 1992, p. 50)

The underlying principle is again one of symptom reduction. It is accepted that negative impacts will occur and that local people will be adversely affected. It is hoped that, by making developments smaller, a more acceptable, sustainable form of tourism will be achieved, the nature of the production, consumption and content of the tourism product not being questioned. This is not to say that this approach is not a step in the right direction and closer to sustainability than a variety of other models in operation. The argument is more whether such approaches earn the titles of 'alternative' and/or 'sustainable'. The above definitions, then, are representative of the reformist approach, one that does not seek or fails to address issues on a fundamental critical level. More deserving of the alternative status is the structural approach adopted by the three destinations on which this research is based. Although structural approaches are inevitably also susceptible to reductionist tendencies, it is argued that, while embryonic, they *have* taken a crucial first step in recognizing that symptom solving, as opposed to curing the ailment, and thrift are unlikely to offer successful long term solutions.

The connection between the culture of consumption and the tourism and leisure industries is paramount (Featherstone, 1990), the experience encountered by the tourist being the commodity to be consumed. In order for the tourism product to exist, environments and host communities are to a varying extent being consumed and exploited, which problematizes De

Kadt's (1992) suggestion that a 'Tourism which does not damage the culture of the host community' (p. 51) is a possibility. Of course the process of host and guest interaction is a symbiotic process. However, the benefit to host communities is often variable and is, to a large extent, usually only economically based. While it would be hard to avoid such factors while the custom of temporarily visiting destinations away from home (Smith, 1989) is practised, it could be argued that the content of such encounters could be reviewed at a deeper, more structural level. Finally here, it is worth briefly examining Meadows and Randers' (1992) response to pollution models. Building on their earlier landmark work, *The Limits to Growth* (Meadows *et al.*, 1972), they suggest there are essentially three responses to pollution signals. They can be simply ignored, they can be addressed in the short (and hopefully long) term by technological 'fixes', or, the way they advocate, those involved can:

> step back and acknowledge that the human socioeconomic system as currently structured is unmanageable, has overshot its limits, and is headed for collapse, and, therefore, *(there is need) to change the structure of the system.*
>
> (Meadows and Randers, 1992, p. 191)

This appears pertinently to frame the above discussion and could be used to summarize the responses of the various tour operators. Tour operators who do not appear to have engaged in any rhetorical discourse regarding sustainability could accord with the first response to pollution or human impact signals. Those expressing interest and concern in adopting sustainable development paradigms could be seen to be adopting the second response. The third response seems to be that which is adopted by the three destinations labelled as structural in approach: the belief that the structure of the system itself has served its purpose and fundamental change is necessary. The three responses could be viewed as a process, the individuals responsible for the three structural outlets being perhaps a result of the move towards a postmaterialist (Inglehart, 1977, 1981), postmodern society, a further development along the time continuum occupied by the alternative commune movements in the 1960s. This claim is justified by collated data pertaining to the staff's and participants' social status at the three researched destinations. Again, a significant majority could be described as middle class in origin and were often highly educated, some having achieved considerable status in their profession. Evidence suggests that a common rationale for attending was the quest for alternative values and a more complex understanding of their existential position and human potential.

Characteristics of a Structural Sustainability Approach

The type of tourism experience offered by the three research destinations could be depicted merely as a 'lifestyle signifier' (Shaw and Williams, 1994) for individuals seeking a compatible product and thus only focusing on sustainability on the basis of its current *vogue* status. This would mean the approach would be reflexive, subject to external influences and responsive to demand and therefore perhaps fluctuating. However, this can be coupled with Krippendorf's (1987) description of the 'critical consumer' or Gordon's (1991) 'inner-directed' lifestyle groups: evolved individuals increasingly coming to demand a more ethical, ecological and socially responsible product. Although the research reported on in this chapter supports such notions, and it should be noted such destinations have arisen in conjunction with other environment sentiments, this is only half the issue. The operators concerned can also be characterized as actively designing experiences in order to facilitate and impart an ecological and ethical ideology resultant from their combined vision and value system. Some participants engage in experiences for purposes of reinforcing existing value judgements, but a variety of the activities offered at the destinations are designed to challenge individuals and present alternatives that might not be easily absorbed or internalized. This is reflected in the advertising literature of Ecologic College, which states: 'the College is offering neither reassuring certainties nor particular belief systems' and 'the issues raised can lead to the questioning of personal values and assumptions' (College Prospectus 1993/94: not paginated), which implies its aim is, through intellectual and personal challenge, to enable participants to deconstruct existing assumptions in order to construct alternative ones. If it were not for the fact that each destination offers what can be described as a 'ready made' ideology for individuals to adopt, this might seem like a contradiction to the above point. However, an alternative value base and ideology can be clearly seen, as within this context the College (again in its Prospectus, 1993/94) explains how it aims to take individuals 'below the surface of the current predicament to scrutinize the assumptions that have given rise to it' and asks: 'what can people do in the face of such political complacency and industrial recklessness?'. Labelling itself a 'centre for ecological and spiritual studies', this implies an explicit interest in criticizing and evaluating the current social, political and economic system, but with the intention of implanting a more environmentally and spiritually harmonious and satisfying one in its place.

In practice, the imparting of a particular ideology manifested itself in a number of situations and activities which in aggregate can be described as attempts to foster an *eco-consciousness* among participants. A guided walk whereby participants were informed about and encouraged to harvest edible, uncultivated vegetation for incorporation into the forthcoming meal inferred a return to 'natural' sustainable food sources, a reminder of the perceived vital, but diminished, partnership between people and their environment. A 'Deep Ecology' workshop, in which participants were taken to nearby undeveloped 'beauty spots' to focus on both the breadth and the minutiae of the habitat and to celebrate the life processes within it, inferred both a spiritual and an intellectual celebration of human involvement in organic processes. Lastly, a dance workshop in which individuals were encouraged to participate bare-footed in a number of unstructured dance modes was apparently to relinquish inhibition and recover or establish a sense of 'connectedness' and 'natural rhythm', echoing the ritual of more primitive cultures (Brown, 1984). In addition to the content of visits, attempts were made to ensure they were local and inexpensive, the majority of which took place on the host's premises. It seems that all three destinations, contrary to the quest for sensational and exotic experiences, apparently often sought by modern tourists (Cohen, 1974; Gorman, 1979), were concerned to provide fulfilling and satisfying experiences through providing a new awareness of 'everyday' aspects, perhaps taken for granted by participants. In this way, attempts were made to transform the tourist's approach to tourism itself, indicating an attempt to alter the structure of tourism activity. Such activities were also generally met with enthusiastic responses, such as individuals stating intentions to begin organic and 'nature' gardens on their return, indicative that the participants were beginning to internalize the sustainable message. It seems the proviso of meaningful, as well as socially responsible, experience is a goal of each of these destinations and this too is reflected in Ecologic College's Prospectus: 'An Ecologic College course, we know ... is one that participants are unlikely to "recover" from!' (1993/94: not paginated). Thus a central aim of each operation is to provide a sustainable product not only in terms of encouraging eco-consciousness, but in terms of actually sustaining the individual after the experience. It seems participants are encouraged to cherish their experience and to aim to apply newly acquired principles on their return home, sustainability being not so much in terms of ensuring the survival of the product, but of sustaining the spiritual and intellectual life of the individual and changing typical patterns of behaviour. This implies the seeking of the 'sacred' (Graburn, 1989) through tourism which can be compared to a mundane, normative existence, which validates the claim that such destinations are in operation firstly to impart a particular structurally challenging ideology and value system, tourism being the vehicle through which this is achieved.

Cohen (1988) suggests authenticity is without an objective quality and is subject to negotiation between host and guest. The research destinations appear to attempt to negate the quest for authenticity (MacCannell, 1976) through a philosophy of encouraging individuals to become physically and emotionally involved in the processes they are witnessing. At each destination there was emphasis on pro-active personal involvement ensuring each experience was encountered in a complex and holistic way by each individual. Examples of this included: a variety of craft workshops whereby individuals with the aid of tuition could create objects to keep as mementoes; the designing of sustainable garden and dwelling plans, to be potentially put into practice when applicable; and individuals being invited to participate in a 'soirée', a social occasion whereby the participants would contribute directly to the entertainment in the form of a song, poem or sketch. Notably, when this idea was introduced to the participants it was accompanied by a staff speech on the holistic benefits of participating in active rather than passive entertainment. A structuralist approach to achieving sustainability in tourism, then, would seem to view the individual as integral to the tourism process, rather than as an external force whose behaviour and demands are fixed. Structuralist sustainable tourism, similar to other forms of sustainable tourism, can be characterized as aiming to provide quality, rather than quantity, products. However a crucial further step is taken in that it aims also, through incorporating the individual more fervently into experiential processes, to produce tourists who are harbingers of sustainability, rather than merely consumers of a sustainable product.

The attempt to draw visitors into the sustainability argument and thus to encourage them to question the principles of consumption and production on which modern economies are largely based, is again illustrated through the localized nature of workshops and visits. This is enforced only partly through limiting the extent of the product and controlling the participants visiting potential. Significantly it is also achieved through encouraging participants to explore their inner, as opposed to external, realities. Visits were often utilized as vehicles to build on intimate and emotional personal experience and stimulus, rather than seeking satisfaction from experiencing, through witnessing, the external environment (Urry, 1990). Social interaction was, to a great extent, confined to within the tourist group and its staff. In this way, consumption was limited and the idea of self containment was introduced. Social, cultural and environmental needs were met from within the individual and localized group, again the example of the soirée illustrating that the participants could create the product as well as experience it, lessening the need to commoditize other cultural groups.

Each of the destinations had attempted to embrace the sustainability message in terms of the operation's physical consumption. Both Recton Farm and Ecologic College operated a policy of vegetarianism and had a stated policy of acquiring organic produce whenever possible. At Alston Manor, all staff members were vegetarian and only a vegetarian diet was offered unless visitors specifically requested otherwise. Each also had a stated policy of buying local produce to support local community outlets. Analogously each, as a matter of course, used only 'eco-friendly' cleaning products and recycled appropriate waste. Notably, a 'hands on' approach had

been adopted at each destination too, whereby staff members were expected
to contribute where needed and participants were allotted chores as part of
the experience, such as washing dishes, cooking and gardening. Although the
participants were paying customers, there seemed to be little complaint, the
policy announcement having been coupled on each occasion with a justification broadly explaining how this formed part of the process of developing
group intimacy and allowing each project to be sustained. Although there is
undoubtedly a complex rationale for such a policy, aspects of which are not
relevant to this discussion, this highlights what can be described as an attempt
to circumvent recent mass tourism trends (especially in foreign travel)
whereby the tourist product may be based on tourists purchasing relative
luxury and indolence for the duration of their stay, such as in the case of some
timeshare organizations (Haylock, 1994). It implies a return to the notion of
symbiotic relationships, host and guest being in partnership on the basis of
mutual advantage and understanding. An important strand in both the
sustainability argument and the values on which alternative tourism is based,
namely, 'positive and worthwhile interaction and shared experiences'
(Eadington and Smith, 1992), this policy suggests this concept is being
applied structurally. For these tourists, the notion of a hierarchical relationship between the visitor and visited is challenged; participants, although
embroiled in economic exchange, are requested to commit other resources to
the experience, physical work in a commitment to aid the provision of
activities. This challenges the assumptions that have underpinned the process
of tourism and infers a new set of relationships within the concept of tourism
activity.

Finally, at each destination a number of 'tips' and skills were imparted
which could be described as sympathetic to the 'self-help' or 'self-sufficiency'
synonymous with the earlier commune movements. It is not within the
context of this paper to analyse this heritage, but it is pertinent to suggest that
such an approach infers a movement away from prevailing trends whereby
transactions are usually solely economically based. At each destination, a
number of peripheral services were offered, some of which were economically orientated, such as Shiatsu massage and hypnotherapy sessions by
qualified therapists. However, the majority of services were integral to the
experience, although not actively structured as part of the product or, for
example, cited in the advertising literature. Examples of these ranged from
impromptu lessons in bread, beer and cider making, to 'hands on' gardening
instruction and meditation instruction. Again, this did not form part of the
formal agenda and the experience participants may have felt they had
purchased; they were informally structured and impromptu and seemed to
derive from the staff and volunteers who wished to impart what they
perceived as beneficial skills and thus, values, to individuals. This again
illustrates how a particular value base and ideology is imparted by such
destinations, the development of an eco-consciousness seeming to be focused

on developing self-sufficiency and encouraging ideas of sustainability that will be retained by the individual. Although beyond the remit of the research, a metamorphosis of values and behaviour is one potential outcome.

Conclusions

This chapter sought briefly to examine sustainable approaches to tourism and to evaluate how these are interpreted. It suggested there is a continuum along which these can be measured, the most fundamental approaches being 'structuralist', the most conformist (that is, in terms of existing development patterns) being 'reformist'. As a tool for investigating this hypothesis, recent ethnographic fieldwork was evaluated. It indicated that the three tourism destinations whose cultural ethos is unorthodox, was structuralist in approach. Although loosely defined as alternative in common with a variety of other forms of tourism, it was shown that these particular types of destination warranted a further, or 'alternative' alternative classification, owing to their offering a product more in keeping with the sustainable approach originated by the Environment Movement and more closely linked to the alternative cultures developed considerably in the 1960s.

The destinations on which this chapter is based were described as experimental in form and, again, possibly resultant from the 1960s commune movement. It was stated that through this their primary goals were focused more on enabling the experiment to continue and succeed, than obtaining profit through tourism. Tourism was seen as a necessary vehicle to enable this to happen. The movement of such organizations into the tourism market can be seen as a continuation of these original alternative currents, although, in a different temporal space, or in different ideological time frames, they can be seen as the survival or burgeoning of earlier lifestyle experiments. The structuralist form of alternative tourism discussed can be depicted as empirical data relating to an alternative theory approach. Although this is unlikely to transform the tourism industry, it nevertheless provides a critique of the values that have underpinned the wider industry' development and behaviour, and the assumptions on which these are based.

Destinations such as those described can be criticized for being on the periphery of the tourism industry and for being merely responsive to a wider developing environmental awareness. While there are undoubtedly grounds for such claims, it can be argued that a truly sustainable product has been created. The destinations themselves attempt to be sustainable in terms of consumption patterns and social responsibility. The way the product is structured encourages participants to contribute and help create and maintain the product; the experience which is provided is designed to sustain the individual beyond the realms of the experience itself and the principles of sustainability are imparted, both wittingly and unwittingly, at every opportunity. Within this, standard exchange and consumption patterns are challenged structurally too. Again, economic exchange, although part of the equation, is not the only foundation on which this model is built. Participants are expected to contribute in a variety of ways to make the experience worthwhile. This also occurs within other forms of tourism activity, but it has emphasis here and is incorporated into the experience with purpose. As a much more fundamental approach that intends to address the consuming nature of tourism practise, the aim is to make experience meaningful, enduring and authentic. Through suggesting an alternative to the prevalent 'been there, done that' (Gorman, 1979) style of tourism, this structuralist approach is concerned to emphasize the process, as opposed to the product. It attempts to apply sustainability structurally and provides a tangible, working model of sustainability in application.

A central strand of this approach is in the imparting of an ecologically aware sustainability ideology. Eco-consciousness is fostered, not only through the structure, but through the discussions and attitudes of staff members and volunteers. The cultural ethos itself appears to be self sustaining, participants frequently returning as staff members. Again, because the hosts of such destinations have a personal interest and empathy with the values they are imparting, the ideology is conveyed through a number of channels. This is tourism with an ideological and cosmological message.

Finally, similar to other alternative value approaches, it seems unlikely that existing structuralist sustainable tourism products will become mainstream in themselves. However, this is not to say that such ideas will not inform the mainstream and may be appropriated to an extent. Again, this approach succeeds in questioning the fundamental assumptions that underpin modern development models. History has proven the difficulty people have in confronting such challenges openly. This approach demonstrates that tourism itself can be used as a tool for education about the need to address such changes and in this way its potential is great; tourists themselves can be encouraged to further contribute constructively to the debate.

References

Brown, C.H. (1984) Tourism and ethnic competition in a ritual form: the fire walkers of Fiji. *Oceania* 54, 223–244.

Bulmer, M. (ed.) (1982) *Social Research Ethics*. Macmillan Press, London.

Cohen, E. (1974) Who is a tourist? A conceptual clarification. *Sociological Review* 22, 527–553.

Cohen, E. (1988) Authenticity and commoditization in tourism. *Annals of Tourism Research* 15, 371–386.

Davis, J. (1991) *Greening Business: Managing for Sustainable Development.* Basil Blackwell, Oxford.

De Kadt, E. (1992) Making the alternative sustainable: lessons from development for tourism. In: Smith, V.L. and Eadington, W.R. (eds) *Tourism Alternatives: Potentials and Problems in the Development of Tourism.* Wiley and Sons, Chichester, pp. 47–75.

Devall, B. and Sessions, G. (1985) *Deep Ecology: Living as if Nature Mattered.* Peregrine Smith Books, Salt Lake City.

Dobson, A. (1990) *Green Political Thought.* Unwin Hyman, London.

Eadington, W.R. and Smith, V.L. (eds) (1992) *Tourism Alternatives*: *Potentials and Problems in the Development of Tourism.* John Wiley and Sons, Chichester.

Eckersley, R. (1989) Green politics and the new class. *Political Studies* 37, 205 –223.

Farrell, B. and Runyan, D. (1991) Ecology and tourism. *Annals of Tourism Research* 18, 26–40.

Featherstone M. (1990) Perspectives on consumer culture. *Sociology* 24, 5–22.

Frommer, A. (1988) *The New World of Travel.* Prentice Hall Press, New York.

Gordon C. (1991) Sustainable leisure. *Ecos* 12, 7–13.

Gorman, B. (1979) Seven days, five countries. *Urban Life* 4, 469–491.

Graburn, N. (1989) Tourism: the sacred journey. In: Smith, V.L. (ed.) *Hosts and Guests: The Anthropology of Tourism*, 2nd edn. University of Pennsylvania Press, Philadelphia.

Hammersley, M. and Atkinson, P. (1983) *Ethnography Principles in Practice.* Tavistock Press, London.

Hanna, N. (1992) Eco-tourism. Unpublished draft paper, *Tourism Concern.* Roehampton Institute, London.

Haylock, R. (1994) Timeshare – the new force in tourism. In: Seaton, A.V. *et al.* (eds) *Tourism the State of the Art.* John Wiley and Sons, Chichester, pp. 230–237.

Heywood, P. (1990) Truth and beauty in landscape – trends in landscape and leisure. *Landscape Australia* 12, 43–47.

Inglehart, R. (1977) *The Silent Revolution: Changing Values and Political Styles Among Western Publics.* Princeton University Press, Princeton, USA.

Inglehart, R. (1981) Post-materialism in an environment of security. *American Political Science Review* 75, 880–900.

Krippendorf, J. (1987) *The Holiday Makers: Understanding the Impact of Leisure and Travel.* Heinemann Professional Publishing, Oxford.

MacCannell, D. (1976) *The Tourist: A New Theory of the Leisure Class.* 2nd edn. Macmillan, London.

Meadows, D.H. and Randers, J. (1992) *Beyond the Limits: Global Collapse or a Sustainable Future.* Earthscan, London.

Meadows, D.H., Randers, J. and Behrens, W. (1972) *The Limits to Growth: A Report on the Club of Rome's Project on the Predicament of Mankind.* Universe Books, New York.

Murphy, P.E. (1992) Data gathering of community-oriented tourism planning: a case study of Vancouver Island, British Columbia. *Leisure Studies* 11, 65–79.

Naess, A. (1989) *Ecology, Community and Lifestyle: Outline of an Ecosophy.* Trans. D. Rothenberg, Cambridge University Press, Cambridge.

Pigram, J.J. (1992) Alternative tourism: tourism and sustainable resource management. In: Eadington, W.R. and Smith, V.L. (eds) *Tourism Alternatives: Potentials and Problems in the Development of Tourism.* John Wiley and Sons, Chichester.

Porter, D.J. (1994) The limits and beyond: global collapse or a sustainable future. *Journal of Sustainable Development* 2, 53–57.

Rigby, A. (1974) *Communes in Britain.* Routledge and Kegan Paul, London.

Sharpley, R.A. (1993) Sustainable tourism in the English countryside – should the user pay? *Journal of Sustainable Development* 1(3), 49–63.

Shaw, G. and Williams, A.M. (1994) *Critical Issues in Tourism: A Geographical Perspective.* Basil Blackwell, London.

Smith, V.L. (ed.) (1989) *Hosts and Guests: The Anthropology of Tourism*, 2nd edn. University of Pennsylvania Press, Philadelphia.

Spradley, J.P. (1980) *Participant Observation.* Holt, Rinehart and Winston, New York.

Urry, J. (1990) *The Tourist Gaze: Leisure and Tourism in Contemporary Societies.* Sage Publications, London.

Valentine, P.S. (1992) Nature based tourism. In: Weiler, B. and Hall, C.M. (eds) *Special Interest Tourism.* Belhaven Press, London, pp. 105–127.

Vinten, G. (1994) The sustainable company: the need for environmental concern. *Journal of Sustainable Development* 2, Summer, 1–8.

Weiler, B. and Hall, C.M. (1992) *Special Interest Tourism.* Belhaven Press, London.

World Commission on Environment and Development (1987) *Our Common Future.* Oxford University Press, Oxford.

Yearley, S. (1991) *The Green Case.* Harper Collins, London.

The Tourism Industry's Response to Sustainability Principles

The common thread running through these six chapters, which tend to concentrate on the objectives of the tourism industry, is that, while the attainment of sustainability is an important goal, there are both business and economic prerequisites which must be met. The chapters are quite strongly industry-centric but they do progressively indicate that cooperation with non-commercial bodies is often desirable in order to maintain the viability of businesses and almost certainly essential to sustain the tourism resource base.

Eaton in the first contribution in this part (Chapter 7) considers a business strategy on sustainable tourism from a largely tour operator and resort provider perspective. He examines the practical applications of sustainable tourism by showing opportunities for suppliers to increase turnover and profits by reducing costs. He accepts sustainable tourism objectives as given and discusses how the industry can capitalize on the ecotourism market by meeting consumer demands. The discussion includes reference to product development; its quality and differentiation; the strategies of suppliers, particularly what is termed 'generic' differentiation and cost leadership as ways of promoting eco-green tourism.

In Chapter 8, Tregear, McLeay and Moxey investigate the relationship between sustainability and tourism marketing to attempt to ascertain whether the two are competitive or complementary. While acknowledging that neither tourism nor marketing has a very secure academic pedigree with respect to sustainability, as both are strongly market driven, it is possible to suggest compatibility if market stakeholders' needs and wants are identified and matched with those of providers. Using a case study in South Wales, in

which a development initiative aimed at encouraging endogenous but
sustainable growth emphasizing local community empowerment is considered, the authors argue that tourism development can be achieved in a
sensitive way. This, it is posited, is possible through appropriate marketing
techniques such as segmentation, branding and marketing mix.

It is clear that in the initiative sustainability is essentially concerned with
securing the long-term growth and survival of particular forms of tourism.
Sustainability, in the sense of safeguarding environments, whether physical,
ecological or socio-cultural, is perceived as being attained by targeting would
be tourists who would be sensitive to community needs and behave
accordingly. It is claimed that marketing also contributes to sustainability by
identifying the changing needs and wants of stakeholders over time.

A pragmatic view of regulation to achieve sustainable tourism is taken by
Middleton in Chapter 9. In a quite provocative way he suggests that not all
tourism pressure is due to long-stay tourists but is caused by recreation by
residents. Not only does he press for both a top-down and bottom-up
approach but he advocates local partnership strategies for visitor management at specific destinations. It is suggested that the problem is essentially
very local. Nevertheless, it is acknowledged that the tourism industry needs
to recognize that it has a joint responsibility for environmental protection.
A sceptical view is taken of the effectiveness of regulators, it being asserted
that the public sector must appreciate and take account of market forces.

A particular aspect of tourism development is examined by Harper in
Chapter 10. The promotion and marketing of farm tourism in a destination
are taken to illustrate the importance of community leaders' involvement in
attaining sustainability and the consequences for public organizations and
the visitor. An initiative to coordinate the operation of eight farm holiday
groups was undertaken by means of a project manager whose function was
to facilitate action within the groups through local influencers or opinion
leaders, rather than from outside via public and tourism agencies. The value
of working from within local community organizations is that the limitations
of such an approach as well as the advantages can be identified at an early
stage. A crucial factor determining success is the retention of 'local
ownership' of the 'product' in order to maintain commitment to the initiative
and its quality. Another important consideration is mediation between public
sector organizations and the community because of the tendency of the
process to be slower and more expensive at the outset than the former would
countenance if they were the initiators.

While it might seem strange to invoke Wordsworth as an early
conservationist, in Chapter 11 McCormick uses his experiences as both a
curator for the Wordsworth Trust and Chairman of the South Lakeland
Tourism Partnership to open up the much more fundamental issue, with
wide-ranging implications, of the cultural, political and social forces which
drive both promotion of a destination and the sustainability movement, and

for whose benefit. A telling point is made about the cultural ethos of the British and how this might impinge on the evolution of tourism, not only in the Lake District but elsewhere. This is the deep inhibition to action and change of the 'Golden Age' sentiments, which is partially social class based because the perpetuation of this myth reinforces the stand against the threats which tourism poses to landowners and land managers.

Against this background, McCormick poses the question as to the feasibility and effectiveness of organizations which combine public sector guardianship with private (business) sector creativity in this conflict between tourism development and the natural (unchanging) order. He considers that there is a danger that organizations ossify and their members devote all their energies to avoid change. He suggests that a systems approach with community involvement is required to avoid a narrow class-ridden sectarian possession of sustainability. He concludes that the task is a difficult one but the spectre of what would occur if the problem is not tackled is the driving force to attempt to resolve the issue.

The twelfth chapter, by the Edingtons, serves two purposes. First, as a cameo review of two projects it reveals some criteria which are relevant to attaining sustainable development while benefiting the local community. Second, it acts as an illustration, from a practitioner's stance, of the issues which any such initiatives engender, albeit virtually unstated explicitly. The almost deliberate 'under-emphasis' by the contributors in some ways makes more impact than perhaps some of the more strident utterances on such tourism developments. For example, with regard to the development of schemes, their small scale, the necessity of involving the local community and the need to conserve the resource base seem obvious requirements, but the predilection for such ecotourism developments to grow out of initiatives by non-commercially orientated enterprises and the need to rely on subsidies for their continued existence do not augur well for their long-term survival as profit-making businesses. Another interesting aspect of the cases is that the provision of facilities and services, seen as crucial for the comfort of visitors and longer-term viability of enterprises, without regard for the impact of eco-tourism on local resources and the demonstration effect on residents, is taken for granted. Such cases, therefore, underline the misgivings voiced elsewhere in this book.

Sustainable Tourism: Industry Responses and Industry Opportunities

7

B. Eaton

Coventry Business School, Priory Street, Coventry CV1 5FB, UK

The Green Tourist: Economic Characteristics

As far back as 1987, the green tourist was described as

> a well informed selective individual from a higher socio-economic group, taking a second or third holiday in a rural area and often reasonably informed, but, nevertheless potentially benefiting from better coordination of provision in the countryside.
>
> (Jones, 1987, pp. 354-356)

It is not the purpose of this chapter to discuss definitions of green or sustainable tourism so this description is here accepted throughout. Other writers, however, differentiate between a narrow idea of ecotourism and a wider view of sustainable tourism, such as Forsyth (1995). Implicit in this description is the idea that the green tourist, because of his or her socio-economic grouping, is a higher spender. This is implied by the reference to a second or third holiday. Indeed, it has been stated by Mintel Leisure Intelligence (1989) that national parks or National Trust land had been visited by 20% of its sample in the previous six months, an increase of one quarter. So, within the UK, a 'greener' aspect of tourism is quantified in that higher percentages of people are visiting national parks. But what sort of people? The Mintel report says that although the overall figure is 20%, the figure rises to 45% for classes AB, so confirming that national parks attract a more affluent and educated visitor. Here then, is a sound business reason as to why leisure operators should regard sustainable tourism as a commercial opportunity. The customers buying in this market segment are higher spenders, and are likely to repeat their purchases more than once a year. As such, there is the scope for multiple sales to the same individuals.

Ansoff (1968) describes a variety of growth possibilities for an organization. That which relates to current product range and existing customers is called market penetration (see Table 7.1). This involves existing customers buying the organization's leisure services more frequently and could, if one accepts the Jones (1987) idea of the green tourist, be a valid growth strategy. For instance, a hotel group could sell a 'short break' at a sister hotel in addition to a main holiday to a particular customer. It should be said, however, that this strategy is usually used as a device for maintaining market share rather than increasing it. In practice, therefore, leisure operators who desire growth are likely also to have to find new products, new customers or else both.

Product development occurs when an organization maintains its knowledge of and strengths in present markets while developing and introducing new products for those markets. For leisure, product development is likely to mean that organizations provide for new and changing needs in their existing customer base as existing leisure offerings fade in popularity and come to the end of their product life cycles. The Rank Organization, for instance, own the Butlins brand name but have changed what it offers considerably over the years, whilst maintaining its traditional customer base as a mass market operator. This may seem an obvious strategy, and one by which Rank could appeal to an increasingly sophisticated (and slightly greener?) tourist. However, those leisure organizations who have used product development successfully are largely those who have invested in organized ongoing customer monitoring schemes which are able to spot changing trends, fashions and needs. This is not as simple as it sounds, and many organizations have made costly mistakes as a result of assuming too much knowledge about customers.

An alternative to the development of new products is to maintain the existing product range and find new customers and markets through a strategy of market development. Of course, there is no reason why a broad based leisure organization should not carry out product development and

Table 7.1. Possible directions of growth for tourism and leisure businesses.

Customers	Product range	
	Existing	New
Existing	a. Withdrawal b. Consolidation c. Market penetration	Product development
New	Market development	a. Related diversification b. Unrelated diversification

Source: Adapted from Ansoff (1968) in Eaton (1996).

market development simultaneously. A further important form of market development is that of taking a niche market and turning it into a far broader based market. Here is an apparent opportunity for green tourism. If green tourism is indeed a niche market, then market development will create a mainstream market, which is potentially more profitable, simply because of its size.

Diversification will always be available to green tourism operators, as will be the opportunity of moving into green tourism for mainstream operators, or even those businesses with no current connection with tourism. The issue of diversification is, however, broad enough to merit separate discussion elsewhere.

Existing customers of green tourism operators are likely to regard the quality of the natural environment that they visit as an indicator of the quality of the holiday that they are on. This will apply whether that natural environment constitutes, as in the United Kingdom for instance, the Lake District National Park or an established overseas destination, such as the Languedoc in France, or a 'new' one, for instance The Gambia. Conversely, the tourists' activity, including their economic activity, will also be influenced by the local quality of the natural environment. It is reasonable to expect higher spending in environments that are of higher quality, perhaps in terms of average daily spending or indeed in terms of a longer length of average stay. This will also have an impact on the amount of spending, which is multiplied by indirect and induced spending through the local economy. Meanwhile, a business organization which has provided, say, the hotel accommodation or holiday package can look forward to increased future expenditure by the tourist as a result of 'trading up' or more regular purchases. The amount of repeat business can, over the medium term, affect profits and pricing strategy because the costs of promotion to first time buyers are avoided.

Quality and Differentiation

In marketing terms, customers buy a series of product benefits. In the case of tourism, the quality of the visitor experience is a key benefit of the holiday. Quality is often closely associated with exclusivity and the idea of doing something a little different from friends or colleagues. These kinds of benefits are not usually associated with mass markets, but certainly are features of the 'green tourist' segment identified by Jones (1987). The relative sophistication of such informed and affluent consumers is likely to have to be met by organizations through the longer term development of relationship market- ing. This form of marketing is particularly appropriate to leisure and tourism services since it is as much to do with retaining customers (by building a positive relationship with them) as with attracting them initially. Over a period of time, effective relationship marketing will enable the development of several distinct tourist niche markets of varying shades of green. Recently, marketing methodology has become more sophisticated to the extent that psychographic, or lifestyle, segmentation can be used successfully in many cases, particularly in the form of relationship marketing, which is geared specifically to the retention of customer loyalty and spending (Christopher *et al.*, 1991).

Already there is a trend towards fragmentation of the industry in the UK. Two examples of this are: (i) the rising number of Association of Independent Tour Operators (AITO) members; and (ii) the introduction of an increasing number of sub-brands by the major operators. Thomson's, for instance, produced 27 separate brochures in 1995.

Generic Strategies

Quality is not the only way by which providers of tourism can compete with each other. Price competition has been keen, particularly in a UK economy in which there was comparatively little confidence in the early 1990s. Non-price competition itself can take various forms, but will almost always be based on differentiation of the offering by one operator from that offered by others. So what are the strategic options for firms? The firm will seek to achieve a position within the industry which is sustainable and credible within both the medium and the longer term. Various bases can be found within Porter's (1980) concept of generic strategies.

Differentiation

Porter (1980) argued that there are three ways by which organizations can compete with each other in the longer term. The first of these is differentiation strategy. This involves something which is unique and which is valued by buyers. This differentiation may be achieved, for example, by convenience within a booking system or a free airport transfer, both of which have been proved to be of value to customers. Other forms may be the differentiation of the holiday itself on the basis of accommodation and/or destination or by the use of overseas couriers.

In the context of this chapter, the point is that environmental features themselves can be built into a product as a form of an additional product benefit and therefore constitute a form of differentiation. So product differentiation can confer a competitive advantage which is sought by businesses offering sustainable tourism as part of their product. Differentiation may well be by way both of sustainability and quality. A case can be made for the need to preserve longer term quality which embodies sustainability as a prerequisite anyway. Sisman (1993) maintains that client loyalty increases, complaints decrease, and repeat trade increases as a result of sustainability.

An often quoted example of tourism development based on sensitivity to the environment which has produced significant economic rewards is that of Center Parcs. Originally founded in Holland in 1967, Center Parcs now consists of 16 sites in five countries. To illustrate the claim with regard to Center Parcs, three different accounts follow, the first by a market research agency, the second from a Government commissioned report and the third by Center Parcs themselves.

> Center Parcs was acquired by Scottish and Newcastle Breweries in 1989. In Sherwood Forest a holiday village was opened in 1987 and contains over seven hundred villas. Special features include a circular golf driving range and a country club. Its Elveden Forest village in Suffolk was opened in 1989. Special features include a covered parc plaza and the provision of studio apartments and executive villas. Center Parcs aims its villages at the ABC1's with adults in the 25 to 44 age range forming the bulk of the market. Smaller, but also important markets include young peer groups aged over 21, and retired people, perhaps as part of an extended family group. Center Parcs have been extremely successful in reporting very high occupancy levels, and surveys have indicated very high levels of customer satisfaction with (positive) response levels of over 90% for the sub-tropical swim paradise and villa comfort.
>
> (Mintel Leisure Intelligence, 1992, p. 13)

> Center Parcs at Sherwood Forest is a good example of a large but sensitively designed visitor attraction, 'hidden' in a forest environment. The attraction is a combination of quality accommodation and leisure facilities which are discretely woven into a natural environment of woodlands and water. The area has been improved by the creation of lakes and streams, the planting of some half a million native trees and bushes, and the seeding of native grasses and wild flowers. These, combined with wildlife management and the restoration of heathland, are all embraced in a management plan designed to increase ecological richness and diversity. Good design helps the environment successfully to absorb people, their cars and the visitor facilities with minimal impact on the wider landscape.
>
> (English Tourist Board/Employment Department Group, 1991, p. 56)

> We will develop our business activities by continuing to invest in those resources which are central to our success - our people and the environment. Center Parcs also pursues environmental policies which seek:
>
> ● to make a positive contribution to the global environment by efforts at a local level
> ● to accept responsibility for the environmental consequences of activities
> ● to conduct all activities in the spirit of being custodians of the villages environment
> ● to enable guests and employees to experience the process of environmental care at first hand
> ● to be acknowledged as setting the standard for the industry by demonstrating that sustainable tourism is achievable.
>
> (Center Parcs, 1994)

These lengthy quotes extol the business performance of the brand, based on quality of provision, as measured by the 90% positive response rates, 71% rebooking rates and highly profitable occupancy rates of 94%. They are also, without any apparent conflict, able to reconcile intensive use with environmental achievements. Traditionally, many writers have suggested that these two objectives are intrinsically opposed. Leslie (1986) claims that this is not necessarily the case and the evidence here seems to support this.

This example of Centre Parcs should lead to the conclusion that individual companies can enhance some or all of revenues, growth and profit over the longer term by using positive environmental policies to improve quality. More generally, the potential of destinations to embrace sustainable tourism in return for economic reward is noted by Stancliffe (1993), who states, 'with flair and imagination, tourism can work hand in hand with conservation and local development interests to the benefit of both' and 'any green tourism initiative must bring in more visitors or create more visitor spend' (both p. 15). This theme has been taken up at destination level by, for instance, the Lake District Tourism and Conservation Partnership (LDTCP) (see Harper – Chapter 10 in this book), which draws together smaller scale providers of accommodation and facilities with public sector bodies such as the English Tourist Board and Rural Development Commission and voluntary sector bodies such as the National Trust. Under the slogan 'Protecting the landscape makes business sense', the partnership claims that businesses can 'gain financially by safeguarding the environment', because it will 'enhance your image, profile and reputation among guests and the industry' (LDTCP, 1996). Businesses donate services, time or money in return for conservation orientated marketing advice.

The competitive environment

New entrants will always be attracted into industries where there seems to be the potential for reasonable profit-making and/or growth. How many new entrants actually occur will depend on the extent of so called barriers to entry. For instance, for many forms of tour operations, the large capital requirements which need to be found will act as an entry deterrent, as would the economies of scale open to larger-scale operators. However, a new entrant to an industry will usually have more chance of being successful if it is able to differentiate its product or services. This is more likely to be possible within leisure industries since leisure is not a standardized item anyway. For example, a new entrant to the 'natural holiday villages' market is that of Rank's 'Oasis' villages. The first two of these will be located at Whin Fell, near Penrith, in Cumbria and at Folkestone in Kent. They will be designed to attract socio-economic groups B and C1 customers to well fitted out accommodation in woodland settings. Rank acknowledges that this market has been pioneered by Center Parcs, but claim that their development can improve on the Center Parcs resorts. In other words, they can positively differentiate their product for a variety of reasons. Firstly, location: there is, simply, no Center Parc either in Cumbria or in Kent. Secondly, Rank regard themselves as specialists of providing UK holidays for UK residents. Since the Center Parcs concept was originally Dutch, Rank feel that they can tailor their own product rather better to the particular market that they are operating in. Thirdly, Rank has carried out extensive research on Center Parcs customers and have identified certain aspects about Center Parcs which are sometimes rated less than satisfactorily by its customers. One of these includes the physical appearance of the chalets themselves in a wooded setting which Rank thinks it can improve upon significantly. Also, Rank challenges Center Parcs policy of ensuring that capacity is utilized at all times, which has, in practice, meant that at certain times of year customers arriving at their centres have had to book almost their entire activities for their stay on arrival. Rank intends to put in extra leisure and sports facilities so as to allow holiday makers to be slightly more spontaneous in at least some of their choice of activities. So the company will seek to gain competitive advantage in this relatively affluent and environmentally sensitive market on the basis of differentiation with respect to the selected location of the villages, accommodation design and the quantity of leisure facilities on offer.

Costs and cost leadership

Porter's second generic strategy is that of cost leadership (but not necessarily price leadership). Essentially, a low cost producer will find and take advantage of any cost savings available. This may well imply an organization carrying a minimum number of staff, contracting out of some activities to specialists in the field and seeking out the lowest cost suppliers. But are there any cost savings associated with embracing sustainable tourism? Advocates of sustainable tourism will certainly suggest that there are savings in terms of reductions in environmental costs (such as pollution) and social costs (such as diverting resources into tourism development, particularly in less developed economies), which should feature in cost–benefit analyses. Here, though, the emphasis is on direct monetary savings, or economic costs. Woodward (1993) refers to Intercontinental hotels' launch of an 'Environmental operating manual' in 1991, requiring the responsible handling of issues such as 'waste management and reduction; recycling; product purchase; energy conservation; noise control; laundry and dry cleaning, etc.' Activities included a purchasing policy based on higher quality durable goods which would need to be replaced less often, buying for energy efficiency and purchasing goods with a recycled component. The purposes of this initiative included fulfilling a 'moral responsibility for environmental protection' but also, importantly, the achievement of 'energy and cost savings'. Other major hotel operators took these ideas of waste reduction on board, coming together to form the International Hotels Environment Initiative (IHEI) in 1994. So the promotion of these aspects of sustainable tourism ran parallel to direct economic cost advantages.

This may not always be the case, however. For instance, in 1992, Eurocamp adopted an explicit and positive environmental policy whereby all their sites have can crushers and recycling facilities. At the end of the season, the tents bought by Eurocamp are also recycled, in recent years to Romania and Albania. Customer monitoring of British customers after 12 months of operating this environmental policy suggested that customers were very neutral about it. The company found this quite surprising, given the relatively high socio-economic profile of Eurocamp customers. On the other hand, their German customers are rather more conscious about green issues. In Germany, the company's brochures are printed on recycled paper, not because it is a legal requirement, but rather because customers expect it. Likewise, tickets are no longer issued (in any country) in plastic wallets, but in fabric pouches. The company no longer buys mahogany furniture for its reception areas at sites. Overall, however, it is calculated that this environmental policy actually costs Eurocamp money rather than producing cost savings.

Brochure production would seem to give an opportunity for cost savings, however. The tourism industry spent an estimated £86m on brochures in Britain in 1995, contributing, on average, an added £20 to the price of each holiday (40% of brochures, some 24,000 tons of paper, were simply thrown away). But reducing such waste may be easier said than done. For example, Guerba expeditions have run into technical problems by using recycled paper, negating any possible cost saving (Hodson, 1996). The evidence from these examples relating to economic cost savings from pursuing sustainable tourism by tour operators is therefore mixed.

Hence it would seem that the pursuit of a cost leadership strategy by the use of sustainable tourism is unlikely on its own to gain a competitive advantage for an operator. In addition, a company such as the Ladbroke group, whilst a member of the IHEI, can claim, according to Eaton (1996), that their 'environmental policy is based on standards of corporate citizenship', which suggests a broader approach based on social criteria as well as cost savings.

Focus strategy

The third of Porter's generic strategies is that of a focus strategy, which involves the targeting of client groups within a particular market and concentrating on the provision of tailored leisure services for them (a strategy closely related to market segmentation). If green tourists themselves represent a market segment, then the discussion of differentiation and cost leadership earlier will by definition be in the context of a focus strategy rather than a mass market.

Recently, Green Horizons Travel entered the travel agency sector with a specific remit,

> to offer a range of holidays with operators who are taking action to increase the benefits for local communities in the destinations they visit and minimise the adverse effects of their operations on the environment. We believe we are the first retail travel agency to offer such a service.
>
> (Green Horizons Travel, 1995)

Although the company suggests that it did not consciously seek a commercial advantage in setting up this sort of business, there is clearly the potential for a differentiation focus strategy. This is enhanced by the fact that the company is currently the only such agent, which means it enjoys 'first mover advantages'. These arise from a certain degree of (at least temporary) monopoly supply and from the differentiation of service based on offering only 'approved' tour operators products, limiting distribution of brochures on the grounds of wishing to avoid unnecessary expense and giving a donation to an environmental charity for each holiday booked.

Conclusions

In the future then, the message for the private sector must be that the objectives should be: maximize income; minimize damage; promote good practice. This can be done by cost reductions and improvements in resource use and efficiencies as well as making better use of capacity than has traditionally been the case. The minimization of damage will pay regard, for instance, to the recyclability of materials, local sourcing, quality and durability of items. These opportunities may also be threats, since 'such a changed climate of thinking would be ignored at the industries peril' (Romeril, 1989, p. 205). Romeril also suggests that

> an aware industry can sustain tourism ... if tourism is sustained significant steps have then been taken towards maintaining environmental integrity. A healthy environmental integrity means the possibility of successful tourism which, when managed properly, becomes a resource in its own right.
>
> (Romeril, 1989, p. 207)

Tourism can certainly be regarded as a conserving industry and capable of effective use of the environment through such activities as reclaiming land and using redundant buildings. This chapter suggests that, alongside this, one of the very tools for achieving sustainable tourism is business strategy itself.

References

Ansoff, H. (1968) *Corporate Strategy.* McGraw Hill, New York.

Center Parcs (1994) Environmental Policy Statement. Nottingham.

Christopher, M., Payne, A. and Ballantyne, D. (1991) *Relationship Marketing: Bringing Quality, Customer Service and Marketing Together.* Butterworth/Heinemann, Oxford.

Eaton, B. (1996) *European Leisure Businesses: Strategies for the Future.* ELM Publications, Huntingdon.

English Tourist Board/Employment Department Group (1991) *Tourism and the Environment: Maintaining the Balance.* ETB, London.

Forsyth, T. (1995) Business attitudes to sustainable tourism. *Journal of Sustainable Tourism* 3(4), 210–231.

Green Horizons Travel (1995) Press release, December.

Hodson, M. (1996) The case against brochures. *Sunday Times,* 14th January, p. 5.11.

Jones, A. (1987) Green tourism. *Tourism Management* 8, 354–356.

Lake District Tourism and Conservation Partnership (1996) Membership leaflet.

Leslie, D. (1986) Tourism and conservation in National Parks. *Tourism Management* 7, 52–56.

Mintel Leisure Intelligence (1989) *The Day Visit Market,* Vol. 2. London.

Mintel Leisure Intelligence (1992) *British on Holiday at Home,* Vol. 1. London.

Porter, M. (1980) *Competitive Strategy.* Free Press, New York.

Romeril, M. (1989) Tourism and the environment – accord or discord? *Tourism Management* 10, 204–208.

Sisman, D. (1993) Sustainable tourism as a business concept. Paper presented at *Tools for Sustainable Tourism,* a conference hosted by the Royal Geographical Society, London, 6th October 1993.

Stancliffe, A. (1993) Green tourism: what's in it for you? *In Focus,* Autumn.

Woodward, D. (1993) Hotels and the environment. Paper presented at *Tools for Sustainable Tourism,* a conference hosted by the Royal Geographical Society, London, 6th October 1993.

Sustainability and Tourism Marketing: Competitive or Complementary?

8

A. Tregear, F. McLeay and A. Moxey

Department of Agricultural Economics and Food Marketing, University of Newcastle upon Tyne, Newcastle upon Tyne NE1 7RU, UK

Introduction

Tourism and marketing are both areas of human activity with long histories, yet relatively short academic pedigrees (Casson, 1974; Grether, 1976; Van Doren *et al.*, 1995). Both are also subject to popular misconceptions. Tourism is often viewed as an industry which promotes mass movements of people with little regard for fragile natural or socio-cultural environments: a misrepresentation which the label 'sustainable tourism' is presumably intended to address (McNulty, 1993; Middleton, 1993; Wheeller, 1995). Marketing is often caricatured as simply selling and advertising, or taking something simple and obvious and sticking it in fancy packaging (Harding and Walton, 1987). Just as tourism is not merely about mass travel, marketing is also a broader area of activity and enquiry which considers the process of matching organizational aspirations and capabilities with the needs of consumers and stakeholders and is concerned with long term relationships: in the words of Davies (1995, p. 59), 'Marketing might be likened to marriage, selling to seduction.' This chapter offers a brief review of marketing and its relevance to sustainable tourism issues and then considers marketing's role within the tourism element of rural development by reference to a case study in South Wales. Some concluding comments on the contribution of marketing to the pursuit of sustainability are offered.

Marketing and Sustainable Tourism

As a field of academic enquiry, marketing gained scholarly acceptance in the post-war period as researchers attempted to understand changes in market behaviour in Western economies (Bartels, 1962). Initially, attention was focused on the economics of markets and market institutions. Gradually, however, this evolved into the study of the process of exchange with particular emphasis upon consumer needs, organizational and managerial capabilities, and buyer–seller relationships (Sheth and Gardner, 1982). Throughout this evolution, ideas and techniques from the behavioural sciences, including psychology, sociology, anthropology, and the political and management sciences, have also been incorporated (Bartels, 1962; Horsky and Sen, 1980; Desphande and Webster, 1989). The outcome of this evolution is an unashamedly multi-disciplinary perspective which recognizes that the complex real world does not respect disciplinary boundaries. Marketing thus seeks to offer a 'boundary-spanning' approach to planning and management. This is of direct relevance to the planning and management of tourism since the multi-disciplinary nature of tourism (Brent Ritchie, 1993) requires an interdisciplinary approach to research (Przclawski, 1993).

The absence of a single dominant disciplinary paradigm means that it is difficult fully to establish conceptual boundaries for marketing (Krapfel, 1982; Ardnt, 1985), which inevitably gives rise to the misconception that marketers merely borrow from other, established disciplines. Such critisism ignores the holistic boundary spanning nature of marketing. It is possible, however, to identify a central marketing concept which holds that an organization can best achieve its objectives by determining the needs and wants of target markets and delivering desired satisfactions efficiently and effectively (Kohli and Jaworski, 1990; Kotler, 1994). One interpretation of this is that, to be successful, organizations need to be market driven in that they need to identify and respond to the needs and wants of 'market stakeholders'. Moreover, as markets are dynamic, organizations must be prepared to anticipate and respond to changes in needs and wants over a given period of time. This temporal dimension is relevant to the sustainability debate. Although definitive definitions of sustainable tourism are elusive (Wheeller, 1995), concern over sustainability implies some long-term view which must encompass not only consideration of the future resource base, but also the consideration of future demands upon resources.

The use of the phrase 'market stakeholders' in the previous paragraph is deliberate. Increasingly, marketers are recognizing that consumers are not the only people with an interest in the process of market exchange and its outcome. In particular, the concept of relationship marketing emphasizes long-term relationships between a series of stakeholders or markets, beyond the traditional basic buyer–seller dyad (Christopher et al., 1991). This emphasis seems particularly appropriate in sectors involving a high service element. This is because services necessitate interaction between providers and consumers. Relationship marketing is also appropriate where the outcome of the exchange process has implications for groups in society other than the providers and consumers (McCort, 1993). In the tourism sector, market stakeholders could include current tourists, future tourists, service providers, other members of the host community and environmental protection agencies. Taking this perspective, a marketer would maintain that sustainable success is most likely to be achieved if market stakeholders' needs and wants are identified and systematic efforts are made to match the capabilities, needs and wants of providers with those of other stakeholder groups.

Implementation of the marketing concept in tourism can be achieved through the use of a variety of marketing tools and techniques (Popadopolous, 1989; Calantone and Mazanec, 1991). Of these, three merit discussion here: market segmentation; branding; and the marketing mix.

Market segmentation can be described as the process of dividing a market into segments of consumers that have similar needs and wants. The segmentation concept suggests that a market is composed of numerous consumers with heterogeneous needs and wants, but that it is possible to use some classification criterion or criteria, such as social class or age or income, to group consumers into more homogeneous market segments. For example, Levine (1975) identifies five groups of consumers in the travel market: culture seekers, pleasure tourists, root seekers, fun lovers and bargain hunters. By segmenting a market, a tourism provider is able to adapt product and service offerings so that they more precisely match the providers' capabilities with the needs and wants of a target segment or segments (Calantone and Mazanec, 1991). Issues and trends in tourism market segmentation research are discussed by Weber (1992).

Branding is the process of creating 'an identity' for a market offering (i.e. good or service). This is employed to improve recognition of the market offering to customers, to differentiate the offering from competitors, and to give confidence to the customer when purchasing. In tourism, branding is often associated with travel companies or hotel chains (Holloway and Plant, 1992), but may also be applied to places or regions. In this case, branding is more appropriately termed 'identity creation' and is becoming increasingly popular amongst UK area tourist boards as they seek to present consumers with a succinctly identifiable offering such as, for example, 'Catherine Cookson Country' (Northumberland), 'Robin Hood Country' (Nottinghamshire) or 'Rob Roy Country' (Stirling and The Trossachs).

The idea behind the marketing mix is that the market offering itself is just one variable which needs to be considered jointly with several other controllable marketing variables (Kotler, 1994). These include, but are not restricted to, communication with consumers, pricing, availability, and mode of delivery. To increase the likelihood of success, all elements of the marketing mix need to be in harmony: there is little point in, for example, offering a high quality product if consumers are unaware of its existence. With respect to tourism, this suggests that attention needs to be paid not only to the provision of quality visitor services, but also to, for example, spatial and temporal availability and promotion of consumer awareness. The notion that marketers need to consider a broad mix of elements associated with a market offering, rather than focusing narrowly on the 'product', mirrors the holistic perspective of the marketing concept.

The marketing concept and various marketing techniques have been employed widely across many sectors, including agriculture, retailing and, increasingly, health and education. However, although marketing may be mentioned in many tourism studies, Calantone and Mazanec (1991) suggest that marketing techniques currently employed in the tourism industry are less advanced than in other sectors, whilst Greenly and Matcham (1990) cite survey data to argue that UK tourist businesses have a relatively low marketing orientation. This is reflected in the sustainable tourism literature where studies which have examined problems and issues relating to marketing tourism in a sustainable manner appear to be limited in number (for example, Wight, 1994; Eccles, 1995).

SPARC: a Case Study in Sustainable Tourism

This case study illustrates the marketing techniques employed by SPARC (South Pembrokeshire Partnership for Action with Rural Communities), a community-led rural development initiative, as part of its sustainable tourism strategy for rural South Pembrokeshire. SPARC is a LEADER group operating under the auspices of the EU Objective 5(b) programme, and, as befits a group of this type, it aims to encourage sustainable, endogenous growth, with an emphasis on the empowerment of people in local communities. The background to this project has been reported on elsewhere (Midmore *et al.*, 1994). From the outset, tourism was considered to be of major importance to the initiative, and the tourism strategy adopted by SPARC sought to build upon the strengths of the region, in terms of both physical and human resources. As well as having the economic objectives of increasing visitor numbers and expenditure in the region, SPARC sought to achieve these objectives in a sensitive and sustainable way. In implementing the strategy, a number of marketing techniques were employed which assisted the initiative in the achievement of its objectives: namely, market segmentation, branding (in terms of the creation of an identity for an area) and the use of a coherent marketing mix.

With respect to tourism strategy development, the area itself presented SPARC coordinators with something of a *carte blanche*. Rural South

Pembrokeshire represented an unknown quantity to the visitor, being underdeveloped and underexplored, and possessed of a limited tourist infrastructure. Potential tourist offerings of the area included a pleasant agricultural landscape, attractive villages and numerous historical sites. Given these strengths, SPARC sought to select the appropriate type of tourist for the area by segmenting the tourist market.

Market segmentation

Markets may be segmented according to a number of criteria. In the case of the SPARC project, a multiple criteria segment was selected. The target tourist segment had the attributes of being older, higher-spending and quality conscious, and in search of off-season short break destinations to complement a main, usually foreign, holiday. Choice of this target segment was based on two premises. First, it was felt that the needs and wants of this kind of tourist would be well matched by the visitor experience which rural South Pembrokeshire offered. Second, the principle of targeting off-season visitors was consistent with the objective of sensitive and sustainable tourism development. A further consideration was the belief that the target visitor would be open to, and appreciative of, a high degree of community involvement in the visitor experience provided to them. Successful market segmentation and targeting involve the appropriate matching of tourist needs with the experience being offered, and, in the case of sustainable tourism, the need for this match is particularly important. The targeting of inappropriate tourists not only jeopardizes business objectives because the needs of inappropriate tourists will not be met, so that resource efficient repeat visits do not ensue, but also detracts from sustainability objectives, since inappropriate tourists may cause environmental damage or act insensitively towards a local culture.

Sustainability objectives were also met by gearing local communities into action to build upon the basic tourist offering of the area. For example, visitors in the target segment were identified as being interested in investigating the history of an area. To meet this need, each parish produced its own local history information leaflet for tourists, researched and written by community residents, following their attendance at historical research training courses provided by SPARC. Local communities undertook to reinstate footpaths to improve access for visitors. The target segment was also identified as being quality conscious, which meant that much of the existing accommodation required upgrading. This was undertaken via an incentive scheme, whereby providers were granted funds to improve their facilities if they contributed an initial amount themselves. The communities themselves decided which of these activities were to be undertaken in their area, thereby ensuring a bottom-up development approach took place. Segmentation was linked to the tourism strategy through the hope that the target visitors would appreciate, and be willing to pay for, the type of experience which came as a result of community involvement in the tourism offering.

Branding

Branding, or the creation of an identity for a product or place, has become an increasingly popular promotional tool amongst tourism strategists. The identity created by SPARC for the South Pembrokeshire area was that of 'the Landsker Borderlands', an identity which referred to the important medieval period in the area's history. A distinctive black and white logo was designed, which featured prominently in the newly allocated Tourist Information Centre. The logo was also used in the design of parish information leaflets, it appeared in press advertisements, and was used by accommodation providers, to identify themselves as participants in the SPARC initiative. Thus, an identifiable 'theme' for the tourism strategy was developed, to enhance visitor awareness and recognition of the area, both of which were important to the achievement of business-related objectives. The logo also provided an umbrella identity under which the diverse tourism-related activities of each parish community could be unified.

A coherent marketing mix

Although market segmentation and branding were important to the overall success of the tourism strategy, they did represent only two elements of the whole tourism offering developed by SPARC. In order to ensure success, it was important that all the elements of the tourism offering were well coordinated and compatible with one another. In other words, the initiative sought to achieve a coherent and well-blended marketing mix. The tourist experience being offered was one of a tranquil, historic, rural retreat, with an opportunity to explore and undertake activities such as walking and cycling. The quality, comfortable, bed and breakfast accommodation on offer complemented the style of the holiday. Accommodation and attractions were priced appropriately, with the area being promoted as a holiday destination in selective media such as *Country Life*. In this way, each aspect of the whole tourism offering was geared to the expectations and needs of the target segment, while also building on the human and physical strengths of the area.

Emergent questions

The marketing techniques employed in the SPARC initiative gave rise to benefits, and assisted the coordinators in the achievement of their objectives. In employing these techniques, some issues arose which merit attention here. First, in relation to segmentation, it was realized that although the target segment was an appropriate one to choose, the 'empty nester' tourist is a popular target customer for other regional tourist boards in the UK, particularly those who aim to extend their visitor season through the promotion of short break packages. Thus, SPARC chose to enter a competitive market with their selection of target visitor. Competitive markets present a particular challenge to those tourism developers who, like SPARC, seek to create awareness and interest in a previously 'unknown' destination, because they are competing against better known and recognized areas.

A second area of consideration is the use of branding for a location. In terms of enhancing visitor recognition and confidence, branding has much to commend it. However, there are cultural questions at stake. For example, is it possible to impose an identity on something as diverse as a place or region? Even relatively small regions like rural South Pembrokeshire can boast considerable diversity in physical and human terms. Also, what are the cultural effects of historical branding in a region? If regional identities are based on historical periods and events to which modern residents bear little relation, what is the effect of this identity on the modern way of life? Hewison (1987) neatly addresses this question in the context of tourist promotions based on historical eras by suggesting that the protection of the past conceals the destruction of the present. In addition, there remains the problem of tourist expectations being disappointed if the promise of the identity, communicated through promotional literature for example, does not match up with the reality of the area, as actually experienced by the visitor. When developing a sustainable tourism strategy, the interest of people, both visitors and residents, need to be taken into account.

Conclusions

Marketing, like tourism, is a relatively young academic discipline suffering from popular misconceptions. This chapter has sought to dispel some of these by outlining the contribution that marketing can make to sustainable tourism. The espousal of marketing may seem little more than support for apparent common-sense approaches. It is surprising then that the marketing concept is apparently often implemented only superficially and that few tourism organizations utilize marketing functions effectively, typically failing to adopt a holistic approach. Moreover, many aspects of the consumption process in relation to space and place are under-researched (Urry, 1995).

As a set of functional tools, marketing can contribute to sustainable tourism by offering means of identifying market segments whose consumption patterns are more likely to be in harmony with the interests of other stakeholder groups, and by offering branding and marketing mix techniques to target these desired market segments. In short, marketing functions facilitate the matching of market offerings with the needs and wants of various stakeholders. It is important to note here that neither the marketing concept, nor the various marketing functions imply that all needs and wants can be met.

As an underlying concept, or philosophy, marketing can also contribute to the sustainability debate by holistically drawing attention to the importance of considering not only the needs and wants of different tourism stakeholder groups but also, perhaps more crucially, the fact that these needs and wants may change over time. In effect, by highlighting the exchange process, the marketing concept implies that sustainable tourism should be tourism as if people mattered, after Schumacher (1973). This means finding out what matters to people (Mulberg, 1995).

References

Ardnt, J. (1985) On making marketing more scientific: role of orientations, paradigms, metaphors and puzzle solving. *Journal of Marketing* 49, 11–23.

Bartels, R. (1962) *The Development of Marketing Thought*. Richard D. Irwin, Homewood, Illinois.

Brent Ritchie, J.R. (1993) Policy and managerial priorities for the 1990s and beyond. In: Pearce, D.G. and Butler, R.W. (eds) *Tourism Research: Critiques and Challenges*. Routledge, London.

Calantone, R.J., and Mazanec, J.A. (1991) Marketing management and tourism. *Annals of Tourism Research* 18(1), 101–119.

Casson, L. (1974) *Travel in the Ancient World*. Allen & Unwin, London.

Christopher, M., Payne, A. and Ballantyne, D. (1991) *Relationship Marketing*. Heinemann, London.

Davies, A. (1995) *The Strategic Role of Marketing: Understanding why Marketing should be Central to your Business Strategy*. McGraw Hill, London.

Desphande, R. and Webster, F. (1989) Organisational culture and marketing: defining the research agenda. *Journal of Marketing* 53(1), 3–15.

Eccles, G. (1995) Marketing, sustainable development and international tourism. *International Journal of Contemporary Hospitality Management* 7(7), 99.

Greenly, G. and Matcham, A. (1990) Market orientation in the service of incoming tourism. *Marketing Intelligence and Planning* 8(2), 35–39.

Grether, E.T. (1976). The first forty years. *Journal of Marketing* 40, 63–69.

Harding, G. and Walton, P. (1987) *Bluff Your Way into Marketing*. Ravette Books, London.

Hewison, R. (1987) *The Heritage Industry.* Methuen, London.

Holloway, J.C. and Plant, R.V. (1992) *Marketing for Tourism*, 2nd edn. Pitman Publishing, London.

Horsky, D. and Sen, S. (1980) Interfaces between marketing and economics: an overview. *Journal of Business* 53(3), s5–s12.

Levine, P. (1975) Locating your customers in a segmented market. *Journal of Marketing* 39, 72–73.

Kohli, A. and Jaworski, B. (1990) Market orientation: the construct, research propositions and managerial implications. *Journal of Marketing* 54, 1–18.

Kotler, P. (1994) *Marketing Management: Analysis, Planning and Control*, 8th edn. Prentice Hall, Englewood Cliffs, New Jersey.

Krapfel, R. (1982) Marketing by mandate. *Journal of Marketing* 46, 79–85.

McCort, J. (1993) A framework for evaluating the relational extent of relationship marketing strategy in non-profit organisations. In: *American Marketing Association Educators Conference Proceedings*, Chicago, Illinois, pp. 409–416.

McNulty, R. (1993) Cultural tourism and sustainable development. In: Go, R. and Frechtling, D. (eds) *World Travel and Tourism Review*, Vol. 3. CAB International, Wallingford.

Middleton, V.T.C. (1993) *Tourism* 78. Tourism Society, London.

Midmore, P., Ray, C. and Tregear, A. (1994) *The South Pembrokeshire LEADER Project: an Evaluation.* Report published by the Department of Agricultural Sciences, University of Wales, Aberystwyth.

Mulberg, J. (1995) *Social Limits to Economic Theory.* Routledge, London.

Popadopoulos, S. (1989) A conceptual tourism marketing planning model: part 1. *European Journal of Marketing* 23, 31–40.

Przeclawski, K. (1993) Tourism as the subject of interdisciplinary research. In: Pearce, D.G. and Butler, R.W. (eds) *Tourism Research: Critiques and Challenges.* Routledge, London.

Schumacher, E.F. (1973) *Small is Beautiful. A Study of Economics as if People Mattered.* Blond Briggs, London.

Sheth, J. and Gardner, D. (1982) History of marketing thought: an update. In: Bush, R. and Hunt, S. (eds) *Marketing Theory: Philosophy of Science Perspectives.* American Marketing Association Proceedings Series, Chicago.

Urry, J. (1995) *Consuming Places.* Routledge, London.

Van Doren, C.S., Koh, Y.K. and McCahill, A. (1995) Tourism research: a state-of-the-art citation analysis (1971–1990). In: Seaton, A.V. *et al.* (eds) *Tourism: The State of the Art.* John Wiley & Sons, Chichester, pp. 308–315.

Weber. S. (1992) Trends in tourism segmentation research. *Marketing and Research Today* 20(2), 116–123.

Wheeller, B (1995) Ecotourism, sustainable tourism and the environment. In: Seaton, A.V. *et al.* (eds) *Tourism: The State of the Art.* John Wiley & Sons, Chichester, pp. 647–654

Wight, P. (1994) Environmentally responsible marketing of tourism. In: *Ecotourism: a Sustainable Option?* Wiley, New York, pp. 39–56.

Sustainable Tourism: A Marketing Perspective

9

V.T.C. Middleton

Independent Tourism Consultant, The Charcoal House, Low Nibthwaite, Ulverston, Cumbria LA12 8DE, and Oxford Brookes University, Oxford, UK

Introduction

Tourism professionals broadly agree that the core resources on which much of the commercial holiday sector trades are comprised of a combination of the natural and built environment of destination areas and their communities and cultures. It is also commonly accepted, at least in principle, that the long run prosperity of the tourism industry depends on these resources being looked after and sustained at a level of development which does not erode their intrinsic value. Common acceptance extends to the view that constructive partnerships between industry, local populations and their representative governments are a necessary condition for sustainability. Therefore, the real issue for sustainability in tourism is not about ends; it is about means. In other words it is not what ought to be done but how best to do it, how to achieve sustainable tourism and development and how to harness the energy and vision to make it happen at destinations.

The dominant view held by most academics is that sustainable tourism can best be achieved through the creation and operation of a legal and regulatory framework establishing the goals and ground rules within which the industry will work. This is an attractive and seemingly logical proposition because there is a process of statutory regulation to be found in all countries at national, regional and local level. Increasingly, for the environment, the pressure for regulation emerges from international conventions and agreements and its application is a national and local responsibility within the wider framework. The UK, for example, is now heavily constrained by environmental regulations emanating from the European

Union. In practice, a regulatory view is essentially a top-down approach which looks to national politicians and civil servants to help define the necessary regulations and enforce their operation. A great deal is expected of local planners in this process.

An alternative view, not popular among academics, is that commercial organizations in the industry are now able and many are willing to shift toward sustainability through a process of self-regulation. For reasons related to their role in marketing, the practical influence of tourism businesses is immense, although diffuse in its operation. Now, and more so in the future, self-regulation is not just another factor but arguably a necessary foundation and continuing input to whatever overall regulatory framework exists. The self-regulation view is essentially a bottom-up approach that looks to industry and trade associations to define sustainable goals which are practical, to help frame relevant regulations, and to alter operational practices to achieve them, using commercially relevant enforcement methods.

In practice, of course, sustainable solutions will always be a balancing act of top-down and bottom-up approaches. As Cairncross (1991) put it succinctly, 'To rely exclusively on the force of the market, however ingeniously harnessed, is as naive as relying solely on government intervention.' The notion of balance is unfortunately another woolly concept and politicians' sound bite. This author believes that exploring the differences between a regulatory and self-regulatory regime is a fertile process likely to lead to wholly practical conclusions and to point towards a workable balance. It is especially necessary in an industry as diverse and a market as heterogeneous as tourism.

The Tourism Context in Cumbria

Cumbria is one of the most precious natural environments in Great Britain, in defence of which the National Trust was established a century ago. It is appropriate, therefore, to set the issues of sustainability within a local context. It is a common belief in Cumbria that tourism is growing, threatens to destroy the places tourists come to visit, and requires sustainable policies based on increasing regulation, especially of cars.

Looking ahead into the early years of the 21st century, it hardly needs a crystal ball to predict on current expectations that the national park will have reached a point at which on many days and in many locations the overall volume of tourism demand will be judged to exceed acceptable levels of capacity (however defined) and planners and other bodies will be required to act. Apart from the usual planning and building regulations designed to contain development for all purposes including tourism, likely actions, many of them currently under active discussion and consultation, will involve a combination of:

- Access controls on specified roads for cars.
- Speed restrictions on specified roads.
- Restrictions on at least some of the most popular fell walks (to help restore erosion).
- Limitations on access for four-wheel drive vehicles and trail bikes and possibly also for mountain bikes.
- Control of use of the lakes (numbers of boats/noise/speed/boat pollution).
- Provision of more public transport linked with park and ride schemes.
- Parking restrictions in selected 'honey pot' areas and more pricing mechanisms to inhibit vehicle use.
- Controls on towed vehicles such as caravans.
- Use of price and taxation measures to curb demand and supply.
- Specific development controls aimed at restricting the growth of tourist accommodation.

It is difficult, on current evidence, to avoid the conclusion that the strategy for tourism will be primarily a negative one of restriction and curtailment of access. Whether such a strategy can possibly achieve what its proponents hope for is another question. There is a real danger that the restrictive measures will actually serve to drive away the most environmentally sustainable forms of tourism leaving Cumbria with the lowest common denominator forms of tourism: those which create the greatest damage in return for the least economic and environmental gains.

In common with most destinations, the real knowledge of who visitors are in Cumbria is sketchy. The Cumbria Tourist Board have done more than most in research terms but the patterns even of staying visitors are not known in sufficient detail for visitor management purposes. The sample on which the annual estimates of tourism are based in Cumbria is derived from a national survey of tourism demand and the number of respondents available for analysis is somewhere between 250 and 350 for all purposes, grossed up to some three million visits. One person more or less in the sample represents around 10,000 tourists and around £1.5 million revenue. This is not conducive to confidence in tourist board trend data.

Over a 12 month period people choose to stay in Cumbria for many reasons. These include business visits and work related trips outside the usual routine, conferences, visiting friends, visiting relatives, taking part in or watching a wide range of sports and active recreation, attending and participating in the wide range of country shows in Cumbria, attending and participating in a wide range of arts and cultural activities, visiting second homes including caravans, and holidays in a range of commercial accommodation extending from five star hotels to homely bed and breakfast houses and farms. Holidays vary from main holidays, through additional holidays, to short breaks; and that is just the staying visitors. In addition there are day

visits from outside Cumbria and, in many ways even more important, there
are residents of the local community whose recreational activities are part of
the visitor volume and pressure at specific destinations. These points are
summarized in Table 9.1 and developed in the next section.

In terms of trends, the available research evidence suggests that the level of
commercial holidays in Cumbria has not increased at all in the last 20 years. At
around 1.5 million visits (Cumbria Tourist Board, 1995), this form of tourism
has actually spread away from the traditional July/August peak as the
incidence of additional holidays and short breaks have altered the patterns. On
rather weaker evidence there is no reason to suggest that the number of day
visits from outside the county, estimated at 9 million in 1992 (ECOTEC, 1992)
is significantly larger than it was 20 years ago. A combination of economic
recession in the early 1980s and 1990s, perceived pressure on incomes, and
growing awareness of the certainty of congestion and frustration on motor-
ways and trunk routes has probably inhibited the use of cars for longer
recreational or 'day-out' journeys. The development of week-end car boot
sales and Sunday trading in refurbished town centres and the newly developed
out-of-town 'retail parks' now serve as attractions for many who might
otherwise drive for five or more hours a day getting to and from the national
park, driving within it, and looking for somewhere to park.

This author, at least, is convinced that the perceptions of continuing
tourism growth in Cumbria have little to do with what many, especially the
wider public, understand as tourism. They are primarily the result of the
recreational use of cars by residents of the local community. There are some
half million residents in Cumbria at the present time. Based on national
estimates of car ownership and use, at least 200 million journeys are
estimated within the county over the year, for all purposes. On sunny
weekends in the summer and in school holidays it is obvious that the local
scenic attractions are a natural target for the enjoyment of residents. Car
ownership and use in Cumbria have probably doubled over the last 20 years
or so, and they are set to increase by another 25% or so by around the turn
of the century. This level of car usage far outweighs any overall pressures that
staying visitors generate in Cumbria; yet it is the latter which have been
specifically targeted for draconian measures in the 1996 traffic management
initiative. The Cumbria Tourist Board has estimated residents' recreation day
visits from home for purposes other than work and regular shopping trips at
around 6 million but this appears to be a significant underestimate having
regard to the range of purposes included within the internationally agreed
definition of tourism.

It is of interest that in its recent traffic flow analyses (Cumbria County
Council, 1995) the County Council appears to overlook and misrepresent
the massive upsurge of recreational traffic by the local community, attribut-
ing the increase in traffic flows over the period since the early 1980s to tourist
traffic. 'The council's report also asserts that there is a prominent peak in

August, the most popular month for tourists visiting the Lake District' (Appendix A3.2), and that the 'peak flows generally occur over weekends. This is to be expected because the majority of tourist traffic will occur then' (A4.2). Furthermore, it is argued that 'flows in August are due to increased tourist traffic' (p. 47). No evidence is provided of what proportion of car usage at busy places and busy times is attributable to residents' recreation activity rather than to tourism. Common sense suggests it is high in most congested places and a principal cause of traffic growth over the last two decades and looking ahead.

What is a resident anyway? Does a resident of Barrow-in-Furness look different from a tourist in the Duddon Valley and going over Hardknott Pass? Do all the residents of Kendal have an automatic right to transport themselves to the Lyth Valley as often as they wish for picnics and walking their dogs? Should priority in Borrowdale be reserved for those who live in Keswick? Of the millions of boots that erode the popular fells and scar the mountain environment, how many are worn by locals, not tourists? Who own most of the four-wheel drive vehicles and trail bikes that scar the bridleways and green lanes? Should not planners at least consider these issues and identify local residents moving around outside their own locality when they are, say, ten or more miles from where they live (or any other arbitrary distance), as a major contribution to what is commonly understood to be tourism pressure?

Beneath the Surface of Lake District Tourism

The central proposition in this chapter is the pointlessness of maintaining the convenient but fallacious view that there is a more or less homogeneous aggregate term 'tourism' for which it is possible to have a policy and strategy at destination level, for example the Lake District. Increasingly it appears that tourism is a meaningless word and tourism industry is a meaningless concept. A sustainable tourism strategy is a notion with little meaning at the local level where it is meant to apply. Whether one takes the broad definition ratified by the United Nations Statistical Commission in 1993 (World Tourism Organization, 1994) or the even broader definition (including turnover and employment implications of investment) promoted by the World Travel and Tourism Council (1995) one ends up with an aggregate called tourism which is attractive to politicians, academics and the media but has little or no practical relevance locally.

According to the World Tourism Organization (1991), 'Tourism comprises the activities of persons travelling to and staying in places outside their usual environment for not more than one consecutive year for leisure, business and other purposes.' To try to unravel tourism and get below the surface, Table 9.1 presents an indicative list of ten segments. Some can be

Table 9.1. Tourism in Cumbria in the mid 1990s – five impact indicators by visitor segments.

	Volume indicator (%)	Economic benefit	Fragile environment impact (location)	Traffic congestion impact	Seasonal impact	Social culture impact
Staying visitors (approx. 15 million nights)						
On business trips	10	5	0.5	1	1	0.5
Short breaks – commercial	15	4	3	3	2	2
Main holidays – commercial	25	3	4	5	5	4
Additional holidays – commercial	20	4	4	4	3	3
Visitors to friends and relatives	20	1	2	3	3	1
Use of holiday homes	10	1	3	3	3	5
Day visitors (approx. 18 million days)						
On business trips	10	4	0.5	1	1	0.5
Visiting friends and relatives	20	1	1	3	3	1
Distant recreation trips	35	2	5	5	5	5
Cumbria resident recreation	35	1	3	4	4	4

Volume estimates are based broadly on Cumbria Tourist Board (CTB) estimates for 1992 (day) and 1994 (stay).
Economic benefits (employment and support for local economy) are positive: 5 is high; 1 is low. All other impacts are negative (e.g. impact on fragile environment): 5 is most negative; 1 is least negative.
© VTCM Author's estimates, Newton Rigg, 1996.

broadly quantified on existing research evidence but for others, for example business visits, it is a matter of using judgement based on experience. Broadly based on Cumbria Tourist Board (CTB) estimates, the table provides a volume indicator for each segment (staying visits and day visits each add to 100) and they are based on some 15 million staying visitor nights (say 3.5 million visitors – CTB estimate of 3.2 million in 1994) and 18 million day visits (CTB estimate of 15 million in 1992). To put these tourism figures into some sort of perspective it is worth at least noting that the half million residents of Cumbria spend around 175 million resident days in the county going about their normal business, many in the same places as those used by visitors, for example when shopping. Because the issue for this conference is sustainability, the table provides indicators for five of the important factors influencing economic value and environmental impact. The scoring system based on 1 to 5 reflects judgement, not research. A system based on 1–10 would be more sensitive but such issues would be for debate.

There are at least two dimensions of difference revealed in Table 9.1. Segments vary on the demand side by differences attributable to:

- Types of people – where they live; their age and stage in the lifecycle; their income.
- Different attitudes – to recreation pursuits; preferences for peace and quiet, for energetic sport, for gregariousness, etc.
- Choice of specific products purchased in the course of tourism activities; choice is wide.
- Choice of time of the year to visit Cumbria.

Segments vary widely in terms of impact and sustainability according to:

- Volume and trends (growth or decline over a decade or so).
- Contribution to the local economy measured in terms of employment and support for locally provided services.
- Impact on the social and cultural environment in Cumbria.
- Impact on fragile environments such as the high fells and lakes.
- Traffic congestion, noise, and pollution.

To attempt to aggregate these segments and their very different motivations as tourism or tourists is counter-productive. It produces sledge-hammers to crack nuts and, at the risk of a pun, is likely to destroy what it seeks to protect. Yet aggregation is invariably the process when county, national park and district planners conceive of tourism policies in Cumbria. Cumbria, however, is not just another tourism destination, but one which has been an attraction for visitors and the basis for a large tourist industry for more than a century. If Cumbria, a county with a century of experience of tourism, in a country with one of the most sophisticated planning mechanisms available at local level in the world, is still at this stage of sophistication of visitor management, how will planners achieve sustainable tourism policies for

Cumbria in the next decade? What hope is there that developing countries around the world will devise realistic, sustainable tourism policies, when they have none of the experience and local government traditions on which to draw that are part of our heritage in Cumbria? These are not popular views but they are rooted in reality and ought to be debated.

Sustainability Defined

Environment means

> The quality of natural resources such as landscape, air, sea water, fresh water, flora and fauna; and the quality of built and cultural resources judged to have intrinsic value and be worthy of conservation.
> (Middleton and Hawkins, 1993)

In addition to those already coined, there will doubtless be many new definitions of sustainable tourism. Whatever the words used one can be confident by now that all will be nuances on a central theme, which is that the cumulative volume of visitor usage of destination – by the full range of its visitor types – should not exceed the point at which the regenerative resources available locally are capable of sustaining the environment. Beyond that point the environment, in its physical, social and cultural dimensions, may not recover but be damaged in the long run, possibly permanently.

It is important to recognize that regenerative resources are part natural and part managed. The natural part means ecosystems, the robustness of which is a function of specific areas as some are more fragile than others. The managed part means the controls, technology, support systems and influences put in place by human intervention of individuals, government and other bodies, of which the aim is to protect and enhance environmental quality. Looking ahead into the next century it is likely that developments in technology will increase the relative importance of the managed part of regenerative resources.

The key word in the broad definition of environment offered in this section is quality. Quality in this context, even for factors such as air and water, is a relative not an absolute condition. Quality is defined by perception and is capable of extensive human intervention. The attractiveness of a destination's environment for its visitors is also a matter of perception and this chapter argues that environmental quality, interpreted as attractiveness for potential visitors, is part of product quality and thus a primary concern for marketing managers as developed in the next section.

As an aspirational theme, sustainability seems splendidly logical and achievable. Theoretically, it is simply a matter of identifying the capacity of natural and built environments to absorb tourism, and the willingness of residents to accept an 'appropriate' level of intrusion in their lives, and then managing tourism around it. In practice of course, capacity turns out to be another of the woolly concepts and it is never that easy. As numerous surveys have discovered since the 1970s, all forms of capacity are hard to define and impossible to measure with precision. Capacities vary, for example:

- Having several different forms ranging from eco-capacity to cultural, economic or road capacity.
- According to local destination circumstances – there are no simple overall guidelines.
- Over time because social and cultural capacities are matters of perception, not of fact.
- According to different types of local community residents whose perceptions will vary.
- According to the level of technology which may be brought to bear to ameliorate the effects of visitor activities which create damage and pollution.
- According to the knowledge and skills applied to the management of visitors at the local destination – including the knowledge of tourism and stage of development of local government.
- According to the types of visitor and activity putting the destination environment under pressure.

The more one considers notions of tourism capacity the more its complexities emerge. The simplest approach available in practice, and because of its simplicity it is the approach most likely to be sought by regulators and planners, is the attempt to define capacity in terms of existing tourism uses of an environment. Appealing in circumstances of perceived crisis, and where lack of research information reduces professionalism to guesswork, the existing use approach is probably the worst, most sterile, backward-looking way of tackling capacity issues. It may worsen rather than alleviate the environmental damage it responds to. It is not difficult for tourism professionals to conceive of potential tourism developments which, through the use of improvements in new technology and better management procedures, could double the existing use capacity of an area and at the same time significantly enhance sustainability. Walt Disney World and Center Parcs are examples. But once the regulatory shutters are up to contain existing uses, the new and better uses are likely to be ignored or rejected. Regulating capacity to existing uses, although intended to protect the environment and ameliorate problems, may actually serve to make the situation worse.

A Marketing Perspective

Because all visitors accommodated in the commercial sector purchase specific products to undertake particular activities in particular places (Middleton,

1994), the commercial marketing of products to targeted groups of pro-spective visitors provides what is often the major influence over what happens with tourism on the ground. Other businesses providing services to day visitors have a lesser but still significant influence over the choice of activities and location. A successful marketing approach is always specific to segments and not products.

Marketing people around the world will appreciate that:

1. travel and tourism in an area such as Cumbria is comprised of hundreds of individual businesses (public and commercial sectors), each in a specific location.
2. marketing skills focus on knowledge of existing and emerging new segments and existing and developing product offers – and how to bring the two together most efficiently.
3. each business has its own target segments for which they tailor and promote product offers.
4. for commercially provided vacations in particular, the environment is often the core component of the quality of the product offer – either explicitly, as in lakes and mountains, most outdoor activity holidays, most short breaks, or implicitly, as in products featuring aspects of the arts/culture and heritage of an area.
5. all customers have choices and their holiday behaviour decisions reflects their view of the importance of the environment identified (explicitly or implicitly) as a product quality expectation and choice.

In marketing terms, sustainability is primarily an issue of product quality. There is no clear evidence in the developed world that more than a small minority of visitors understand concepts of sustainability and environmental good practice and draw on them when choosing products, although travellers from countries such as Germany, Holland and Scandinavia appear to be further ahead in this respect than the UK. There is even less evidence that the great majority of visitors are willing to pay premium prices for the products of tourism businesses operating to high environmental standards. But there is convincing evidence that customers turn away from what they consider to be overcrowded, polluted destinations which have allowed their environ-mental quality to become eroded through overdevelopment. This is espe-cially true where health risks, as from air and water pollution, are perceived as problems. There is also convincing evidence that customers generally in the 1990s are more experienced in travel, more demanding, and searching for a combination of quality and good value for money which they are increasingly able to recognize.

Product quality in Cumbria is directly associated with scenic values, freshness and tranquillity associated with mountains and lakes and the heritage attractions of towns and villages. From a marketing perspective, sustainability comes down in practice to an issue of product design and development, product quality reflecting customer expectations and presentation to target market segments. Tourism in Cumbria is an aggregate of dozens of individual segments and thousands of specific product offers made by hundreds of business operations. Sustainability or degradation through tourism at particular localities emerges collectively from the individual marketing processes of business which begin and end with customer choices and expectations of the quality of product offers. It can only be sensibly tackled with this essential insight.

Leading businesses in tourism destinations increasingly recognize this. Smaller operators within the local community are much more likely to make their business decisions based on short-term opportunism and it is likely to be an increasing role for tourist boards and trade associations of small businesses in the next decade to restrict membership and its benefits to those who observe agreed codes of good practice. Such codes, for example that now operated by the British Home Parks and Holidays Association, already include the environment.

A Regulatory Perspective

As noted earlier, sustainability always involves the issue of regulation versus self-regulation. Regulation to protect the environment and secure sustainability through a form of environment planning and policing is superficially attractive. It has many armchair devotees. It seldom works in practice and only in specific destinations such as Bermuda which have particular advantages. The problem with regulation is that it requires regulators. Regulators are typically not tourism professionals but politicians, public servants and lawyers with responsibility for but no direct experience of the industry. They are expected to grapple with tourism and the tourism industry as meaningless abstracts.

Regulators typically lack adequate research data even at national level, and at local level are often faced by a complete absence of usable information for management decisions. They have no means of assessing the relative value of segments as set out in Table 9.1 and determining which segments deserve expansion and which ought to be discouraged where possible. They are typically required to work with data and ideas at least five years out of date and are expected to regulate for a rapidly changing industry looking ten years hence. At best, regulators may understand the techniques and controls available in the public sector for influencing tourism but few, if any, are likely to have a knowledge of tourism marketing in their backgrounds.

If the realities of achieving sustainability in tourism are still the subject of debate and emerging concepts among tourism professionals, one should not be surprised if regulators do not understand the issues. In addition to the problems already noted, they often move so fast between jobs that they leave tourism before their learning curve begins to reach the point where they could make a real contribution. In the best of circumstances, with the best of intentions, regulators will always lack the knowledge and the speed of response to deal sensitively with an industry comprising so many different facets.

There remains an important role for regulators to establish and enforce controls for particular tourism destinations, especially to define the limits to tourism development based on flexible, changing approaches to tourism capacity and technology, and to control pollution, waste disposal, use of water, and other environmental impacts imposed by tourism businesses. Regulation is also likely to include fiscal measures and incentives designed to achieve agreed ends. To achieve success in this difficult role it seems essential that regulators create systems to draw on the expertise and local knowledge of the important stakeholders in the local economy, particularly tourism businesses. It is equally important and in their own self-interest that the local businesses and their representative associations do all they can to facilitate the process and participate in it.

Conclusions

For those who thought about the issues it has been obvious since the 1970s that tourism and the environment – social, cultural and especially physical – would never achieve a harmonious balance solely through any benign influences inherent in market forces. The principal reason for this appears to be less a comment on weaknesses inherent in market forces but more in the extraordinary complexity of the tourism industry and the predominance of small businesses involved in it. In different ways, invariably using the dreadful cliché about tourists destroying what they come to see, every part of the world can produce a catalogue of environmental degradation said to be due to tourism. Regulation is inevitable.

Equally obviously in the 1990s, such harmony as may be identified cannot be achieved and sustained without harnessing the power and the revenue contribution associated with market forces generally, and marketing management in particular. Are market forces and environmental sustainability in tourism irreconcilable, as conventional wisdom has it, or are they logical partners? Therein lies a paradox for the 1990s and beyond. Sustainable development has become a fashionable global notion since the Brundtland Report (World Commission on Environment and Development, 1987) and the Rio Earth Summit (United Nations Conference on Environment and Development, 1992). Sustainable tourism has become a mantra for tourist boards and government tourism departments in the 1990s, at least in their public pronouncements. In practice, they are far more dependent on market forces and marketing skills, which they do not control, than is generally recognized.

The best prospects for achieving sustainability in tourism appear to lie with local partnership strategies for visitor management at specific destinations. That avoids the meaningless national aggregates of tourism and tourism industry noted earlier. The partners are regulators and planners, mostly within the public sector at local level on the one hand, and commercial sector bodies and trade associations on the other. Partnership is based on recognition that there is a joint responsibility for the quality of the environment, for its intrinsic value and as the core element of products, which generates and sustains vital economic activity.

Visitor management means devising locally achievable strategies and programmes to balance the eternal triangle of visitor segments, destination resources, and residents. Management embraces an understanding of market forces and the power of tourism marketing to influence the behaviour of visitors and businesses as the core tool of strategic action. Effective marketing management for destinations, especially segmentation and related product design and quality controls, draws on the contribution of all the partners and offers the best prospect of achieving long run sustainable goals at specific tourism destinations. This is a positive, pro-active approach open to new ideas, new technology, new markets and new uses for environmental resources.

Of the two parties, planners and local politicians are directly involved not only or perhaps even primarily because they control the local planning and regulatory process and often also take on a destination marketing role. They are primarily involved because responsibility for and ownership of much of the natural and built environment at destinations is typically invested in local government. Their responsibilities include a duty of care for the quality of the environment resource. They also have responsibility to support economic growth and employment in their areas.

Private businesses are directly involved in influencing and helping to shape the regulatory process because none of them can insulate themselves from the potentially damaging activities of others at the destination, and also because their long run prosperity and return on investment is directly linked to maintaining and enhancing product quality in a fiercely competitive world. Tourism businesses, either directly if they are large enough or through their representative trade associations, have a major role to play in partnerships in interpreting and communicating to regulators the processes of continuous change at work in the market in terms of growth markets and product development.

References

The concepts and practice expressed in this chapter derive from:
Middleton, V.T.C. and Hawkins, R. (1997) *Sustainable Tourism: A Marketing Perspective*. Butterworth-Heinemann, Oxford.

Specific references

Cairncross, F. (1991) *Costing the Earth*. Business Books Ltd, London.

Cumbria County Council (1995) *Traffic Flow and Accident Monitoring Report*.

Cumbria Tourist Board (1995) *Facts of Tourism*. CTB, Windermere.

ECOTEC (1992) *Tourism in Cumbria in the 1990s*. Report for Cumbria Tourist Board.

Middleton, V.T.C. (1994) *Marketing in Travel and Tourism*, 2nd edn. Butterworth-Heinemann, Oxford.

Middleton, V.T.C. and Hawkins, R. (1993) Practical environmental policies in travel and tourism. *Travel and Tourism Analyst* no. 6, 63–76.

United Nations Conference on Environment and Development (1992) *Agenda 21: A Guide to the United Nations Conference on Environment and Development*. UN Publications Service, Geneva.

World Commission on Environment and Development (1987) *Our Common Future*. Oxford University Press, Oxford.

World Tourism Organization (1991) *Tourism Trends Worldwide and in Europe*. WTO, Madrid.

World Tourism Organization (1994) *Recommendations on Tourism Statistics*. WTO, Madrid.

World Travel and Tourism Council (1995) *Travel and Tourism's Economic Perspective*. WTTC, Brussels.

The Importance of Community Involvement in Sustainable Tourism Development

10

P. Harper

ADAS, Agricola House, Gilwilly Trading Estate, Penrith, Cumbria CA11 9BN, UK

Introduction

The Agricultural Development Advisory Service (ADAS) was, until its privatization on 1 April 1997, the UK national consultancy agency of the Ministry of Agriculture. Over the past 45 years its staff have been given the task of helping to improve the technical efficiency of the agricultural sector and have been the main organization responsible for the dramatic improvements that have taken place. Within the rural tourism sector ADAS has a long history of involvement, having established most of the 70 Farm Holiday Groups in England and Wales, several farm attraction groups and the Farm Holiday Bureau, which is the national association responsible for promoting farm accommodation in the UK.

The Cumbria Farm Tourism Initiative is an example of a local rural tourism development project. This chapter will examine the importance of community involvement in creating sustainability and the consequences of this approach for public organizations, local communities and the visitor. It uses the example of a particular task that was carried out by the project manager to show the methodology used to mobilize opinion, gain local ownership and manage change.

How the Cumbria Farm Tourism Initiative Started

In 1992, ADAS presented a paper on the future development of farm tourism to the Cumbria Tourist Board and the Rural Development Commission which raised several obstacles to future development, highlighted new opportunities and suggested that extra public funding was needed if the potential was to be realized. These three organizations agreed to work together and start a collaborative public and private sector project. The work of the project was to follow the good example of initiating change in the farming community that had been adopted by ADAS since its inception.

After undertaking desk research, a draft strategic plan was produced, which formed the basis of several grant applications, which were approved in August 1993. The project obtained initial core funding of £32,000 from the Cumbria Tourist Board, Cumbria Training and Enterprise Council and the Rural Development Commission. Whilst funding was only for one year it was intended to be a three year project. The main tasks to be undertaken included product and marketing research, preparation of a strategic plan, conducting free business advice, offering specialist training, the rationalization and improvement of the promotion and marketing of farm accommodation within the county and to generate an extra £4000 extra income.

The Task

The example which best illustrates the recommended principles to follow is the element of the work programme relating to the improvement of the marketing of farm accommodation within the county. The objective was to encourage existing operators to promote themselves together in one publication covering all Cumbria, with at least 70% of existing group members participating.

Farm Accommodation in Context

In order to appreciate the circumstances and context in which the project manager had to operate, it is important that the characteristics of the farm accommodation sector within Cumbria at the time the project started are understood. Local research had shown there were between 1000 and 1200 farmers in the county offering farm based holiday accommodation. Most of these farmers operated tourism enterprises with turnover of often less than £10,000 with the income generated being very much of a secondary nature to that from the main farming enterprises. These tourism enterprises had evolved over many years with little history of the farmers taking any external advice on tourism matters. Attitudes were mainly fixed with no thought that any 'professional outsiders' could do anything to help. Within the farm accommodation sector there were eight existing Farm Holiday Groups all of which were (and still are) fiercely independent with, for instance, different rules for joining, different opinions about marketing and different costs of membership. Most were producing their own group brochures, which yielded few bookings and in some cases costing £100 per head for group members to enter.

Research had shown the sector had good opportunities for growth but there were many barriers in the way of change, particularly if suggestions came from 'professionals', rather than from within the community from someone who was respected. It was recognized by the funding agencies that, if the good potential was to be realized, change was needed, particularly in the marketing of farm accommodation and in the organization and structure of the existing groups, which had to be encouraged to work more closely together. It was acknowledged that, if the project was to be sustainable, the different communities had to take ownership of the work of the project, particularly in the marketing sector.

Methodology

Motivating local people

'Local influencers', defined as the prominent peers, were identified in each farm holiday group and they were interviewed to discover existing attitudes to change, prejudices, potential personality clashes and so on. The broad objectives and principles of the project were presented to these people in such a way that they could relate to and agree with. Above all, the local influencers were given the impression that their views were important and that they would be listened to. Most agreed that their cooperative marketing could be improved but there was disagreement on how best to proceed. It became apparent that there could be a consensus to start a county-wide promotional campaign which the public agencies would also be happy to support, but not everyone agreed on its detail. It was recognized that if the public agencies were to support the action required it had to be shown that there was a common goal which would have widespread and increased benefits.

The next step was to hold a joint meeting between farmers and funding agencies with the objective of acquainting all concerned with the value of the common goal and to allow farmers to voice any disagreements and to listen and take note of these and be prepared to change if necessary. Great thought was given as to how this meeting was to be conducted. Local influencers with particularly strong views which did not concur with a likely common goal (previously identified from the initial interviews) were placed in discussion groups with other local influencers from other parts of the county with equally strong views in favour of the common goal. Each discussion group was chaired by a person who was aware of these attitudes beforehand. In addition, a nationally recognized expert in farm tourism research, Richard

Denman, who had recently carried out some consumer research, was asked to present a paper at the meeting to keep all concerned market focused. By adopting this approach, agreement was reached to produce and promote a county-wide farm accommodation brochure, channelling the enthusiasm generated into a momentum which launched this sector of the work programme.

Maintaining the momentum and overcoming objections

Within any project of this type, whilst the principles or common goal may be agreed upon, great problems can arise when trying to reach agreement on the detail, particularly where there is such a wide range of very individual operators with differing views. To overcome this problem, the method adopted was to form a working party of farm accommodation operators which comprised 'local influencers' from each farm holiday group plus representatives from farmer operators who were not in any existing farm holiday group.

The first meeting, chaired by the project manager, identified areas both of common agreement and of disagreement. Where there were differing views, these were discussed and either they were resolved by agreement of the farmers or additional research was carried out to enable the decisions to be taken to be based on better information. In the event of an impasse the project manager as chairman was empowered to take the casting decision as an impartial person. This requirement arose only occasionally, mainly for minor issues, for example on design of the brochure. After only two meetings, broad agreement was reached on the production and promotion of a single county farm accommodation brochure with all farm holiday groups in agreement that they would promote themselves together. The ultimate goal was that each group would give up its brochure in the long term and put the money saved into the county publication.

The best example of how well this approach worked in overcoming objections was in connection with the minimum standards of accommodation that would be accepted. At the first meeting of the working party, all the farmers who were in farm holiday groups were in agreement that only farms that were inspected under the English Tourist Board scheme should be allowed to enter. The farmers who did not belong to any group were not so adamant and it was suspected that if this rule could be relaxed many more farmers would be encouraged to enter. It was agreed to defer the decision until some extra research had been carried out to discover the likely difference in membership if entry into the brochure was dependent or not on having an inspection carried out.

At the second meeting the research results showed that if a policy including only those subject to inspection was adhered to, the potential membership would be at least 30% lower than if this rule were to be relaxed. After learning of these results, the farmers completely changed their original views and it was unanimously agreed to relax the 'inspection only' criteria for the first year. Thus this major problem had suddenly disappeared and harmony returned. Therefore, the principle which subsequently was applied was that the working party resolved major disagreements themselves, facilitated by the project manager, rather than having solutions imposed by him. In general, it is apparent that if farmer members are presented with all relevant facts, a consensus often emerges and they will come to the correct decision. It is surprising how attitudes can change if the communication process is properly followed; openness with information is fundamental as part of this process.

The Lessons Learned

If the above approach is adopted, the role of project manager or the person responsible for initiating change is critical if the changes are to be sustained with local communities. The person must have several attributes, with the requisite skills to be able to work within rural areas and communicate effectively with the independent operators as well as officers from public funding agencies. Very often these two groups find it very difficult to communicate effectively between each other so that the project manager can facilitate this work. The project manager must, above all, listen and be prepared to adapt the role as a result of input from the local community. The local influencers in the community must be involved at the start of the project in order that they can initiate and influence change in the community through peer pressure. Throughout the life of the project, the project manager can gradually take on a more subservient role to the local community, making sure that individual groups are not dominated by particular individuals who can detract from and move the direction of these groups away from the wishes of the majority. Local opinions above all must be respected and it is important to realize that local influencers have much to offer as they have probably thought about the problems to be tackled for far longer than any of the 'professionals' that are often imposed on the community from outside. The project manager must also gain the trust of the local community which means that information must be open to all, arguments must be thoroughly backed by fact and the whole partnership must be encouraged to have open minds to appreciate each other's opinions.

The approach which is advocated to ensure change is sustained in rural communities does require some change of attitude by those in public organizations. Too often the attitude is 'we know best' and new development ideas are imposed. This creates local antagonism and mutual distrust even if their objectives may be the same as those of the local community. It is their method of implementing policies which can often be completely inappropriate.

An Appraisal of the Local Community Approach

There are many benefits of following the suggested approach identified above. Firstly, the social and cultural limits of development options are easily identified at an early stage. For example it was soon discovered that one of the original ideas from the public funding agencies was the provision of a vacancy advisory service and it became clear that this would not be acceptable to the farmers in the first year.

The approach does engender good local ownership which ensures increased commitment from the farming community. A good example of this was that by involving the local influencers on the working party for the production of a new county brochure, this resulted in attitudes completely changing from initial scepticism to wholehearted support. It was directly responsible for the project achieving its original objective by attracting 71% of existing farm holiday group members into the first brochure. It also results in a high level of understanding of the work of the project by the key local influencers so that when things go wrong within their sphere of influence, the group is much more tolerant of mistakes and problems which occur, compared with projects where they have no direct influence.

Direct involvement of the farmer operators has also led to a much higher quality and more effective product than would have been the case if they were not involved. By listening to the farmer operators much extra knowledge was gained from their experience, which helped with subsequent promotion and helped the project achieve over 12,000 bed nights for 88 entrants in the first year.

Some of the disadvantages of the approach are, firstly, that direct involvement of the local influencers within the work of the project results in them becoming very frustrated in being closer to the bureaucratic decisions and procedures that necessarily have to be taken in dealing with the public funding agencies. Also, because the approach recommended is rather slower than the implementation method, it is difficult to maintain the commitment of some of the public sector organizations. It also places high demands on the communication skills and level of commitment of the project manager.

The approach described shows the role of the project manager to be one of a facilitator, who guides rather than directs. Openness and hard work are essential to gain respect among farmers. Acceptance by the farmers helps to instil confidence in the project manager, which is vital when trying to obtain funding from the public sector organizations. However, the option is relatively expensive initially compared with the alternatives of a low level of community involvement and/or higher level of imposition from outside agencies. Some public organizations have yet to grasp the fact that the approach can take several years but in the long term is the cheapest as it can be sustained. Using public funds in other approaches for short term benefits often leads to the collapse of schemes when public funding is withdrawn.

Conclusions

Sustainable tourism in the context of working with rural communities is like any other rural development project which involves change. In order for the change process to be sustainable it is vital to involve the local community. To influence change within the local community is a slow process which requires a thorough understanding of its social make-up. It is considered that the best approach to encourage change is to identify the local opinion leaders in the community, understand them, listen to them and work with them so that the change process is driven by these 'local influencers' rather than be imposed from outside.

A collaborative project of this type where the project manager or development worker acts as a facilitator is much more likely to lead to the required change, which can be sustained to achieve common objectives. The speed of change may be slower than that desired by the public funding agencies but it is considered the best approach in the long term.

Wordsworth, Sustainable Tourism and the Private Sector

11

T. McCormick

Heritage and Tourism Consultant, 17 High Fieldside, Grasmere, Cumbria LA22 9QQ, UK

A Pioneer of Deep Ecology

It is assumed that there is widespread agreement about Wordsworth's impact as a pioneer deep ecologist. His poetry of place with its exploration of the then *terra incognita* of mind and nature left a new map for a new territory which has guided, for better or for worse, subsequent journeys. He is a poet who, above all others, rescued and rejuvenated an encrusted 'spirit of place', giving it a modern meaning which we can all understand and work with: 'a living ecological relationship between an observer and an environment, a person and a place' (Cobb, 1970, p. 125). Those messages appealing to all tourists to 'leave only footprints' can be tracked directly back to Wordsworth's championing of the discrete (sometimes disappearing) self merging into the natural world:

> ...Verily I think
> Such place to me is sometimes like a dream,
> Or map of the whole world: thoughts, link by link,
> Enter through ears and eyesight, with such gleam
> Of all things, that at last in fear I shrink,
> And leap at once from the delicious stream.
> (Hayden, 1977, I, p. 566)

What is perhaps not so widely accepted is the poet's transference of this vision into an argument that such moments of merging, such 'spots of time', depended upon a beauty which was only possible when the 'economy of nature', Wordsworth's 1823 equivalent of the term 'ecology' (Owen and

Smyser, 1974, II, p. 185), was functioning appropriately as it did, for him, in
one of his favourite places, Loughrigg Tarn:

> Having spoken of Lakes I must not omit to mention, as a kindred feature
> of this country, those bodies of still water called Tarns. In the economy of
> nature these are useful, as auxiliars to Lakes ... Of this class of miniature
> lakes, Loughrigg Tarn, near Grasmere, is the most beautiful example. It has
> a margin of green firm meadows, of rocks, and rocky woods, a few reeds
> here, a little company of water-lilies there, with beds of gravel or stone
> beyond; a tiny stream issuing neither briskly nor sluggishly out of it; but its
> feeding rills, from the shortness of their course, so small as to be scarcely
> visible. Five or six cottages are reflected in its peaceful bosom; rocky
> and barren steeps rise up above the hanging enclosures; and the solemn
> pikes of Langdale overlook, from a distance, the low cultivated ridge of
> land that forms the northern boundary of this small, quiet, and fertile
> domain.
>
> (Owen and Smyser, 1974, II, p. 187)

Here, the 'economy of nature' fulfils Wordsworth's definition of beauty: 'a
multiplicity of symmetrical parts uniting in a consistent whole' (Owen and
Smyser, 1974, II, p. 181). Romantic aesthetics are a template for and from a
healthy ecology.

Wordsworth's insight, his vision, were worked out within a unique and
unrepeatable cluster of political circumstances which both fostered and
oppressed his ambition. By the early years of the nineteenth century his
youthful (and once treasonable) republicanism had been confined within the
boundaries of his native region:

> Towards the head of these Dales was found a perfect Republic of
> Shepherds and Agriculturists ... Neither high-born nobleman, knight, nor
> esquire, was here; but many of these humble sons of the hills had a
> consciousness that the land, which they walked over and tilled, had for
> more than five hundred years been possessed by men of their name and
> blood; and venerable was the transition, when a curious traveller,
> descending from the heart of the mountains, had come to some ancient
> manorial residence in the more open parts of the Vales, which, through
> the rights attached to its proprietor, connected the almost visionary
> mountain republic he had been contemplating with the substantial frame
> of society as existing in the laws and constitution of a mighty empire.
>
> (Owen and Smyser, 1974, II, pp. 206-207)

For Wordsworth, by 1810, the republic needs the 'laws and constitution
of a mighty empire' to protect it and most politics is local. His *Guide through
the District of the Lakes* (first published anonymously in 1810 and then in
Wordsworth's name in 1820 as *Topographical Description of the County of
the Lakes in the North of England*) represents his first bout of activism
directed, as the title-page makes clear, towards 'Tourists *and* Residents'
(emphasis added). In this text Wordsworth makes a powerful case for the link
between sensitive tourism and ecological awareness:

> ... Nature will be far more bountiful in granting what they (visitors) severally
> look for if he aspires at something beyond a superficial entertainment of
> the eye. The Soul of objects must be communicated with, and that

> intercourse can only be realized by some degree of the divine influence
> of a religious imagination.
>
> (Owen and Smyser, 1974, II, p. 306)

This Wordsworth sets against the impulse 'to hurry forward with the lightminded train from one distinguished spot to another', which 'is nothing more than the prolongation of the labour of vanity, invigorated and refreshed by a change of scene' (Owen and Smyser, 1974, II, p. 305). But the focus of his attention and anxiety throughout his *Guide* is not tourism but the emergence of a new resident community which was not only supplanting the 500 year old republic of shepherds and agriculturists but was also a threat to 'the native beauty of this delightful district, because still further changes in its appearance must inevitably follow from the change of inhabitants and owners which is rapidly taking place' (Owen and Smyser, 1974, II, p. 223). This orientation (so pertinent, it seems, to the current place/visitor/host community axis in this region) actually underpins Wordsworth's seed proposal that the Lake District should be, effectively, a national park. He calls upon the support of visitors (tourists) to claim the region with a new kind of democratic, open ownership:

> It is then much to be wished, that a better taste should prevail among
> these new proprietors; and, as they cannot be expected to leave things
> to themselves, that skill and knowledge should prevent unnecessary
> deviations from that path of simplicity and beauty along which, without
> design and unconsciously, their humble predecessors have moved. In this
> wish the author will be joined by persons of pure taste throughout the
> whole island, who, *by their visits (often repeated)* (emphasis added) to
> the Lakes in the North of England, testify that they deem the district a sort
> of national property, in which every man has a right and interest who has
> an eye to perceive and a heart to enjoy.
>
> (Owen and Smyser, 1974, II, pp. 224-225)

In the context of a revolutionary poetry which showed that 'men who do not wear fine cloaths can feel deeply' (Merchant, 1970, p. 839), Wordsworth's intention here in proposing a national park is to protect the region from new settlers through forming an alliance between a native community rooted in place and a visiting community whose chief virtue is that it seeks to cherish that place without inhabiting it.

Marketing Wordsworth in the Tourism Economy

Within the heritage sector, the word 'marketing', with its direct equivalence to 'sell', is too limited a description of work which is actually 'audience development', encompassing visitor management and educational activity alongside marketing. As Curator for the Wordsworth Trust from the early 1980s, there was no way one could bury oneself in collections care and avoid the responsibility of audience development; and one did not want to! If one were to set the work out in phases it would run as follows:

1. training to ensure a sensitive individual experience within a high volume of visitors in a small site with clear conservation imperatives (the Wordsworth Trust won a Lady Inglewood Award for Training in 1987).

2. visitor services to identify and meet a range of visitor needs and develop techniques for managing a wide variety of visitor volumes (the Wordsworth Trust was cited as an example of good practice in *Tourism: the Environment Visitor Management Case Studies*, 1991).

3. marketing to ensure that skills and techniques from the commercial sector were learnt and applied to develop and maintain national and international visitor loyalty.

These three areas overlapped all the time, of course, but all underpinned the trust's commitment to sharing its treasures and knowledge with as many people as possible while strengthening its economic viability. The validation in all this work is the actual experience of a place which recalls extraordinary creativity almost 200 years ago. Most of the trust's visitors, and there are currently 75,000–80,000 a year (over one million since Dove Cottage first opened to the public in 1890), are, in some way, seeking to make contact with this creativity in this small house in Grasmere Vale. The adjacent museum then provides opportunities to inform and develop that experience through its presentation of manuscripts, books, and fine art treasures.

Audience development for the Wordsworth Trust is inextricably bound up with a wider promotion of the destination: the Lake District. The trust worked closely with the Cumbria Tourist Board and joined the district tourism association, then known as the South Lakes Organization for Tourism. What was discovered in the early 1990s, as the recession began to bite, was a sectarian tourism culture in which conflict and parochial distraction seemed to be the norm. It rapidly became clear in the trust's bit of the region that there was no consensus about values and objectives and, therefore, no spirit and practice of cooperation and partnership which could encourage planning for projects with direct, identifiable benefits. The public and private sector representation of place, visitor, and host community had not come together in a triangle; more often than not it was a self-defeating wrestle to defend the territory of one set of interests against another. The resulting lose–lose situation forced the author and his fellow marketeer, Sylvia Wordsworth, into a choice: go it alone and seek collaboration where the benefits are clear or attempt to change the context and try to build a culture of creativity and partnership. The fact that the trust was already in dialogue with active private sector players and district council officers who wanted to make a commitment to the second option made the choice straightforward. Another factor for the Wordsworth Trust was a recent, problematic, history of misunderstandings with its immediate community, particularly in connection with planning issues. The second course could help to break down the local perception of the trust as an alien organization, committed to its own agenda, remote from the needs of its neighbours. And so it was resolved that the organizations would act together to try to build a partnership which would bring benefits to the resident and visiting communities, and the place they inhabited, within a programme which supported economic development.

It is worthwhile, at this point, noting some key facts (dating from 1991) about the tourism economy of South Lakeland which distinguish it from other districts in Cumbria and place it in the same category as, for example, Stratford upon Avon, the centre of Cambridge, and parts of the Peak District.

1. Out of a total annual visitor spending of £812m (direct, indirect, and induced) £400m was spent in South Lakeland (49.3%).
2. Out of a total direct spend of £195.5m in South Lakeland, £41.5m (21.4%) was spent by day visitors.
3. Out of a total working population of 37,410 people, 17,015 (45.5%) are working in tourism; this represents 40% of all tourism jobs in Cumbria, a county which has one in six of its workforce involved in tourism.

These figures are exaggerated in the national park portion of the district, where approximately 70% of the population are involved in tourism. Any discussion of the tourism economy must, of course, acknowledge the facts of seasonal work and low wages: average earnings for men and women in hotels and catering were, in 1995, approximately three-quarters of the average UK adult employee rate (Labour Party, 1995). South Lakeland does not have immunity from these facts, but it is also irrefutable that the district has a vibrant tourism economy underpinning a community with high scores on all the quality of life indicators. Complementing, and often conflicting with, this culture is a resident community with an age profile, certainly for Windermere (South Lakeland Tourism Partnership, 1994, p. 12), which bears out the area's appeal as a retirement zone for those tourists whose 'fancies', as Wordsworth put it 180 years ago, 'were smitten so deeply that they became settlers' (Owen and Smyser, 1974, II, p. 208) (Table 11.1).

A small group of members within the South Lakes Organization for Tourism, concluding that it would be more effective to realign the core objectives and re-form the structure of the existing tourism association rather

Table 11.1. Age distribution of Windemere and Cumbria.

Age group	Windermere	Cumbria
Children	12.7%	17.8%
Over 70s	17.3%	11.9%
Over 60s	29.8%	23%

Source: 1991 Census.

than create another organization, agreed and proposed a new mission to:

strengthen the economic base of the South Lakeland District through
sustainable tourism which benefits the resident community, the natural
environment and the visitor.
(South Lakeland Tourism Partnership, 1992)

A range of initiatives to support this objective was developed over a number
of meetings and discussions. South Lakeland District Council offered a
£3000 a year grant to pay for an administrator along with the provision of an
office and facilities. In June 1992 the author was elected Chairman and a
committed group of private sector activists joined with officers from the
district council, the tourist board and the national park to form a management group which was to meet monthly to oversee an action programme.
The South Lakeland Tourism Partnership came into existence. The record, in
terms of new initiatives, over a two year period as Chairman, was as follows:

For the resident community:
- conceiving and establishing, with South Lakeland District Council, a
  Residents Open Week: almost 40 attractions involved with 5000 passes
  issued (6000 in 1995).
- establishing a visitor attractions preview, with the National Park
  Authority, for local guest house owners and hotel managers to learn
  about and promote 'heritage on your doorstep'.
- supporting the district council and the National Park Authority in
  holding a Craft Fair; giving an opportunity for local craftspeople to
  promote their work.
- developing and presenting, with consultation, submissions to the Cumbria and Lake District Structure Plan and the Local Government Review
  Commission.

For the natural environment:
- 'The Lakes and Beyond'; a collaborative marketing project with British
  Rail to encourage increased use of trains.
- 'Hilltop to the Lake' footpath scheme; a fundraising venture to finance
  the building of an off-the-road footpath for residents and visitors from
  Beatrix Potter's house to the shores of Windermere.

For the visitor:
- special local offers for purchasers of British Rail Intercity tickets from
  Euston to Windermere.
- development of itineraries for the travel trade.
- integrated presence at British Travel Trade Fairs in 1992 and 1993, with
  awards for stand and promotion.
- integrated presence at World Travel Market from 1992 onwards.
- facilitating an 'Action Kendal' initiative to enable and encourage
  partnership work in the town.

Behind these initiatives there was a great deal of unrecorded work by individual members and groups within the partnership which, through active communication and mediation, often helped to resolve potential conflicts.

The key outcomes were as follows:

1. over the two years 1992–1994, the membership increased from an inherited 44 to 103 (70% representation from businesses operating within the national park and with a roughly equal division between hotels, services, and attractions).

2. there was an increase in private sector voluntary time and effort, particularly from the management team; this was calculated in terms of loss of potential earnings to be £60,000 p.a.

3. the private sector was in closer working collaboration with three public agencies (district council, national park, tourist board).

4. communication and rapport with resident community not directly involved with tourism were enhanced.

5. energy and investment from the private sector in natural environment increased through the footpath project.

6. core commercial activity was strengthened.

7. for the Wordsworth Trust, its audience development work was strengthened through partnership work and networking: a balance was maintained.

During the last months of chairmanship one began to reflect upon the constraints which were preventing further development of this initiative. There was an identity issue in that the tourism partnership was tackling challenges that were unique to the national park and only 30% of its members lived and worked in a large area of South Lakeland District beyond the park. Its title was not accurate. The logical next step would be to build a park-wide public agency/private sector forum but this would involve three other district councils and a great deal of political graft, which, at that time (all public agencies defending their patches under the review of local government), was just not feasible. Unlike, say, the Stratford upon Avon Visitor Management Action Programme or the Peak Tourism Partnership, the South Lakeland Tourism Partnership was (apart from the catalysing district council contribution) private sector and voluntary led and yet, although its virtues were private sector (a 'can do' approach with rapid-response creativity), it was run, despite the intentions, as if it was a public agency, with time-consuming accountability procedures.

The sense, in 1994, was that there had to be a different kind of paradoxical network, a sort of intermediate organization which embodied the best of private sector creativity and speedy responsiveness working to a rigorous values-based forward plan which would be supported by tactical long-term funding from public agencies. Such a network could not flourish without a clear regional and community identity, which would have to be genuinely inclusive and, probably, Cumbria-wide. But, in 1994, these half-formed (some would say half-baked) notions were being conceived in the dark; there were, apparently, no models to refer to. The question remains now as it did then; is this feasible? The experience since 1994, during which time intense involvement with the region's tourism as Chairman of Cumbria Tourist Board's commercial members had occurred, revealed that the answer to this question will have a practical, long term bearing upon the evolution of tourism in this region, and, insofar as Cumbria can be seen as a case-study, possibly beyond.

Golden Age Politics: Some Conclusions

But before one can begin to answer the question about feasibility, one has to acknowledge and work round and through the power of certain perceptions which are actually deep inhibitors within the British approach to tourism and the environment. From a green political perspective, the inclusive, contrary, consciousness that is ecology does not work productively with widespread golden age sentiments which appear to link people together but, in fact, prevent concerted action. John Barrell got at this paradox when he focused on the paralysing strength of the golden age myth in UK culture, which ensures that there will be no concerted attempt to change the way the countryside is owned and managed:

> That myth is of a 'natural' order, before politics, even before history: and as long as that remains our ideal, it seems incapable of being realised by political action as it is of being realised without it ...
>
> (Barrell, 1984, p. 89)

Without this political action, those who own and manage the land on the nation's behalf defend golden age sentiments against all threats, and tourism is seen to be one of the most serious of these. And because, in UK culture, the ownership and management of land have conserved a class structure, class-enthralled attitudes often permeate perspectives and policies towards tourism. This is acknowledged by the Oxford English Dictionary in its definition of tourism: 'The theory and practice of touring; travelling for pleasure; usually *depreciatory*' (emphasis added). Wordsworth is frequently invoked as the champion of golden age pastoralism. A recent text on sustainable tourism supplements a chapter title with that favourite con-servationist quotation, 'Is then no nook of English ground secure/From rash assault?' and sub-titles this with 'Wordsworth on Tourists' (Croall, 1995, p. 39). In fact this is Wordsworth, old and tired in 1845, campaigning against the extension of the railway from Kendal to Windermere so that a certain *class* of tourists would not be encouraged to visit the Lake District. These are not the tourists he rallied in his conceiving of the national park 35 years earlier, but the new emergent working class whose potential power in the land he had resisted through his opposition to the 1832 reform act. It is a cause of melancholy that his gift of communicating a deep ecology vision is used, in 1845, with such profligacy to support social division. Sustainable tourism is susceptible to, and can be a mask for, the insidious hypnotism of class prejudice. It was left to the civility of the Board of Trade to rebuke the poet laureate's offensiveness:

> We must therefore state that an argument which goes to deprive the artisan of the offered means of occasionally changing his narrow abode, his crowded streets, his wearisome task and unwholesome toil, for the fresh air, and the healthful holiday which sends him back to his work refreshed and invigorated simply that individuals who object on the grounds above stated may retain to themselves the exclusive enjoyment of scenes which should be open alike to all, provided the enjoyment of them shall not involve the infringement of private rights, appears to us to be an argument wholly untenable; and we are of opinion that there are no public grounds which ought to be decisive against the Kendal and Windermere railway receiving the sanction of parliament.
>
> (Owen and Smyser, 1974, III, p. 334)

This episode underlines the paradox of Wordsworth's legacy; in terms of the individual his work is inspirational; in terms of the community it is cautionary, arriving with a large label 'Handle with Care'. If the inspirational is used to over-ride the cautionary then Wordsworth will be press-ganged into support for the *myth* of a natural order and ensure that its practical realization, certainly in this part of the world, will be prevented.

Returning to the question of the feasibility of a new kind of intermediate organization, combining the best of private sector creativity and public sector guardianship, nothing is feasible without a strong steadily evolving economic culture within which tourism, set to be the world's largest industry by the arrival of the millennium, is managed to serve all needs. The issue of livelihood and social equity is the hub around which that awkward, bumpy, triangle of place/visitor/host community revolves. As Raymond Williams put it during the mid 1980s frenzy of free marketeering:

> The most hopeful social and political movement of our time is ... 'green socialism', within which ecology and economics can become, as they should be, a single science and source of values, leading on to a new politics of equitable livelihood. There is still very much to be done in clarifying and extending the movement and in defining it, practically and specifically, in the many diverse solutions and resolutions, where it must take root and grow. But here, at least, is a sense of direction, born in the experiences between city and country and looking to move beyond both to a new social and natural order.
>
> (Williams, 1984, p. 219)

In its own extremely modest way (and probably against the party political loyalties of many of its participants) the South Lakeland Tourism Partnership was (and in some respects still remains) a stab at one of these diverse 'solutions and resolutions'.

The outstanding matter is the character and nature of the organizations which are required to embody and promote the new social and natural order which is fundamental to sustainable tourism. When does a network ('A system of units, for example, buildings, agencies, groups of people, consisting of a widely spread organization and having a common purpose') go through that mysterious precipitation in which the original common purpose becomes so completely identified with the structure and survival of the organization that bureaucracy replaces effective administration? Usually, it is when the common purpose changes but the people in the network or organization do not! For this precipitation to be avoided one of the basic tenets of ecology – recycling – has to be given divine status. When the common purpose of a network changes or ceases to be, so should the network change or cease to be. If the network has become a traditional organization, there will be a resource-devouring protracted delay while the organization devotes its energy to avoiding change and mortality. In matters of tourism and the environment, this delay cannot be afforded because it alienates the private sector and the communities it supports.

If Wordsworth is to be embraced as a key representative of golden age politics (and it seems to happen often enough) the way to include what he represents and move beyond it is to recognize the template between 'economy of nature' (ecology) and romantic aesthetics (beauty as a 'multiplicity of symmetrical parts uniting in a consistent whole') but then add another template of application: the network which embodies the principles it promotes. This is the organization rooted in 'systems thinking' (and, more than incidentally, supported by the permanent evolution in information and communication technology):

> The new vision of reality we have been talking about is based on the awareness of the essential interrelatedness and interdependence of all phenomena — physical, biological, psychological, social, and cultural. It transcends current disciplinary and conceptual boundaries and will be pursued within new institutions. At present there is no well established framework, either conceptual or institutional, that would accommodate the formulation of the new paradigm, but the outlines of such a framework are already being shaped by many individuals, communities, and networks that are developing new ways of thinking and organizing themselves according to new principles.
>
> (Capra, 1982, p. 285)

The systems approach, as all self-respecting new-agers know, is not new; its advocacy dates from the early 1980s. What would be new would be the application of this approach to the 'management' of tourism and the environment because it requires the participation of the whole community working to forward plans which have been thoroughly developed in close consultation with the whole community. A model for how this could be done has recently been outlined by Hanna Hoffmann (1995) on behalf of the World Travel and Tourism Research Centre, and, it seems, copyrighted or patented for commercial purposes. Such exclusive possession is impossible to claim beyond a form of words; inclusivity is the precondition of success in the systems approach and models and case studies must be shared. It will be problematic enough developing networks which do not support a regional or national status quo maintained by public agencies which will fight to ensure their possession of responsibility. This is more critical and more feasible for Cumbria than for many other regions because the county is a crucible for all these processes and, with the status and glamour of the Lake District, has a clear national and international profile.

The etymology of sustenance and nurturing in 'sustainable' *should* deny sectarian possession of it: nurturing is for wholeness in the person, the community, the place. And yes, most will agree that, in cosmological terms, possession is illusory. But one also knows that real politics is about possession of agendas, of language, of resource. This region, Cumbria, has experienced many possessions. Wordsworth's Duddon Sonnet on Hardknot Roman Fort charts colonizations by 'Druids', 'Scandinivians' and 'Romans', all of which have been included and survived. Above the fort flies the golden eagle. Its appeal for the poet is clear: '... behold the Eagle upon the wing, retaining her empire when that ambitious People who adopted her image for its standard, have for ages been but a name' (Owen and Smyser, 1974, II, p. 345). But painful too because the eagle's survival in this region during his own lifetime was under constant threat. It is known now what is required to try to keep the eagle alive in these mountains, less than 20 miles from here, but at this moment it is not known if this will be the third season when an egg does not hatch. There will be no flinching from the difficulty despite doubts about the feasibility of this project. The inclusive, systems approach and the political movement which is its context are fraught with difficulty and their success cannot be guaranteed. But if, like endeavours with the Lake District eagle, one seeks to enable places and their communities and the people who visit them to be the foundation for a vibrant pattern of growth rather than a symbol recollecting a past golden age, then the work which has already begun must be developed as a grand project for the new century.

References

Barrell, J. (1984) The golden age of labour. In: Clifford, S., King, A. and Mabey, R. (eds) *Second Nature.* Jonathan Cape, London.

Capra, F. (1982) *The Turning Point: Science, Society and the Rising Culture.* Fontana, London.

Cobb, E. (1970) The ecology of imagination in childhood. In: McKinley, D. and Shepard, P. (eds) *The Subversive Science.* Houghton Mifflin, Boston.

Croall, J. (1995) *Preserve or Destroy: Tourism and the Environment.* Calouste Gulbenkian Foundation, London.

Hayden, J.O. (ed.) (1977) *William Wordsworth: The Poems.* Penguin, Harmondsworth.

Hoffmann, H. (1995) *Integrated Total Quality Tourism Management, A SEEing*

Approach: An Emerald Isle Case Study. Executive Summary, World Travel and Tourism Environment Research Centre, Oxford.

Labour Party (1995) *The Tourism Economy*. The Labour Party, London.

Merchant, W.M. (ed.) (1970) *Wordsworth: Poetry and Prose*. Reynard Library, London.

Owen, W.J.B. and Smyser, J.W. (eds) (1974) *The Prose Works of William Wordsworth*. Clarendon Press, Oxford.

South Lakeland Tourism Partnership (1992) *Membership Brochure*. Windermere.

South Lakeland Tourism Partnership (1994) *Windermere Inquiry Statement 1994*. Windermere.

Williams, R. (1984) Between country and city. In: Clifford, S., King, A. and Mabey, R. (eds) *Second Nature*. Jonathan Cape, London.

Tropical Forest Ecotourism: Two Promising Projects in Belize

12

J.M. Edington and M.A. Edington

3 Elm Grove Road, Whitchurch, Cardiff CF4 2BW, UK

By general consent the success of ecotourism projects can be judged by three criteria:

- whether they provide the customer with a rewarding experience.
- whether they contribute to environmental conservation.
- whether they bring economic benefits to the receiving communities without also causing cultural disruption.

These issues are explored in relation to two ecotourism projects developed in the tropical forests of the Central American country of Belize.

The Community `Baboon' Sanctuary

The first project for consideration is the so-called baboon sanctuary at Bermudian Landing in the north of the country (Fig. 12.1). Actually baboons do not occur in South America, but the term is used by local Creole people, who are of African origin, to refer to the black howler monkey (*Alouatta pigra*). This particular species has a very limited distribution. It is confined to Mexico, Guatemala and Belize and is constantly under threat from deforestation and from being hunted for food (Emmons and Feer, 1990). That it should feature in an ecotourism project that affords it protection is therefore very much to be welcomed.

The sanctuary at Bermudian Landing is an elongated area extending along the Belize River and encompassing 40 square kilometres. It contains a population of approximately 1000 monkeys. Local farmers, of whom there

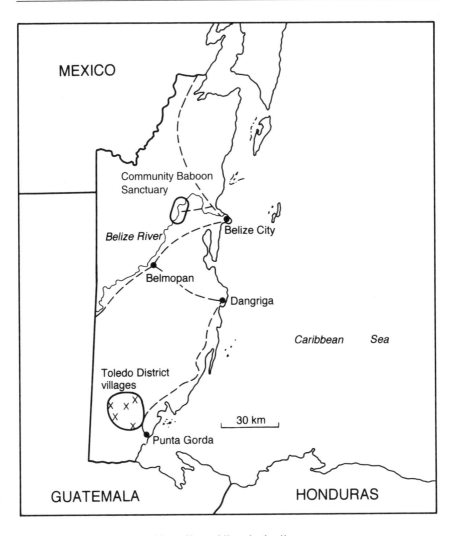

Fig. 12.1. Map of Belize and location of the study sites.

are over a hundred, have committed themselves to managing their land in a manner which is favourable to the monkeys (Horwich, 1990). Farming in the area takes the form of shifting cultivation, whereby patches of forest are periodically cleared prior to the cultivation of rice and beans. When the fertility of a particular plot declines, forest vegetation is allowed to re-invade it to act as a rejuvenating bush fallow. Within this regime, the farmers make concessions to the monkeys' interests by preserving wild fruit trees in otherwise cleared areas and by leaving continuous corridors of trees along the river banks and as connecting links between adjacent patches of forest.

These latter measures serve to facilitate the free movement of the monkey troops which, like many other forest animals, are reluctant to cross open spaces. Rope bridges have been constructed over the main road for the same reason.

From the farmers' standpoint, protecting the monkeys and their habitat has the obvious merit of attracting visitors likely to pay for the services of local guides and for overnight accommodation. Currently 22 families from seven villages are involved with visitors in this way. It should be noted, however, that enthusiasm for the project is not limited to individuals deriving economic benefit from it. A recent social survey has revealed that most local people consider that the monkeys should be protected irrespective of whether they generate revenue (Hartup, 1994).

Tourists visiting the sanctuary, even on a day visit, find much to interest them. They can observe the behaviour of the monkeys in their natural habitat, they can watch birds (of which 200 species have been recorded in the area), they can visit the site museum (containing natural history and ethnographic displays) and they can be guided along trails designed to introduce them to notable plant and animal species. Two trees of particular historic interest are the mahogany (*Swietenia macrophylla*) and the dye-producing, logwood tree (*Haematoxylon campechianum*). Both contributed significantly to the early economic development of the country. Visitors undertaking overnight stays have the added interest of living with a local family and sampling Creole cookery.

The original initiative for the venture came from an American zoologist, Robert Horwich, and his botanist collaborator, Jon Lyon. They made the first surveys of the area and started the necessary negotiations with local farmers in 1984. They subsequently produced a comprehensive account of the forest in a handbook which serves admirably as a visitors' guide (Horwich and Lyon, 1993). Day-to-day management of the sanctuary is now the responsibility of a local manager, working under the general guidance of the Belize Audubon Society.

Factors contributing to the success of the operation as an ecotourism venture include: (i) proximity to Belize City (43 km away by road); (ii) a clear cut conservation strategy; (iii) first class interpretative facilities; and (iv) enthusiastic community participation. The only substantial criticism levelled against the scheme is that the accommodation provided for overnight visitors has not always been regarded as satisfactory in terms of comfort and health protection. If such reservations persist, this could undermine the economic viability of the project, particularly now that the funding initially available for establishing the central facilities has come to an end.

The Toledo Ecotourism Project

Another ecotourism project based on rainforest habitats is to be found in the south of the country. This is the work of the Toledo Ecotourism Association, a consortium formed by five Mayan Indian Villages. In each village, a guest house has been constructed to accommodate visitors. This resembles a village house in overall form, but internally is subdivided into two sleeping areas, each catering for four people. There is an adjacent shower and toilet block.

This scheme has different strengths and weaknesses from the baboon sanctuary project. Particular attention has been given to protecting the health of guests. Beds are fitted with mosquito nets and the use of insect repellents is strongly recommended. These are important precautions in view of the presence in the area of malarial mosquitoes and of sandflies capable of transmitting cutaneous leishmaniasis (Williams, 1970). Meals are taken successively in different houses and each housewife providing this service has received instruction in food hygiene. Illness in a family causes that household to be temporarily withdrawn from the system. Ample supplies of washing water and boiled drinking water are supplied to the guest house to reduce further the risk of enteric infections.

By contrast, attempts at environmental interpretation are much less well developed than at the baboon sanctuary. The main activity offered to visitors is to be taken on a conducted forest walk to some local feature of interest such as a cave, waterfall or temple ruin. One of the opportunities missed is to interpret everyday village activities in terms of available forest resources. For example, visitors would surely be interested to know that the roof of their guest house (and of all the other village huts) is made from plaited leaves of the cohune palms (*Orbigyna cohune*) which surround the village, or that the strange pasta-like material offered for lunch is actually derived from leaf strips of the *jipijapa* tree (*Carludovica insignis*) stained red with seed pulp from the ornamental *annata* tree (*Bixa orellana*) cultivated in most hut gardens.

In addition to the provision of meals and accommodation, income is also derived from dance performances and the sale of handicrafts, the latter consisting mainly of woven bracelets and purses decorated with traditional motifs, and small baskets made from *jipijapa* fibres. A typical 24-hour village stay costs about US $35 per person according to the services requested. This money is paid to the village association and 80% of it is distributed directly to the families providing the meals and other services. The remainder is deposited in an accumulating village fund used to finance health and training projects. Tourist payments are estimated to have raised village incomes by approximately 25%. The conservation element in the project is somewhat vaguely defined and is based on the premise that money earned from tourism should reduce the villagers' need to clear forest for the cultivation of rice as a cash crop.

As tourist destinations, the villages have the disadvantage of being somewhat inaccessible. They are served only by a weekly bus service from the coastal town of Punta Gorda. Otherwise visitors need to charter a local taxi to reach them, an expensive option. Punta Gorda is doubtfully accessible by road from Belize City during the rainy season, but can be reached by air throughout the year using the local air service.

No marketing strategy has been developed for the village project, even the possibility of distributing leaflets to hotels and tourist offices seems to have been ignored. Whilst the danger of swamping the villages with visitors must always be kept in mind, there is equally a risk that if facilities such as the guest houses are manifestly underused they are likely to be converted back into village houses.

As at the baboon sanctuary, the catalyst for the development of the Toledo project has been an outsider, this time in the person of Chet Schmidt, an expatriate American. It seems rarely to be the case that the impetus for ecotourism projects is indigenous in origin, presumably because village communities find it difficult to imagine that visitors would be interested in their daily lives and local surroundings, and be prepared to pay for the privilege of becoming acquainted with them.

Conclusions

Here are two interesting and promising projects. Neither is completely free of difficulties, but considered in combination they provide some useful guidelines for the design of successful ecotourism operations.

From the visitor's perspective, attractions need to be accessible, authentic and well interpreted. Visitors should also be able to feel that adequate attention is being given to safeguarding their health. Particularly in tropical areas, this is a matter which rarely receives the attention it deserves.

From the receiving community's perspective, projects need to be designed to generate significant economic benefits, without causing undue cultural disturbance. On this last score, one approach is to define, and adhere to, a maximum permissible ratio of visitors to residents, say 1:50. On the economic side, experience suggests that the viable projects are those which receive some initial non-repayable funding to finance capital works.

Finally to meet conservation criteria, the anticipated benefits to wild species and their habitats should be clearly defined at the outset of the project and kept continuously under review as it develops.

References

Emmons, L.H. and Feer, F. (1990) *Neotropical Rainforest Mammals*. University of Chicago Press, Chicago.

Hartup, B.K. (1994) Community conservation in Belize: demography, resource use, and attitudes of participating landowners. *Biological Conservation* 69, 235–241.

Horwich, R.H. (1990) How to develop a community sanctuary – an experimental approach to the conservation of private lands. *Oryx* 24, 95–102.

Horwich, R.H. and Lyon, J. (1993) *A Belizean Rain Forest. The Community Baboon Sanctuary*. Orang-utan Press, Gay Mills, Wisconsin.

Williams, P. (1970) Phlebotomine sandflies and leishmaniasis in British Honduras (Belize). *Transactions of the Royal Society of Tropical Medicine and Hygiene* 64, 317–368.

Minimizing the Environmental Impact: Alternative Forms of Tourism

Chapters 13 to 18 develop more strongly the necessity of considering the needs of communities in destinations of any tourism development, thereby taking a rather more dispassionate view of the industry. The perspective is a much more socio-cultural one, which is not surprising given that two of the authors, Wilson (Chapter 13) and Cole (Chapter 15) are practising anthropologists. The first four chapters concern the problems of securing sustainable tourism in developing countries and raise ethical issues with respect to the behaviour of tourists, the industry and the indigenous population, touching also on political matters. The last two (Chapters 17 and 18) examine tourism in developed nations where the pressures of current levels and patterns of tourism threaten not only the viability of the industry but also the sustainability of environments. Indeed, in one case it is shown that there is an urgent need to introduce changes to offset the decline of traditional types of tourism. All the chapters, in different ways, indicate the means by which the impact of tourism might be minimized.

Wilson takes the term 'up-market' and raises the ethical issue of the extent to which it is an elite form of tourism. He compares the apparent control of development of tourism in Seychelles, under a totalitarian regime, with the uncontrolled form in Goa. The former 'officially' aims to prescribe the number of beds, location and siting of accommodation, restaurants, bars and other facilities, but de facto this has not necessarily occurred. Paradoxically the idea of 'up-market' tourism being more sustainable is shown to be invalid as the call on resources, for example water, food, and infrastructural inputs, is so much greater than in Goa where 'traveller' and low-cost domestic tourism has much less impact.

An ironic ethical point arises in that 'travellers', while they have a less marked environmental effect, indulge in unacceptable practices, such as alcohol and drug consumption and casual sex. So there is a moral issue, including in origin countries whose citizens give a bad impression abroad; Israel, for instance, is conscious of the double standards of its nationals on holiday in Goa. Another acute ethical issue is the corruption in the destination where environmental regulations are flouted in the interests of tourism development, which is of economic importance. Consequently, there is continual conflict between the moral high ground and the reliance on tourism as a vehicle of economic survival in many communities. The result is that environmental degradation occurs, with the distinct possibility of neither the Seychelles nor Goa being able to achieve sustainable tourism. The cases also serve to point up the more general issue of the social and political systems and structures and the mores of 'baksheesh', which undermine international and national desires and actions in the form of aid to protect the environment to achieve sustainable development/sustainable tourism.

Chapter 14, by Hamzah, in describing the evolution of small-scale tourism in Malaysia against the changing pattern of both domestic and foreign tourist demand, echoes the problems identified by Wilson in Chapter 13 of the implications of largely uncontrolled development for both natural and socio-cultural environments. The author examines the supply response to the 'drifter' sub-culture of foreign tourism which has concentrated in certain destinations, and has also served an emerging budget domestic tourism, leading to confusion in host communities as to which tourist image they should promote. This is beginning to change as rapid growth in domestic tourism is forcing providers to offer better quality accommodation and facilities as Malaysia enters a mass-tourism phase. As a consequence, the 'drifters' (alternative tourism) move on to new destinations because established sites have begun to invest to upgrade the tourism capital. This rapid development has put severe pressure on natural environments and threatens to overwhelm the traditional Malaysian culture.

From this seemingly depressing and familiar tale of the catastrophic impact of tourism overdevelopment, such as degradation of ecosystems, economic leakages, community displacement, resource shortages and loss of traditional values, there are apparent opportunities for a distinct and unique Malaysian identity and image and therefore tourism product. The author considers technical and institutional proposals for planned small-scale development which can complement the mass market and evolve along sustainable tourism lines. Among these proposals are involvement of local communities, more stringent planning controls, greater use of local inputs, a tighter cooperation of central government, tourism agencies and local organizations and utilization of higher educational expertise. The chapter is concluded on a guarded optimistic note that local business people in destinations are showing greater entrepreneurial and organizational skills and seeking professional advice and in doing so are creating a more sustainable but Malaysian form of small-scale tourism.

The role of anthropologists in establishing whether specific communities are suitable in which to develop cultural/ethnic tourism is explored by Cole in Chapter 15 through two case studies in Indonesia. Cole indicates, by neatly contrasting one community which possessed characteristics conducive to tourism development, such as a varied and attractive culture, an outgoing and friendly disposition and a willingness to serve, with another which did not have these prerequisites. It is argued that there has been too much emphasis on the physical environment as the main concern regarding tourism development and sustainability. More attention needs to be paid to the hosts, particularly because of language problems and cultural differences. The anthropologist can contribute to overcoming such problems by mediating between hosts and visitors and acting as a tour leader, particularly in the early stages of tourism development. Of especial concern is to educate tourists as to the customs and mores of the hosts to minimize any offence which may be committed through ignorance. Through such mediation it is possible for long-term benefits to accrue to communities, perhaps facilitated by initiatives triggered by tourists, for example the provision of water supply to a village through donations into a 'community pot'.

It behoves organizations attempting to promote sustainable tourism projects to appoint appropriate leaders. Cole gives an instance, notwithstanding that the community in question was antagonistic towards receiving tourists, of project leaders making matters worse through a lack of understanding of the culture and language.

The chapter by the Sharpleys considers the debate as to whether tourists need to modify their behaviour to meet the needs of attaining a sustainable product (the environmental, cultural and social sustainability of destinations) or whether to acknowledge that consumers are unlikely to want to forego the quality they expect and therefore to adapt the product. The chapter adopts a largely neutral stance in presenting a reasonably balanced review of the position currently and what the implications are for tourism development, by attempting to suggest the benefits of tourism as well as its costs. The changing patterns of tourism consumption are nicely traced, which reminds the reader of what the ultimate motivations and aims of tourists are and the impact of their behaviour on host communities, including the demonstration effect of their lifestyle and status.

The illustrative study of a development in The Gambia is given to indicate the way in which 'Western' tourists' needs and demands can be met in an almost sustainable way and how the resort confers benefits on the local community in the form of such gains as fresh water, electric power, purchase of local goods and generation of employment. The authors conclude that changes in consumer culture can be accommodated in a redefined sustainable tourism context by offering a quality product which conforms to the meeting of sustainable tourism objectives.

Though gently done, the chapter raises issues which are controversial and likely to be hotly contested by deep-green ecologists/conservationists. It does, however, represent a pragmatic approach which acknowledges both the nature of demand by tourists and the likely response by the tourism industry.

Ireland (Chapter 17) analyses, at both a corporate and individual level, to what extent socially responsible tourism management strategies can be devised to secure sustainability. In a study of the situation in Cornwall in the UK, by means of the review of documents which might show the nature and extent of social responsibility with respect to the environment, he sets out an agenda for its practice. In doing so, a number of business ethics questions are identified and examined.

Chapter 18, by Hunter-Jones, Hughes, Eastwood and Morrison, reviews the decline in tourism to mainland Spain and identifies some factors explaining it, showing that the causes are more complex than simply demand changes. The prerequisites for a reversal of the trend are outlined and policy proposals for tourism and the environment in Spain, the Balearics and the Canaries are examined and the implications of strategies which include extension (spatially) and improved quality are considered.

Strategies for Sustainability: Lessons from Goa and the Seychelles

13

D. Wilson

Department of Social Anthropology, The Queen's University of Belfast, Belfast BT7 1NN, UK

Introduction

Up-market tourism is a commonly understood everyday expression although there are surprisingly few references to it in the academic literature (try looking it up in the index of any publication on tourism). How can such professional reticence be accounted for? One possibility is that up-market tourism is indeed frequently discussed in the literature, but under a variety of other headings. Thus much 'alternative', 'special interest', 'green' or 'sustainable' tourism may be 'elite' tourism in another guise, as authorities as diverse as Wheeller (1994) and Albuquerque and McElroy (1995) have noted. But in that case why avoid such a frequently used term? Perhaps it is simply too vague for academic purposes. Another explanation, not quite so obvious, is that it is avoided because of awkward ethical problems embedded in the concept. It may be ethically problematic because up-market tourism is an elite form of tourism which it is not politically correct to advocate in an age of democratized travel and tourism, even though it may appear to bring benefits which are desirable in terms of sustainable tourism objectives, such as higher earnings from smaller numbers. Governments, however, are not as hesitant as academics and often openly embrace up-market tourism as a strategy to achieve sustainable tourism development. It is uncritically promoted by the governments in both Goa and the Seychelles as a prime policy objective. This chapter contrasts their dramatically different developmental trajectories and assesses the success or otherwise of their respective drives to establish an up-market tourist industry.

© CAB INTERNATIONAL 1997. *Tourism and Sustainability*
(edited by M.J. Stabler)

Tourism in the Seychelles

Tourism in the Seychelles has, from its inception, been a government-sponsored, tightly-controlled, internationally-focused, up-market industry. The first of a succession of development plans advocating tourism development was published in 1959 at a time when the Seychelles was still an impoverished British colony suffering from declining copra production while experiencing a rapidly rising population. There were only a handful of visitors to the islands prior to the opening of the international airport in 1971, which sparked off a construction boom as a number of international hotels were built along the edge of the coastal plateaux. These hotels were built on prime coconut plantation land which was available for sale, their owners being glad to make a windfall profit on their decaying estates. Such space enabled the hotels to extend outwards instead of upwards, a practice which was enforced by government legislation requiring that hotels be set back from the beaches and conform to a maximum height of two or three storeys (the height of the surrounding palms). From the very beginning, the government has exercised strict control over all tourism-related building and construction. Thus the spacing out of new hotels around the main island of Mahe prevented the emergence of enclaves and created new employment opportunities in the outlying areas. With regard to conservation, the 1960s and 1970s saw the establishment of extensive nature reserves and marine parks, the emergence of Aldabara as one of the natural history wonders of the world, the outlawing of spear fishing and a ban on the killing of green turtles, which, along with other environmental legislation, established an international reputation for conservation on the islands (Wilson, 1979).

On his return to the islands in 1991, the author discovered a thriving and successful tourist industry, notwithstanding some dramatic fluctuations in numbers and changes of image during the intervening years since the first visit (Wilson, 1994a,b). There were no obvious signs of discontent among the Seychellois and the majority of tourists enthused positively about their experiences. For example, many remarked how safe they felt on the islands, expressed surprise at the absence of any begging culture and the lack of any hassle from beach vendors or shopkeepers. It can be suggested that a fortuitous conjunction of factors was responsible for the favourable response shown by the Seychellois towards the tourists in their midst and which included their cultural familiarity with European ways, a high level of racial tolerance, a reasonable standard of living, low unemployment, and a generally held belief that the benefits of tourism were being fairly spread throughout the population as a whole (this latter in part a result of the health, education and welfare policies of the revolutionary government). Nevertheless, there were some shadows on the horizon for the industry and these included low occupancy rates, the increasing dependency of the economy on tourism, growing environmental problems, and an uncertain political future. President René's Marxist-based, one-party system was breaking down at the time of the visit and a timetable had been set for the reintroduction of democratic multi-party elections. It was concluded, pessimistically, that the enthusiasm for tourism shown by the poorer Seychellois could rapidly evaporate if a more right-wing government were to gain power and place greater emphasis on market forces and the pursuit of profit (Wilson, 1994a).

Recent Developments in the Seychelles (1995)

In contrast to the generally favourable image of tourism received from Seychellois and visitors alike in 1991, the impressions gained during a second return visit to the islands in 1995 were far more negative. A growing number of complaints, reviewed below, suggest that the government's hitherto generally successful up-market tourism strategy may yet again be in jeopardy.

Rising prices

Prices had risen steeply since the general election in 1993. This included many everyday items as important to the Seychellois as to the tourist, such as fish, beer and vegetables. Each increase had its own local explanation – fish prices had gone up because of the greed of the fishermen, alcohol because of a recent doubling of tax by the government, and vegetables because they were imported and their price had been inflated by the foreign exchange crisis. However, such explanations offered no consolation. As one landlady said succinctly: 'It is all right for ministers with their cars and houses, but how are ordinary Seychellois going to survive?' As tourist numbers were low at the time of the visit, excessive demand from the hotels could not be responsible for the high prices, nor did one hear any such accusations. Tourists were also very critical of these high costs. Many said that they preferred nearby Mauritius, where prices for the above commodities were about a third of those in the Seychelles.

One reason why the Seychelles was becoming so expensive was said to be the acute shortage of foreign exchange. Because of the long delays in getting foreign exchange through the official 'forex pipeline' (whereby foreign exchange was allocated on a first-come first-served basis by the Central Bank of Seychelles in an attempt to ensure that everyone had equal access to foreign currency), overseas suppliers were charging interest on the delay (sometimes up to nine months) in receiving payment, which was then passed on to the customer by the importer. Alternatively, because merchants were having to pay more on the unofficial 'parallel market' (whereby foreign exchange was bought and sold privately at an inflated rate) for the foreign exchange they needed to purchase their imports, they were again passing these additional costs on to the consumer. Not only did this explain the escalating prices for imported goods, the long delays involved also accounted for their growing shortage as well. The shortage of foreign exchange was also said to be responsible for a stagnating economy, declining manufacturing and construction (reflecting the non-availability of imported components), and manpower lay-offs and redundancies.

Crime and safety

The escalating cost of living was seen by many as contributing to the rising crime rates, especially for theft (against both Seychellois and tourists). 'The poor are finding it harder to make ends meet and this is why theft is increasing, especially on Mahe, and the government are trying to conceal its extent so as not to frighten the tourists away,' said another landlady. People also expressed similar concerns about the growing danger of assault and, unlike 1991, the advice was not to walk along the beach at Beau Vallon after dark, although this advice was ignored by everybody with no adverse consequences. In response to the concerns, the government was about to establish a new 'tourist police' to patrol the beaches, as thefts and assaults on tourists were seen as a real threat to the industry.

Tourist growth

Numbers had fallen from a peak in the early 1990s. Visual evidence in 1995 suggested very low occupancy rates in several of the main hotels and two of the smaller hotels visited had no guests at all during the week prior to Easter. Seychellois also felt that business had slumped that year, a decline which many blamed on the high prices and alleged rising crime. The deteriorating condition of many of the hotels was also pointed out by informed observers as another reason why the Seychelles was losing its reputation as a good up-market destination. With regard to the provision of new facilities, slow but steady growth would best describe the general situation in the islands. There were plans for new hotels on St Anne Island and Frégate Island, and another at Beau Vallon. The most noticeable amount of development since 1991 was to be found on the small island of La Digue, with at least eight new guest houses crowding alongside those previously established. Local inhabitants were beginning to complain that there were now too many tourists in their midst. The government had been steadily selling off its own hotel holdings (at one point it controlled about 25% of the bed capacity of the islands) and by 1995 only one was still owned and managed outright by the government. The main problem with state ownership in the hotel industry was that the government had neither the requisite managerial skills nor the financial means to refurbish its deteriorating stock and so had embarked first on a strategy of private contract management and then on a policy of full privatization. In fact, there seemed to be considerable activity in the industry, with several major hotels in the process of changing hands and a number of smaller privately-owned hotels up for sale. This could indicate growing economic difficulties faced by the tourist industry in the Seychelles or alternatively reflect a jockeying for ownership and control during a period of post-revolutionary liberalization and the anticipation of richer pickings ahead. At least three of the large hotels were about to embark on massive re-investment and refurbishing programmes.

Poor service

Unlike 1991, many complaints were heard about poor service and the unfriendliness of the Seychellois, who were variously described as offhand, casual, indifferent, uncaring, slow, surly and careless. Some local business-men blamed the revolutionary government for this because of its ambivalent attitude towards tourism and the education given to Seychellois youth during their service in the YHS (National Youth Service), where they had been taught that Seychellois were no longer slaves to colonial masters, and that tourism was a form of neo-colonial exploitation. Most young people working in bars, hotels and restaurants today would have been through the YHS. Yet the government was aware of this growing image problem and in 1993 had instituted a 'You-First Campaign' to encourage Seychellois to be more welcoming towards tourists and give them better service. The campaign was designed to sensitize Seychellois to the problem by rewarding examples of good practice (rather than punishing bad examples) with certificates to the most outstanding individuals and extensive publicity in the local press. One opposition newspaper suggested other possible solutions:

> Broad smiles and tropical charm have been replaced by poor service and foolish pride. Soaring prices and low standards are becoming our trademark. The U-first campaign is a step in the right direction but other changes are also needed such as a tourism standards agency with statutory powers (and) quality training and compulsory periodic re-training.
>
> (*Regar,* 1995, vol. 4, no. 1, p. 4)

Cruise tourism

The growing presence of cruise ships, especially those anchored between Praslin and La Digue, annoyed many stopover visitors and locals alike. The number of cruise passengers was approaching 10,000 in 1993 (Migration and Tourism Statistics, 1993) and the number of ships had increased dramatically from three in 1990 to 50 or 60 in 1994 (Jaggi, 1994). 'They swamp the place' was one typical response from a Seychellois living on the small island of La Digue. However, the government has been actively encouraging cruise tourism and several ships were now starting or finishing their Indian Ocean cruises in the Seychelles, their passengers either arriving or departing by air. On Mahe, this additional influx of tourists was not so noticeable, but on the smaller islands they were highly visible and could cause disruption. For example, on Praslin local buses were used to take cruise passengers to the beaches and to the Vallai de Mai, thus depriving Seychellois of their own regular services. This could happen several times a week during the season. Some wondered just what the Seychelles gained from these cruise tourists. One local businessman was very sceptical, suggesting that all they got was a few extra taxi fares and a small increase in beer sales. Nor did the cruise ships pay any head tax, although the government charged port fees, pilot fees and lighterage fees, and also argued that cruise ships stimulated the local arts and craft industries. Even though cruise tourism was a seasonal activity (from November to April during the calmer north-west monsoon) its popularity seems set to increase further, reflecting the drawing power of the Seychelles' safe anchorages, renowned diving and other marine excursions, attractive natural history, famous beaches and convenient mid-oceanic location. However, a number of questions can be asked about this rapid growth of cruise tourism. Could the strict government control which has previously characterized tourism development in the Seychelles be weakening? Does cruise tourism provide a timely loop-hole for a cash-strapped government to exploit by enabling it to increase tourist numbers and revenue without increasing the number of beds (thus covertly breaking their own slow-growth policy)? Was cruise tourism welcomed by the government because it was seen as another form of up-market tourism which did nothing to detract from their exclusive marketing image?

Environmental issues

Despite the fact that national parks and nature reserves cover over 40% of the land area, the Seychelles environmental record was not without its critics. Although individual projects were winning awards for their environmental sensitivity, such as the redevelopment plan for Bird Island or the network of trails which had been laid out in the nature reserves on Mahe with attractive and informative guide books to accompany them, other projects had raised fears about conservation. For example, the plans for a five-star hotel on Frégate has aroused concern over the future of the magpie robin and had been criticized for its exclusivity as access to the island would be for hotel guests only, thus excluding Seychellois from yet another of their islands. The government had continued to take action to control the coco-de-mer market and had introduced legislation which required the official registration of all coco-de-mer nuts and artefacts in either private or commercial ownership. Any nuts or products which had not been declared by April 1995 were considered to be illegally possessed and their owners liable for prosecution, although many doubted the effect these regulations would have on those intent on trafficking in coco-de-mer. Similarly, the government had introduced legislation in 1994 forbidding the killing of turtles, the possession of their eggs, the selling of turtle products, and required the registration of all hawksbill shells in private or commercial ownership, yet critics had again pointed out that there was no effective enforcement of the legislation pertaining to live turtles and their eggs and that there had been many reported violations. The government had also withdrawn unexpectedly in 1994 from the International Whaling Commission, in which it had played a significant role in calling for whale sanctuaries. The official reason was to avoid being 'caught between two opposing groups' whereas critics said the government had been bought off by £8 million in grant aid to the fisheries sector by Japan (Jaggi, 1994). The most visible environmental change to be observed in the Seychelles over the last decade was the east coast land reclamation project. It must be remembered that the Seychelles exists primarily on its coastal zone, practically all economic activity is to be found there, and conflicts can occur between the varied interests of those who depend on the zone – tourism operators, coastal home owners, fishermen, farmers, factories, shipping and many others (Shah, 1991). Environmental problems here included preserving coastal habitats, controlling the erosion of the shoreline and dealing with the increasing deposits of silt in the lagoons created by the reclamation. In 1995, the comprehensive new Environmental Protection Act came into force aimed at dealing with such issues and among its requirements were environmental impact assessments for any new hotel or marina projects.

Marketing strategies

An awareness in the early 1990s that the promotion of the Seychelles as a sun, sand and sea destination was losing its uniqueness had prompted the relaunch of the islands as 'Nature's own Paradise' with a greener image and an increased emphasis on their other natural assets. However, the small scale of the industry meant that resources for marketing were severely limited, especially in comparison to some rival destinations. The government saw the pooling of marketing resources with other Indian Ocean destinations, especially those which offered two-centre holidays with the Seychelles, as one way of increasing the effectiveness of its advertising (National Development Plan, 1990–1994). Joint marketing arrangements had been made with other destination countries in the region, for example with Mauritius and

Kenya. However, one problem here was the unfavourable comparison many tourists made with Mauritius (lower prices and better value for money), which could create a wave of adverse publicity seriously damaging to the image of the islands abroad.

Reflections on the Seychelles

The impact of tourism on a host community cannot be understood without firmly locating tourism in its local political, economic and cultural contexts. Regarding politics, the first multi-party general election for 17 years, held in July 1993, was unexpectedly won by the ex-dictator Albert René, perhaps because the ordinary Seychellois felt he better represented their interests than the more right-wing opposition parties. Whatever the reason, he has since embarked upon his own policy of economic liberalization and privatization. The islands seem to have entered a period of relative political stability with Sir James Mancham, leader of the main opposition party, espousing a policy of reconciliation rather than recrimination. Economically, however, the islands were experiencing a difficult phase. With the end of the cold war, bilateral aid from international powers seeking influence in the strategic geopolitical arena occupied by the Seychelles has plummeted and the American Satellite Tracking Station located on the islands (which has long been one of the main foreign exchange earners) is due to be closed by October 1997, losing the Seychelles US$4.5 million in rent and the loss of 150 jobs for Seychellois employed in the services sector (*Indian Ocean Newsletter*, No. 686, 23rd September 1995). Such factors are as much responsible for René's conversion to free-market economics as any political change of heart, and, with 3000 new school leavers looking for jobs each year, he and his government have their work cut out to generate new investment and employment opportunities. The government is trying hard to develop a third strand to the economy, alongside tourism and fishing, and, with the continued failure of oil exploration to find the deposits everyone in the Seychelles believes are out there somewhere, a lot of effort has been put into trying to establish the islands as an offshore service centre. However, the government has received considerable adverse publicity in the international press as a result of legislation it introduced in 1955 which seemed to be actively encouraging the use of the islands for laundering drug money. It did this by:

> ... offering foreign investors immunity from prosecution by any police force in the world in return for a payment of $10 million. This 'subscription' would give them diplomatic status and Seychelles citizenship. Further, the government would guarantee to protect their assets on the islands if foreign authorities tried to seize them ... In short, millionaire criminal gangs were being offered carte blanche to conduct their business in the Seychelles.
>
> (*The Sunday Times*, 14th January 1996, section 1, p. 12)

Visitor numbers have fluctuated over the last few years and from an all time high in 1990 of 103,770 they have oscillated as follows: 90,050 in 1991 (reflecting the adverse effect of the nearby Gulf War), 98,547 in 1992, 116,180 in 1993, and 109,901 in 1994 (Statistical Bulletin, 1995). This most recent slump is said to be partly due to increased competition from other long-haul destinations, such as Mauritius and the Maldives. The number of beds available has continued to increase slowly but steadily from 3680 in 1991 to 4240 in 1994. Bed numbers have been limited to 4000 on the islands of Mahe, Praslin and La Digue, and to 500 on the outer islands, in order to help with 'preserving the reputation of Seychelles as an exclusive and unspoilt holiday destination' (National Development Plan, 1990–1994). The occupancy rate has fallen from a 1990 high of 67% to 54% in 1994, mainly due to this increasing capacity (Statistical Bulletin, 1995).

So can the Seychelles' up-market strategy still be described as successful and an outstanding example for other small island microstates to follow in spite of the current hiccups, or is the start of an inevitable decline being witnessed? Government strategy has been to target 'the middle to higher echelons', although it is aware that 'service and facilities must improve considerably if this strategy is to succeed and be sustainable' (National Development Plan, 1990–1994, Vol. 2). In 1994 the government issued a memorandum outlining the goals for which the tourist industry should aim. These included commitments to limit the number of large hotels, encourage the development of smaller establishments, promote the development of island-hopping holidays and the expansion of facilities in the largely uninhabited outlying islands, support the continued improvement of sports facilities (such as yachting, golf, boat charter, diving and fishing), increase its calendar of special events (especially during the low season), and open up new places of interest in the national parks and other areas associated with special interest and alternative holidays (Ministry of Tourism and Transport, 1994). All these policies were designed to increase regional competitiveness and broaden market appeal whilst maintaining the Seychelles' status as an up-market resort. However, pressure to keep expanding tourism in an economy now largely dependent on it for jobs and income, especially in a reformed political system in which votes are now at stake, remains a contradiction at the heart of tourism development in the Seychelles.

One recent report has attempted to assess how close the Seychelles has come in practice to achieving its objectives. Writing specifically about La Digue, Karkut (1995) notes four major 'discontinuities' between sustainable principles and actual practice. The first concerns poor channels of communication between government and many members of the local community who feel excluded from the decision-making process. The second relates to the existence of politically-based patron–client networks which by-passed the official regulations, supported vested interests on the island, and prevented an equitable distribution of the benefits created by tourism. The third appertains to an inadequate system for training local supervisory and management personnel, thus reinforcing the heavy reliance on the expertise of expatriates. Finally, insensitive and heavy-handed policing, although aimed at reducing thefts against tourists, had the opposite effect of making tourists feel less secure due to the unexplained presence of armed police and troops on the beaches (Karkut, 1995). Notwithstanding these difficulties, he considers that the trend on La Digue towards small-scale projects was in keeping with recent government policy, best suited to the needs of the island, as well as being the form of development preferred by the islanders themselves. He concludes:

> The bedrock for sustainable practice has been established. Legislation and statutes are in place to oversee thoughtful tourism policies such as the limiting of bed numbers, dictating the pace of development through a restrained allocation of operational licenses, and crucially in the case of La Digue, acknowledging that small-scale development does seem to be the most appropriate and sustainable option for the island.
>
> (Karkut, 1995, p. 63)

Groote and Molderez (1993) also find that there is still room for improvement and recommend, amongst other things, that the Seychelles government must try to: encourage economic diversification away from a total reliance on tourism; improve the seasonal spread of tourism (perhaps by lower prices during off-peak months, or selectively marketing activities appropriate for the season); encourage other categories than 'rich tourists' to visit the Seychelles as it is not at all obvious that rich people are any more respectful to local people or sensitive to the local environment than any other tourists; improve both service and facilities; promote more ecotourism, not forgetting that the education of the tourist is as important as educating the guides who look after them; and broaden the marketing image away from that of merely a sun–sand–sea–paradise destination (Groote and Molderez, 1993).

At a conference on the concept of 'quality tourism' held in the Seychelles in 1991 (another academic euphemism for up-market tourism?), two government officers (Shah, the Director of the Environment Department, and Chetty, the Director of the Tourism Department) also made suggestions for future development. Shah (1991) was concerned that the present 'beach tourism' focus made the industry especially vulnerable to pollution disasters. He advocated a broader 'nature tourism' approach. However, nature tourism is not automatically synonymous with 'ecotourism' which 'directly benefits the wildlife and protected areas on which it is based' because in the Seychelles most of the revenue accruing from tourism in national parks and reserves 'is not directly used for protection of the environment'. The Seychelles must actively promote ecotourism (Shah, 1991, pp. 101–102). Chetty and St Jorre (1991) assume somewhat optimistically, given the considerable criticism by tourists of the high cost of a Seychelles holiday, that increasing quality will 'provide for increases in revenue through higher tariffs and rates charged to the customers' (p. 106). They also reveal another means whereby the government can increase the number of beds beyond the allotted limit, as 'accommodation capacity provided for domestic holiday-makers will be excluded from the limit' (p. 106). Taken as a whole, these reports also indicate that the Seychelles continues to make great efforts to monitor the progress of its up-market strategy and modify its objectives. However, sustainability is ultimately about viability, and viability in the Seychelles today is about maintaining the relatively high standards of living which were in the past funded to a considerable extent by generous aid from rival superpowers. This economic viability now depends on tourism and the ability to protect the environment which attracts tourists to the islands. To this extent, and unlike Goa, as indicated below, the Seychelles seems to have little option other than to continue along its present trajectory.

Tourism in Goa

In contrast to the Seychelles, international tourism in Goa has emerged as an unplanned ad hoc response to growing numbers of arrivals and is at present undergoing rapid transformation into a major package holiday destination. There have been tourists in Goa since the 1960s. Initially, the area was renowned as a haven for Western hippies and low-budget travellers, although, in recent years, the state government has encouraged the expansion of up-market tourism as Goa acquires a reputation as the Riviera of the Indian subcontinent. There is a well-organized local anti-tourism lobby which argues that a more appropriate comparison is to Benidorm. The lobby has been remarkably successful in publicizing the destructive nature of Goan tourism (see Srisang, 1987; Jagrut Goenkaranchi Fauz, 1991; Bailancho Saad, 1993; Ecoforum, 1993; Lea, 1993; Noronha, 1994; Häusler, 1995). The horrors of Goan tourism have also been disseminated by Tourism Concern (Badger, 1993; Almeida, 1995) and the Ecumenical Council on Third World Tourism (O'Grady, 1990), ventilated in the British Press (Nicholson-Lord, 1993, 1995; Lees, 1995) and exposed on British television (Anderson, 1995; BBC, 1995). The main accusations concern the:

- overdevelopment of the coastal strip.
- flaunting of planning regulations (especially those forbidding any construction within 200 metres of the high tide line).
- danger of exhausting the underground aquifers through overuse.
- threat of salination from wells located too close to the sea.
- preferential access to scarce water supplies by the hotels.
- environmental degradation (including destruction of sand dunes, pollution of sea water by hotel sewage, dumping of garbage and removal of beach sand for construction purposes).

- restricted access for local inhabitants to beaches.
- intimidation of villagers out of their homes as property developers attempt to buy up the remaining coastal strips.
- objections to nudity.
- drugs and moonlight parties held by the 'hippies' (a term still used in Goa to refer to low-budget travellers and young backpackers).
- allegations of increasing prostitution and paedophile activity.
- corrosion of local cultural values.

Taken as a whole, this adds up to one of the most powerful indictments of international tourism to be found in the literature, and for convenience it is referred to below simply as 'the critique'.

The author's first visit to Goa was made in 1994 and the following comments reflect some initial observations (Wilson, 1997). Whilst development had been haphazard and uncontrolled, the construction boom had not led to a rash of high-rise beachfront developments, such as are found in Spain and many other European resorts. Goa was not yet perceived to be a concrete jungle. Commentators' comparisons with Benidorm and similar resorts were grossly inaccurate and misleading. The beaches were not overcrowded by European standards and few hotels or guest houses could actually be seen from the beaches, the majority being small in size and modest in height. Nor was violence or prostitution rampant, the beaches were safe to walk at night, women did not feel threatened or harassed in any way and most tourists enthused about the welcome they received in Goa. It would appear that the critique has exaggerated these problems whilst ignoring other issues, such as government and police corruption and the impact of domestic tourism, which has been estimated to account for 90% of the state's tourists.

However, the rapidly accelerating rate of tourism development was transforming the structure of tourism in Goa and the traditional blend of domestic and low-budget tourism, with its relatively low impact on the local environment, was being replaced by both cheap package tourism and a growing number of up-market hotels. It can be argued that there were positive aspects to the tourism industry, as it had been traditionally structured, which are unappreciated in the critique. These advantages include the wide local ownership of resources and the broad distribution of benefits throughout the community, a view which would appear to be endorsed by the general welcome extended to tourists in Goa. In a survey of tourism development along the 14 km stretch of palm-fringed beaches and bays between Aguada Fort and Chapora Fort, the main tourist area to the north of the capital Panaji, it was estimated that there were a total of 149 hotels and guest houses (including two five-star international hotels), of which 31 were used by the British package tour operators who control 80% of the charter business, the remainder being used by domestic tourists and low-budget Western travellers. In addition, many local people rented out rooms to these travellers and an indication of the size of this market could be gained from the presence of no less than 86 cafes in the Anjuna-Chapora area (the main destination for young people looking for the beach party and rave scene) alongside a mere 16 hotels and guest houses. Finally, along the entire strip, there were approximately 208 cafes and restaurants, 119 beach shacks (serving food and drinks), 169 shops and roadside stalls, 245 beach vendors, and innumerable rickshaw, car and motorcycle taxis. The vast majority of these diverse enterprises were small, independently-owned businesses run by Goans. Most of the remainder were owned by Indian migrants from Kashmir, Rajasthan and Karnataka.

Recent Developments in Goa (1995)

As with the Seychelles, a return visit was made to Goa in 1995, although in this case only a year after the first. Some of the most contentious issues encountered in the later visit concerned conflicts of interest between different market sectors in a situation of uncontrolled growth. However, it is instructive, first, to offer some comments on tourist perceptions.

Tourist perceptions

Unlike the Seychelles, most tourists were generally enthusiastic about Goa, praising its climate (during the peak New Year holiday period), the friendliness of the people, the relaxed easy-going atmosphere, and the low cost of goods and services. However, many first-timers were initially shocked by their first sight of Goa, especially when reaching the area covered in the survey. This did not resemble any European resort – crowded dusty roads with no pavements, teeming with people going about their daily business, the obvious poverty – but, after they had discovered the beach and settled in, their attitudes would change dramatically and many expressed a desire to return again the following year. A number remarked on the hospitality they had received in the beach shacks (some getting invitations to visit local people's homes) and others had got to know and like their 'regular' beach vendors. Not all were so pleased, however, and the few that did complain were often regular returnees who commented on how much the local people had changed in recent years, getting greedy, charging exhorbitant prices, and becoming less friendly towards tourists as well. But they still returned year after year without fail! Tourists as well as locals have their myths of a lost golden age, and never more so than in Goa. One or two package tourists encountered had also found it all a little too 'basic' and would not be returning, although most of these said they had enjoyed the experience.

Uncontrolled growth

Construction

Construction was increasing at an alarming rate throughout Goa. There were at least 61 new medium or larger buildings (either apartments, guest houses or hotels) under construction between Aguada and Chapora alone. One recent guidebook has described the impact of the construction boom around Calangute (the most developed part of the strip) in the following terms:

> As the rash of construction sites around the outskirts blossom into larger resort complexes, what little charm Calangute retains looks set to disappear. Without adequate provision for sewage treatment or increased water consumption, it's only a matter of time before the town starts to stew in its own juices, putting off the very tourists the developers are trying to attract.
>
> (Abram, 1995, p. 114)

Officially, there had been no relaxation of the ban on any new construction within 500 metres of the high-tide line and the Chief Minister had called for a review of all constructions within this area, how these constructions had come to be permitted and what remedial action should be taken against them. At the beginning of 1996:

> The state government has ordered that all constructions, including repairs and renovations, undertaken without the approval of Goa state committee for coastal environments, within 500 metres from the high tide line, be demolished within a month, failing which the government would initiate action for their demolition and expenses incurred would be recovered from them.
>
> (Navhind Times, 4th January 1996, p. 3)

Notwithstanding this, it was still the view of the same local paper that:

> there are disturbing signs of disintegration ... Development in Goa is going on in haphazard manner without any sort of proper planning ... with reckless speed without much thought for basic amenities.
>
> (Navhind Times, 31st December 1995, p. 5)

In spite of laudable government statements of intent, nobody expected it to take action, except perhaps for a token demolition or two. Time to let the fuss subside and baksheesh in the hands of the necessary officials and police were all that was needed before building could be resumed, or so one was told.

Dabolim airport

Another example of the planning chaos was the government's failure to enlarge the handling capacity of the airport, even though the number of flights was rising rapidly. Three times out of the four the author passed through Dabolim there were delays because the apron was full; most tourists had similar stories to tell. The airport terminal held a maximum of two jets and one had either to circle overhead or park on the military side of the airport waiting for space to become available on the apron;

passengers had to endure a long wait in a crowded lounge. The Indian defence forces, which had seized control of the airport during the 1961 invasion, had never completely returned it to civilian hands (an omission which caused great resentment in Goa), allowing only part of it to be used for civilian traffic. Although the terminal building was being extended to handle more passengers, there were no signs of similar activity out on the apron! Unconfirmed reports alleged that the number of international charter flights had doubled over the previous two years; the number of domestic flights had increased as well. Clearly, Dabolim airport has reached its capacity and, unless the apron is enlarged, conditions look set to deteriorate even further. The government had contentious plans for a new international airport in Goa and had two sites shortlisted for the project (Mopa in Pernem and the Siolim plateau) but the conservation lobby was fighting hard against both of them on the grounds of noise pollution and environmental degradation.

Taxi wars

The unregulated way in which tourism development proceeds in Goa was exemplified by the ongoing dispute between taxi drivers and the coaches hired by the package companies for airport/hotel transfers and for sightseeing tours. This threatened the livelihood of the taxi drivers by diverting what they saw as their traditional business (airport transfers) onto the increasing number of coaches. Tactics employed by the taxi drivers included the blocking of roads used by the buses, or surrounding the buses themselves to prevent their departure from the hotels. The taxi drivers had petitioned the government not to permit the coaches to run on certain days. In reply, Chief Minister de Souza had assured the coach operators that they would continue to be allowed to take tourists to and from the airport to the hotels and run tours to the main attractions, such as the churches of Old Goa. The problem was that the number of taxis (said to be around 5000) had been allowed to escalate out of all proportion to the available tourist business for them by the government's inability to restrict the issuing of licences, in spite of assurances that this would be done. The numbers of taxis had also increased rapidly because taxi driving was seen as a way to get rich quick. It was said on the coastal strip that if a person drove somebody else's taxi for a year, they could afford to buy their own the next, and that each subsequent year or two would add another taxi to their fleet.

Beach wars

In the early 1990s there had been an explosion in the number of beach bars or 'shacks' as they were locally known and, as in the case of the taxis, there seemed no control over the issuing of licences. The central role of licences in Goa was referred to by some as the 'licence raj' because it was baksheesh which secured licences not the token official fee, and because it was such a substantial source of extra income for the issuing authorities that ways around the official regulations would always be found. Beach shacks were constructed out of matted palm fronds attached to a framework of poles and all served hot food as well as cold drinks. Licences had to be obtained from the Tourism Department, otherwise shacks could be dismantled by the police. In 1995 the licences had been withheld, but most of the shacks were still standing, doing brisk business, with their owners up in arms against the government's attempt to abolish them. Groups of shack owners had even distributed pamphlets to tourists asking them to boycott Goa until the shacks issue was resolved. The government had also announced that no food was to be cooked in beach shacks, but without exception they all still did! It was said that the government was responding to pressure put on them by hotel and restaurant owners whose businesses had suffered badly with the growth of the shacks (most tourists preferred to stay on the beach rather than eat in the often dusty and noisy roadside restaurants). As one correspondent to a local newspaper put it:

> The very fact that the government shows no inclination whatsoever to rigidly impose the regulations framed by themselves is a clear indication of the existing lack of authority, due to political games being played by honourable ministers for their personal gains and interest.
> (*Herald*, 30th December 1995, p. 6)

This unwillingness or inability of the authorities to enforce their own legislation in such a visible issue as beach shacks exemplifies the extreme difficulty in Goa of enforcing any regulation and points to the near impossibility of dealing with more complex planning irregularities where much larger sums of money are at stake.

Up-market tourism

Up-market hotels have borne the brunt of the critique of tourism in Goa. However, the first observation to make about them is their relatively small number and their wide dispersal around the beaches. The establishment advocates continued expansion of this sector and Chief Minister de Souza has emphasized the need for high quality tourism and for new golf courses to further attract these high-class tourists (*Herald Illustrated Review*, vol. 96, no. 2, p. 18). In 1995, for the first time, Goa participated independently of the Indian delegation in the London World Travel Market Fair (the world's biggest fair for the tourism trade) in 'a major step to attract up-market tourists from Europe to Goa' (Almeida, 1995). The owner of the prestigious Leela Beach Hotel offered his own analysis in inimitable style:

> ... more 'quality' tourists should be encouraged to Goa, 5 star hotels should be tax exempt on the grounds that they bring prosperity to the state — each 5 star hotel employing around 1,000 people ... Two-star

tourism should be discouraged because it clutters the beaches and does not provide sufficient income.

(Herald Illustrated Review, vol. 96, no 2, p. 9)

Water shortages, blamed in part on the rapacious consumption of water by the big hotels, were said to be getting worse, although only towards the end of the season, and the government had banned tube wells in an attempt to help preserve stocks. Luxury hotels required up to '66,000 gallons of water per day for their swimming pools, lush green lawns and 24-hour water supply' (Almeida, 1995, p. 6) and supplemented their reserves by buying water from inland villages with surplus supplies, bringing it in by tanker and pumping it into their own cisterns. The government was supposed to be building a supply dam but, as usual, no one seemed to know when it would be finished.

Low-budget tourism

There seemed to be more beer-swilling British in Goa in 1995 than the previous year, especially around Calangute, where rowdy groups of drunken tourists were encountered shouting and singing as they staggered home along the road at night. This increase can be attributed to the rapid expansion of the cheap package end of the market, but must not be confused with that other group of low-budget tourists, the backpackers and overland travellers, who descend on Goa in their thousands during the winter months. It is these visitors with whom this section is concerned. Their scene was a different one, based on drugs, rave music and alternative lifestyles. They made an important contribution to the local economy and the income of many poor families in Goa was based on them. Such families often started by renting out a room, then adding an extension onto the back, then providing food, then purchasing motorcycles to rent out to their guests, then building a new house, perhaps with a cafe attached to it, as their business acumen improved. In fact, the only way for many young men to acquire their own motorcycles was by renting them out to these budget travellers (another largely illegal activity which the police chose not to enforce, except occasionally to extract baksheesh from the tourists). For example, the author rented a Honda Kinetic which cost the owner 25,000 rupees on a bank loan, which had to be repaid at 750 rupees a month. His own salary at a two star hotel was 450 rupees a month plus tips for a 10 hour day. Tips during December and January were in excess of 1000 rupees a month, but throughout the remainder of the season he got just a few hundred. He could thus only pay off the loan by hiring his bike out to tourists at the standard rate of 150 rupees a day (£1 = 50 rupees).

However, this form of tourism, which had long worked to the mutual advantage of coastal villagers and long-stay travellers, was under threat at the end of 1995. The problem went to the very heart of what makes Goa attractive for these young people, namely parties. The 'moonlight party' scene in Goa had ceased to be a word-of-mouth affair and was becoming well publicized in Western youth magazines. Such publicity was bound to draw international censure sooner or later and, somewhat unexpectedly, it had come from Israel. There were many young Israeli long-stay travellers in north Goa, relaxing after their two years of national service, when the Israeli press ran a sensational exposé of the New Year's Eve rave at Anjuna, with headlines decrying the decadent, immoral, drug-induced behaviour of Israeli youth. The Israeli government had been forced to take action and had put pressure on the government in Goa to ban raves and no more had been held at least up until Easter 1996.

These parties were of course another example of the licence raj, for they could only be held after a licence for a 'sound system' had been 'obtained' from the police. Once a licence had been issued, the police would stay well clear of the rave (a regular location for one was actually behind a police station!) but if a licence had not been obtained the party would be broken up, sometimes by force. The police were not the only beneficiaries of these raves, which could attract up to several thousand young Westerners. A whole infrastructure would develop around them, with hundreds of local people involved in the provision of petty goods and services to facilitate the event and ensure its success. Moonlight parties clearly provided a much welcomed opportunity to supplement meagre incomes and they appear to be regarded as a sort of windfall crop to be harvested whenever the opportunity allowed. If the end of the rave scene is nigh and they are permanently banned, this could also mean the end of a form of tourism which has been of great economic benefit to many coastal villages in north Goa. Whilst the local press generally waged a campaign against such parties, either because it offended their sensibilities or because they thought it would obstruct the development of up-market tourism, the odd dissenting voice could still be heard which seemed a more accurate reflection of the local view:

> Anjuna and some coastal villagers depend on low-budget tourism ... The villagers of Chapora, Vagator and Anjuna and even small house owners cannot get the cream of tourism or the five-star variety and depend on low budget tourists ... (the parties) do not harm locals ... the coastal area of Bardez has survived because of low-budget tourism, banning this kind of tourism has only increased corruption ... banning beach parties in Goa will only benefit the neighbouring states as these parties are organised even in other states of India.
>
> (*Gomantak Times*, 4th January 1996, p. 1)

With an even greater number of young Westerners likely to be flocking to Goa in search of these legendary parties, it seems inevitable that there will be more conflict ahead as competing styles of tourism in Goa and the vested interests that support them continue to clash.

Reflections on Goa

Statistics are far harder to find for Goa than for the Seychelles, but some can be located. In 1980–1981 there were estimated to be around 400,000 domestic tourists and 40,000 foreign tourists in Goa (Angle, 1983). By 1994, the total number had risen to 1,886,000 (*Herald*, 9th January 1995, p. 5), of which 8% were said to be foreign (*Herald*, 3rd December 1994, p. 1), giving a total of around 150,880 overseas tourists. Hayit, presumably writing about the situation in 1992, records that:

> accommodation capacity has grown to over 270 tourist complexes with over 11,000 beds ... However, only 14 of these complexes ... meet the high standards set by western sun worshippers. Only five of them can really be considered luxury holiday domiciles.
>
> (Hayit, 1992, p. 10)

By 1995, there were about ten large international hotels and this number was set to double by the end of the century. The official line on tourism development in Goa was that it could continue to expand. A report commissioned by the government recommended that it could easily grow up to 46,000 tourist beds and projected that 'some 4.1 million tourist arrivals could be received at the optimum development' (World Tourism Organization, 1994, p. 95). As already noted, the government was also encouraging the development of more up-market tourism, and the Director of Tourism told one journalist that:

> Luxury tourism was the way forward. Hippies and backpackers do not bring in enough money. Package tourists might, but only if you allow in millions of them. International high-rollers bring in the money in low numbers.
>
> (Anderson, 1995, p. 10)

Whilst superficially attractive, it seems that this focus on up-market tourism is out of keeping with the traditional structure of the tourist industry in Goa, which is mainly low-budget and served by a multitude of small hotels, guest houses, rented rooms and a host of ancillary services, as has been shown in the survey above. Even though it restricts numbers, the critique has argued that it is up-market tourism which puts disproportionate pressure on both the socio-economic and environmental fabric of the local community. At the same time, the rapid expansion of cheap package tourism also carries problems for Goa. A British travel company representative has warned of the potential consequences:

> ...if mass tourism is allowed to develop unchecked in Goa, prospects are very bleak indeed ... Local government and developers remain bent on making a fast buck, without investing adequately in the future.
>
> (*Herald Illustrated Review,* vol. 96, no. 2, p. 13)

Like its up-market form, package tourism can be far less sensitive to local people and culture than the travellers and backpackers who accept without complaint the 'pig toilets' in the villages of north Goa.

There has been low-budget, relatively low-impact tourism for 30 years in Goa. There is little evidence of hostility and resentment among local people against the majority of such tourists and it has been suggested above that this is in part because the economic benefits they bring have been spread widely throughout the local community. Häusler (1995) argues that it is the higher castes who most resent tourism because it threatens their traditional status within the community. Not only do low-budget tourists require little additional infrastructure to support and accommodate them, they put the least pressure on scarce water resources and bring other less easily quantified benefits. Richter's comments are as applicable to Goa as to Bangladesh, about which she was writing:

> Net economic gains are often higher from tourism pegged at these groups with their long stays and their simple lifestyles than those requiring capital intensive facilities ... the social and political costs, although harder to measure, are likely to be less in terms of relative deprivation, conspicuous consumption, political payoffs, kickbacks, and political grandstanding.
>
> (Richter, 1989, p. 153)

Smith (1992), writing about Boracay in the Philippines, suggests that one problem with low-budget tourism is its fickle nature and that the back-packers may desert Boracay at any moment because of overuse, rising prices or pollution through poor sanitation, leaving in their wake 'a sadly disappointed host population for whom the cash benefits have been short term' (p. 157). If anything, it is the other way around in Goa. It is the backpackers who are gradually being displaced by the new hotels and apartment blocks being built for the package tourists and the increasing numbers of domestic tourists. They are being forced to move further north towards Arambol, one of the few remaining undeveloped sections of coast in the state.

For tourism to be sustainable in the social and cultural sense, it must be wanted by the local inhabitants themselves and it must be perceived as benefiting the majority of local people, not just an elite handful. It must be something which provides employment for the skilled as well as the unskilled and which generates opportunities for social and economic advancement. The jobs up-market tourism provides in Goa are mainly for menial low-paid jobs, such as waiters, room boys and ground staff, which do not allow local people to participate as petty entrepreneurs, establish their own competitive businesses and profit directly from the tourist dollar. Unlike in the Seychelles, up-market tourism in Goa is not wanted by many local people, not just the environmentalists, and this is another reason why continued large scale expansion of this sector would be unsuitable.

In Goa the real battle is over what sort of tourism will dominate the future and what its consequences will be. The government's present policy of advocating a massive expansion of tourism appears shortsighted and more likely to exacerbate rather than solve the existing problems. Sadly, the critique is extremely negative towards tourism in any shape or form and has no positive recommendations to offer. Instead, it has been argued here that a modified form of low-budget tourism might be the least destructive path to follow. Also, if the rate of growth could be slowed down, if planning regulations could be strictly enforced, and if suitable measures could be introduced to alleviate some of the other problems, there seems no reason why an acceptable balance between the costs and benefits of tourism could not be achieved in Goa, although on present evidence it seems unlikely that such a path will be followed.

Conclusions

Both the scale of tourism development and the speed at which it takes place are critical factors to be considered in any analysis of its impacts. Furthermore, as Jenkins has noted:

> Most countries have sought, as a policy objective, to maximise the numbers of tourists arriving in the country. This objective has often brought with it considerable problems ... A more controlled and lower level of arrivals is now aimed at in some countries ... It is now fashionable to describe small-scale tourism as *alternative tourism*. Whether such an approach is desirable, feasible, and possible will again require detailed examination not only of governments' objectives but also of the many factors which constitute a tourist destination.
>
> (Jenkins, 1991, p. 68)

Up-market tourism is one type of such alternative tourism and this chapter has considered the success of attempts to establish it in two long-haul destinations which are both enthusiastic advocates of this particular strategy for sustainable development. However, its perceived impact in the two resorts could not be more different. Put crudely, the strategy seems to work in the Seychelles but not in Goa. The arguments which have been presented can be summarized as follows.

Up-market tourism strategy works in the Seychelles because there is:

- slow planned growth with effective planning controls over hotel location, construction and building specifications.
- all year tourism which creates full time rather than seasonal employment.
- control over type of tourists arriving through pricing structures such as bans on cheap charter flights and strict hotel licensing system.
- strong environment legislation in place and a local population reasonably sensitive towards environmental issues, although there are doubts about the government's ability to police its regulations.
- a majority of Seychellois are in favour of up-market tourism which fits

in with their own self-image of the Seychelles as a high class resort and which also reflects the success of decades of government propaganda directed against young low-budget travellers, who are portrayed as immoral, drug-pushing, political subversives who must be kept out of the country at all costs.

- a general consensus that the benefits of tourism are spread throughout the community, although a changing political situation and rising prices might threaten this perception.
- no large scale influx of migrant traders or hotel workers who have been accused of displacing others, such as in Goa, where there are campaigns for Goan-only employment policies.
- no obvious detrimental impact on the long-term interests of the community caused by corruption.

However, success has to be sustained and the problems reviewed earlier show that success can never be taken for granted. The Seychelles is clearly facing another of its periodic crises, the outcome of which, as always, seems uncertain. The main difficulty with their strategy for up-market tourism is that they are not perceived as offering sufficient value for money in a highly competitive international marketplace, a perception which could precipitate a serious decline in visitor numbers. On the other hand, they are also approaching their self-set limits to growth of 4500 beds and must be under increasing pressure to revise these limits upwards should visitor numbers continue to rise. Indeed, plans for several new hotels are already in the pipeline, thus threatening the present balance between tourism and environmental protection.

Up-market tourism strategy does not work in Goa because there is:

- rapid uncontrolled growth of all types of tourism, which creates many environmental, physical and social problems.
- only a small part played by up-market tourism in a much larger tourism industry and any attempt to curtail these other competing types of tourism would create hostility, unemployment and political instability.
- an inability to control numbers and types of tourists entering and leaving; for example, most low-budget travellers and domestic tourists enter Goa overland from elsewhere in India.
- no clear development plan or policy; instead decisions are usually taken on an ad hoc basis and then rarely enforced effectively or for long.
- limited and ineffective environmental legislation and control.
- endemic corruption, especially in the form of the licence raj, which advances the pursuit of short-term individual profits but which is detrimental to the long-term interests of the community as a whole.
- a plausible case made in the critique that some up-market hotels have ignored the interests of the local community, especially in their drive to acquire development land, and that they impose a disproportionate

strain on scarce resources such as water supplies.

- widespread resentment against up-market tourism among most sections of the local population on the grounds that the big hotels control everything around them, pay extremely low wages and deny opportunities for small independent entrepreneurs and traders to make a living from tourism.
- seasonal tourism which sucks in migrants from other parts of India, causes considerable resentment among Goans and places further pressure on the local infrastructure.

Present day options are determined to a large extent by past decisions and once embarked on a particular trajectory it becomes increasingly hard to change direction. Thus Goa, with its headlong rush to expand all its different types of tourism simultaneously, is finding it almost impossible to regulate and control any single one of them. The elitist way in which up-market tourism is so favourably and uncritically contrasted with the despised low-budget alternative by the government is also not conducive to sound development planning. In sum, up-market tourism as a strategy for sustainable development depends for its success on a favourable conjunction of factors, which include the physical and cultural environment in which it is established as well as the ability of local authorities to implement and control appropriate development plans and policies. Put simply, these exist in the Seychelles but not in Goa. Up-market tourism is not an automatic formula for achieving sustainability.

Acknowledgements

The author would like to thank Margaret Ogg, Declan Quigley and Emil Wendel for their valuable assistance in the project on which this chapter is based.

References

References to local newspapers have been included in the text.

Abram, D. (1995) *Goa: The Rough Guide*. The Rough Guides, London.
Albuquerque, K. de, and McElroy, J.L. (1995) Alternative tourism and sustainability. In: Conlin, M.V. and Baum, T. (eds) *Island Tourism: Management Principles and Practice*. John Wiley and Son, Chichester.
Almeida, A. (1995) Tourism, water and women. *Tourism in Focus* Autumn, no. 17, 6–7.
Anderson, C. (1995) *Our Man in ...* BBC Books, London.
Angle, P.A. (1983) *Goa: An Economic Review*. The Goa Hindu Association Kala Vibhag, Bombay.

Badger, A. (1993) *Sweet Poison?* Press release issued by Tourism Concern, London, 26th January 1993.

Bailancho Saad (1993) Press release issued on 26th September 1993 (obtained through Tourism Concern, London).

BBC (British Broadcasting Corporation) (1995) *Our Man in ... Goa.* Documentary broadcast on 17th February 1995

Chetty, M. and de St Jorre, M.E. (1991) The concept of quality tourism in Seychelles. *AIEST (Association Internationale d'Experts Scientifiquess du Tourisme)*, 33, 103–120.

Ecoforum (1993) *Fish, Curry and Rice: A Citizens' Report on the Goan Environment.* The Other India Press, Mudra.

Groote, P. de, and I. Molderez (1993) A tourism development model based on sustainable tourism: the Seychelles, a unique archipelago in the tropics. Paper presented at the International Conference on Sustainable Tourism in Islands and Small States, held at the Foundation for International Studies, Malta, 18th–20th November 1993.

Häusler, N. (1995) The snake in 'Paradise'!? – Tourism and acculturation in Goa. In: Häusler, N. (ed.) *Retracing the Track of Tourism: Studies on Travels, Tourists and Development.* Verlag Für Entwicklungspolitik Breitenbach GmbFL, Saarbrücken.

Hayit (1992) *Practical Guide A to Z: Goa (India).* Hayit Publishing, London.

Jaggi, M. (1994) Special Report on the Seychelles. *Guardian*, 15th November, pp. 17–20.

Jagrut Goenkaranchi Fauz (1991) Tourism and prostitution: its implications for Goa today. *Contours* 5(2), 15–16.

Jenkins, C.L. (1991) Tourism development strategies. In: Lickorish, L.J. (ed.) *Developing Tourism Destinations.* Longman Group, Harlow, Essex.

Karkut, J. (1995) La Digue: a paradigm in paradise? An ethnographic study of small island tourism development and sustainability. Unpublished MA dissertation, Roehampton Institute, London.

Lea, J.P. (1993) Tourism development ethics in the third world. *Annals of Tourism Research* 20, 701–715.

Lees, C. (1995) Sex-hungry Britons ruin Indian paradise. *Sunday Times*, 19th February, p. 23.

Migration and Tourism Statistics (1993) *Republic of Seychelles Migration and Tourism Statistics.* Management and Information Systems Division, Victoria, Seychelles.

Ministry of Tourism and Transport (1994) *Tourism in Seychelles: A New Vision for Excellence (A Statement of Seychelles Tourism Policy 1995–2000).* Victoria, Seychelles.

National Development Plan (1990–1994) *Republic of Seychelles National Development Plan 1990–1994* Published by the Ministry of Planning and External Relations on behalf of the Government of the Republic of Seychelles, April 1990, Victoria, Seychelles.

Nicholson-Lord, D. (1993) Mass tourism is blamed for paradise lost in Goa. *Independent*, 27 January, p. 9.

Nicholson-Lord, D. (1995) Mass tourism 'is poisoning a paradise'. *Independent on Sunday*, 12th February, p. 6.

Noronha, F. (1994) Trouble in paradise: an update on Goa. *The Eye* 11(4), 44–45.

O'Grady, A. (1990) *The Challenge of Tourism*. The Ecumenical Coalition on Third World Tourism, Bangkok.

Richter, L.K. (1989) *The Politics of Tourism in Asia*. The University of Hawaii Press, Honolulu.

Shah, N.J. (1991) Nature and tourism in Seychelles. *AIEST (Association Internationale d'Experts Scientifiquess du Tourisme)* 33, 89–102.

Smith, V.L. (1992) Boracay, Philippines: a case study in 'alternative' tourism. In: Smith, V.L. and Eadington, W.R. (eds) *Tourism Alternatives: Potentials and Problems in the Development of Tourism*. University of Pennsylvania Press, Philadelphia.

Srisang, K. (1987) A Goan army against five-star tourism. *Contours* 3(3), 9–13.

Statistical Bulletin (1995) *Republic of Seychelles Statistical Bulletin: Visitor Statistics*. Management and Information Systems Division, Victoria, Seychelles.

Wheeller, B. (1994) Egotourism, sustainable tourism and the environment – a symbiotic, symbolic or shambolic relationship. In: Seaton, A.V. *et al.* (ed.) *Tourism – The State of the Art*. John Wiley and Son, Chichester.

Wilson, D. (1979) The early effects of tourism in Seychelles. In: de Kadt, E. (ed.) *Tourism – Passport to Development?* Oxford University Press, Oxford, pp. 205–236.

Wilson, D. (1994a) Unique by a thousand miles: Seychelles tourism revisited. *Annals of Tourism Research* 21, 20–45.

Wilson, D. (1994b) Probably as close as you can get to paradise: tourism and the changing image of the Seychelles Islands. In: Seaton, A.V. *et al.* (eds) *Tourism – The State of the Art*. John Wiley and Sons, Chichester, pp. 764–775.

Wilson, D. (1997) Paradoxes of tourism in Goa. *Annals of Tourism Research* 24, 52–75.

World Tourism Organization (1994) *National and Regional Tourism Planning: Methodologies and Case Studies*. Routledge, London.

The Evolution of Small-scale Tourism in Malaysia: Problems, Opportunities and Implications for Sustainability

14

A. Hamzah

School of Environmental Sciences, University of East Anglia, Norwich NR4 7TJ, UK (from April 1997: Department of Urban and Regional Planning, Faculty of Built Environment, Universiti Teknologi Malaysia, Karung Berkunci 791, 80990 Johor Bahru, Malaysia

Introduction

Small-scale tourism development in Malaysia began as 'drifter enclaves' alongside picturesque fishing villages (*kampung*), either along the mainland beaches or on the offshore islands. As the hippy trail extended into Malaysia in the early 1970s, economically-depressed fishing villages such as Cherating, Teluk Bahang and Pulau Tioman were gradually transformed into *kampung* tourism areas through the development of homestays, chalets and A-frame huts, owned and operated by the local people. From the mid-1980s, small-scale tourism development began to mushroom at a rapid rate and the growth was phenomenal ever since the success of the Visit Malaysia Year (VMY) campaign in 1990.

At Cherating, for example, there were only 19 huts/chalets by the middle of the 1970s but, by 1990, the total number of tourist accommodation had increased to 189. By 1994, there were 356 huts/chalets at Cherating with 192 more either undergoing construction or waiting for planning approval, as indicated by data collected by the town planning section (Majlis Perbandaran Kuantan, 1994). On the offshore islands of Johor, tourist accommodation increased by more than 500% over a two-year period between 1988 and 1990

(Hamzah *et al.*, 1991). Rapid growth was also experienced both at the established *kampung* tourism areas such as Pulau Pangkor and Pulau Tioman, as well as new destinations such as Pulau Perhentian, Pulau Tinggi, Pulau Besar, Pulau Sibu, Marang, and Merang (Fig. 14.1). The largest concentration of small-scale tourism development is located at Pulau Tioman, where 1435 rooms are operated by 263 different owners. In comparison, there is only one resort development on the island which operates 349 rooms (Unit Perancang Ekonomi Negeri Pahang, 1995).

Initially, this form of 'alternative tourism' was frowned upon by tourism

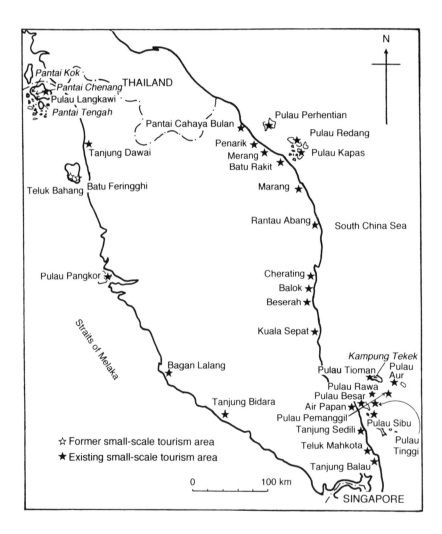

Fig. 14.1. Location of small-scale tourism areas in Peninsular Malaysia.

managers in Malaysia, it being feared that the proliferation of drug abuse and social problems associated with the influx of 'backpackers' (Hong, 1978; Tourist Development Corporation Malaysia/Singapore Tourist Promotion Board, 1983). In the last few years, however, this form of entreprenuership has been encouraged and promoted at all three levels of government (federal, state and local) in the form of financial assistance and relaxed enforcement of site planning guidelines/standards (Jabatan Kerajaan Tempatan, 1993). Far from being considered as an undesirable appendage of the tourism industry, small-scale tourism development is now regarded as a catalyst for improving the income of the rural population as well as boosting domestic tourism. This will help reduce the demand for outbound travel, and subsequently reduce foreign exchange outflow (Government of Malaysia, 1991; Ministry of Culture, Art and Tourism, Malaysia, undated).

The switch to tourism as the major source of employment has increased significantly the income of the local population. In 1983, for example, the average income of the local population of Pulau Tioman, who were mainly traditional fishermen and petty farmers, was below the national poverty line (income of below RM 370 per month for a household size of five). By 1992, 90% of the locals had become directly or indirectly involved in tourism and the average monthly income level rose to more than RM 850 per month (Voon, 1994). By 1994, the lowest income at Pulau Tioman was RM 882 per month, while 21% of the population was earning more than RM 3000 per month (Unit Perancang Ekonomi Negeri Pahang, 1995). At Pulau Perhentian the average income of the chalet operators is currently in excess of RM 1000 per month, compared to less than RM 325 per month when they were employed as full-time traditional fishermen (Jabatan Perancangan Bandar and Wilayah, 1994).

The aims of this chapter are: (i) to describe the evolution of small-scale tourism development in Malaysia *vis-à-vis* the changing tourist motivation/ demand; (ii) to highlight the problems caused by the uncontrolled development of such tourism and the implications on the sustainability of the fragile physical and social environment; and (iii) to recommend institutional and technical measures necessary to ensure the future growth of small-scale tourism development in addition to the planned tourism product complementing conventional mass tourism, as well as evolving along the principles of sustainable tourism development.

The Evolution of Small-scale Tourism Development in Malaysia

By the time unspoilt and 'non-touristic' destinations in Malaysia were being discovered by the early 'drifters', the 'drifting culture' had become a subculture, operating separately but parallel to mass tourism (Cohen, 1973).

Destination areas such as Cherating, Teluk Bahang, Pulau Tioman and Pulau Pangkor were merely an extension of a network of cheap lodgings, coffee shops, restaurants, souvenir shops, and sundry shops offering services such as money changing and international telephone call facilities. These facilities were connected to each other by a cheap transportation system, and informal information boards at every destination helped to promote outlets at the other destinations within this network. As the pioneering *kampung* tourism areas expanded and began to attract more and more tourists, the 'drifters' moved on to new destinations such as Marang, Pulau Perhentian, Pulau Pemanggil, Pulau Aur and so forth. By the mid 1980s, there were about 17 small-scale tourism development areas in Malaysia, mainly along the east coast. In the last few years, new destinations have sprung up, such as Merang, Penarik, Tanjung Sepat and Batu Rakit (Fig. 14.1).

Towards the end of the 1980s there was a rapid growth in domestic tourism, which was propagated by the economic boom and the rise in disposable income, especially among the emerging middle class. Until recently, the government has encouraged the provision of mainly high-class accommodation and facilities to attract international tourists and generate revenues in foreign exchange. Due to the dearth in affordable alternatives to luxury hotels, domestic tourists were drawn to the small-scale tourism development areas, which, when they were first started, attracted only foreign tourists. In the last few years, however, the proportion of domestic tourists visiting these areas had risen by between 20% and 30%, fluctuating to about 80% during the public/school holidays (Hamzah *et al.*, 1991; Unit Perancang Ekonomi Negeri Pahang, 1995).

What has happened at Cherating has been the 'intrusion' of domestic tourists into the tourist space (Mansfield, 1990) of the budget travellers, which has forced the foreign tourists to 'retreat' to the more remote destinations within their network. This process can be expected to continue, with more and more domestic tourists encroaching into the former 'drifter enclaves', facilitated by improvements in transportation. The presence of two distinct groups of tourist in the same tourist space has resulted in an 'image conflict' (Murphy, 1985), which in many ways has confused the hosts about which tourist image they should portray and its function within the overall tourist system.

Methodology

Data on small-scale tourism development in Malaysia is almost non-existent. In order to collect baseline data on the subject, field inventories were carried out at each and every small-scale tourism development area, between 1988 and 1993. In the case of Cherating, data collection began as far back as 1984, the techniques used during the inventories being observation, informal interviews with tourists and tourism managers, and lengthy conversations with the local community. In 1994, two study areas were identified for a detailed field survey: Cherating and Pulau Langkawi.

During the field survey, the techniques used were a combination of field observation, structured interviews and questionnaire survey. Included in the questionnaire were questions on the nature of travel, travel arrangements, travel behaviour and pattern, before and after tourist imagery, and AIO (activities, interests, opinions). Due to the cultural differences between foreign tourists and domestic tourists, it was assumed that there would be significant differences in their motivation and behaviour. Hence, the sample was divided into two main tourist groups, namely, foreign tourists and domestic tourists.

The survey at Cherating was carried out over a two-week period in April/May 1994, the latter half of which coincided with the school holidays. A total of 139 responses were collected (N=139) and, out of this total, 55% (N=77) were made up of foreign tourists, while the remaining 45% (N=62) were domestic tourists. The survey at Pulau Langkawi was conducted in two phases. The first phase was carried out over a 10-day period in October 1994, followed by a 14-day second phase in November. The latter coincided with the school holidays. Altogether, a sample total of 422 was collected (N=422), of which 82% (N=347) were domestic tourists and 28% foreign tourists (N=75).

The changing demand for small-scale tourism development

In predicting tourist demand, it is essential to understand tourist motivation since, together with other factors such as destination features and marketing, it contributes greatly towards shaping tourist demand. In trying to avoid the pitfall of asking tourists direct questions about their reasons for travelling (Krippendorf, 1987), their actual motivation was deduced from interviewing them about their travel pattern, travel behaviour, and perception of tourist images, which were later compared against their demographic profile and observed behaviour pattern. Throughout the duration of the fieldwork, it became clear that there were major differences between the travel behaviour of foreign tourists as compared with domestic tourists. Thus, the following interpretation of the survey results will deal with these two groups separately.

Foreign tourists
As opposed to the overall composition of foreign tourists to Malaysia, those who stayed at the small-scale tourism development areas were mainly from Western Europe, with the exception of Pulau Tioman and the offshore islands of Johor, which attracted mainly Singaporeans. In terms of age,

foreign tourists attracted to these areas were considerably younger than the norm but they were not necessarily classed as the 'youth travellers'. At Cherating and Pulau Langkawi, there were more tourists above the age of 31 years (61% and 55% respectively) than those 30 years old and below (39% and 45%). The age distribution also reflected why there were significantly fewer numbers of students as compared to the proportion of professionals and skilled/semi-skilled workers found at both sites (1:3 ratio).

The demographic profile supports the view that Cherating has become 'mainstream' and thus is being increasingly avoided by 'youth travellers', who have moved on to the 'less touristic' areas such as Marang and Pulau Perhentian, although there are no available data to support this view (personal communication, Unit Perancang Ekonomi Negeri Pahang official, 1994; chalet operators, 1994). Field visits to these areas supported this explanation but it should be added that 'youth travellers' were confined to the spartan establishments on Pulau Perhentian Kechil while the not-so-young/middle aged professionals have also 'encroached', albeit mainly staying at the smarter chalets on Pulau Perhentian Besar. Pulau Langkawi as a whole is more well known (and well promoted) for its duty-free port status and its aspiration to become an international class island resort rather than a budget destination, despite the fact that three of its best beaches are fronted by small-scale tourism development. The travel patterns of the foreign tourists too were different from the average Western tourist, because they stayed longer in Malaysia and longer at a particular destination. They rarely travelled in a large group and they visited more destinations, mainly in the rural areas. In addition, they were attracted to socio-cultural images such as 'village life', 'the natural village environment', and 'the local architecture', notwithstanding the fact that 'the beach' was their main 'pull factor'. However, despite indicating interest in the socio-cultural aspects of the destination area, the activities that the tourists carried out were mainly seeking relaxation and solitude, away from their work/studies and daily routine. Instead of participating in the local way of life, they preferred their own company, carrying out activities that reflected this disposition, i.e. reading, sunbathing and swimming. It is obvious that they could not be described as Cohen's (1972) original 'drifters' as these small-scale tourism development areas had too much comfort (even the spartan accommodation on Pulau Perhentian Kechil) and too many tourists to please the 'anarchistic and hedonistic' soul. Instead, their behaviour bore resemblance to Cohen's description of the 'youth tourists' at Phuket and Kok Samui in Thailand (Cohen, 1982) and Riley's (1988) 'budget travellers'.

In line with Riley's (1988) findings, it can be deduced that the main 'push factor' which had influenced the budget travellers' decision to travel in the first place was 'to escape from routine everyday life' and, to a lesser extent, 'to achieve status or ego-enhancement'. The 'pull factors' on the other hand were the 'nice beach', 'village life', 'natural environment', 'natural beauty', and 'local architecture', attributes that coincide with the tourists' pre-visit images of their destinations. It should be stressed, however, that these were the perceived images and not the real situation at Cherating/Langkawi. In fact, village life had long disappeared before the arrival of tourists to Cherating (Din, 1993, personal communication). Over the last few years, outsiders have bought/leased land belonging to the local residents and introduced commercial tourism developments such as upmarket chalets, batik-painting outlets, seafood restaurants and bars. During the field survey, it was disclosed that the population of Cherating had dwindled to 40 persons, from the official population of 100 persons in 1988 (Majlis Perbandaran Kuantan, 1994). Moreover, most of the residents had moved to the other side of the trunk road, separating them from the tourist accommodation area. In actual fact, the only remaining locals at Cherating were the elderly, who hardly communicated with the tourists due to the language barrier. The 'friendly locals' that the tourists usually referred to were mostly 'imported' workers and the beach boys, whose appearance, mannerism and behaviour were copied from the tourists. As opposed to the exoticism of the natives of the 'Fourth World' (Graburn, 1976) the local population at former fishing villages such as Cherating could be considered as ordinary people. As they were no longer fishermen but had become small business entrepreneurs, they did not represent a 'pull' factor, although their family members and employees were essential 'actors', performing in the background as 'symbols of authenticity' (Cohen, 1987). Their presence in the 'front' regions of the tourist setting (MacCannell, 1989) was not as the sole 'tourist gaze' (Urry, 1990), but as a component of the overall tourist image of Cherating, i.e. 'village life' (Dearden and Harron, 1994).

In the pre-visit images of Cherating, foreign tourists ranked 'nice beach' above 'village life', but there was soon a change in the perception during their stay at Cherating. Not only was 'village life' considered more important than 'nice beach', but the other components of 'village life', such as 'local architecture' and 'batik painting' were also revealed as important during-visit images. This confirms Dearden and Harron's (1994) view that tourists' images of authenticity are adaptable and that the meaning of authenticity was not as inflexible as the interpretation suggested by MacCannell (1989). Nonetheless, negative images were also formed during their stay at Cherating, reflecting the real situation, for example, erosion, too many chalets/restaurants, pollution and noise. There were also clear differences between the real situation and the other pre-visit images, such as Cherating's 'natural environment' and Langkawi's 'natural beauty'. In actual fact, both Cherating and Langkawi have been environmentally degraded in many places, resulting in the loss of mangroves, beach erosion, eroded fill materials and dead trees. The only similarity between the perceived images and reality was regarding the 'nice beach', notwithstanding the presence of groynes constructed by the local chalet operator at the eastern corner of the site to minimize erosion.

Nonetheless, foreign tourists will continue to go to Cherating/Langkawi as long as the perceived positive images of the place are maintained, but it may cease to remain popular once the perceived images are as bad the real situation.

Domestic tourists

The motivation of domestic tourists is more difficult to identify mainly due to the lack of baseline data. The practice of taking a holiday is new to most Malaysians and, in the past, most of the trips made during weekends or holidays have been back to one's hometown or village or parents. This particular trip is called *balik kampung*, which literally means 'going back to the village' (Masri, 1992; Ministry of Culture, Art and Tourism, Malaysia, undated). The extended family system and rapid migration of the rural population into the cities and big towns have created a sense of longing or nostalgia for the *kampung* life which the new breed of urban dwellers seek to escape to, at any given opportunity. The travel motives behind the *balik kampung* trips can be deduced as 'to escape an alien way of life' and 'to return (albeit, temporarily) to familiar surroundings'. Ironically, the somewhat overemphasis on the *balik kampung* phenomenon has obscured the fact that the more conventional forms of domestic tourism have increased fourfold between 1977 and 1987. In 1977, the size of the domestic market was estimated at 2.33 million person-trips, which increased to 4.57 million person-trips in 1987 and domestic travel expenditure increased fivefold over the same period (from RM 251 million in 1977 to RM 1.28 billion in 1987) (Masri, 1992). As the country's population is becoming increasingly affluent, domestic tourism trips and expenditure are expected to grow at the rate of between 10% and 12% annually until the year 2000 (Ministry of Culture, Art and Tourism, Malaysia, undated).

Holidaying, especially overseas, has become a more common practice with the emergence of a new middle class. To the middle class, holidaying became a status symbol, specifically overseas holidays, which 'enhanced one's social standing' (Masri, 1992, p.1). Outbound travel figures show that overseas holidays have increased from 14.9 million in 1990 to 18.8 million in 1994, with a growth rate of between 10% and 15% per year (Malaysia Tourism Promotion Board, 1991, 1992, 1993, 1994). The Frank Small and Associates (1986a,b) study also revealed that 45% of Malaysians went overseas in their recent trip, and, among those who have not travelled in Malaysia, one-third stated that they were not interested in local travel.

Some of the characteristics of domestic tourists, as surmised by the Tourist Development Corporation Malaysia/Institut Teknologi Mara (1988) study, were: (i) few travelled as part of an organized tour (14%); (ii) the majority (76%) knew about their destination from friends or relatives; and (iii) their preferred holiday destinations were beach resorts/seaside, followed by hill resorts, islands, national parks, historical sites, and cultural attractions. It should be stressed, however, that the TDCM/ITM study covered essentially the middle class Malaysians living in the major urban areas and that it was carried out during the global economic recession that also affected Malaysia. The increase in the number of domestic tourists being attracted to small-scale tourism development areas started in 1989, coinciding with the start of the current economic boom. Rapid urbanization and industrialization have resulted in the now familiar need 'to get away from it all', and, due to the dearth of medium-price hotels (Ministry of Culture, Art and Tourism, Malaysia, undated), they flocked to the small-scale tourism development areas. Such areas offered amenities that suited the preference of domestic tourists in that they were mostly located along a 'nice' beach, they were cheap, and the setting reminded them of their *kampung* (personal communication, domestic tourists at Cherating, 1987, 1988, 1989, 1990). Thus the *kampung* tourism areas were able to satisfy their need 'to get away from it all' as well as act as a familiar substitute for their *kampung* environment.

Over the past few years, new trends have been observed in the travel motives and travel behaviour of the domestic tourists. With the improvement in accessibility to these places, the middle class has shown a willingness to stay at small-scale tourism development areas provided that the chalets are comfortable, clean, and fitted with air-conditioners. They will not forego their overseas holiday but will also take frequent short breaks, staying at establishments that do not lower their status or reputation. It was observed from the survey at Cherating that the middle class would avoid the budget accommodation and preferred the more pricey establishments, which were owned and operated by owners from Kuala Lumpur. Likewise, at Langkawi the budget accommodation at Pantai Kok was mostly avoided for the more comfortable chalets at Pantai Chenang and Pantai Tengah. It should be added that the popular establishments among the middle class were those that were able to 'stage' a rustic *kampung* ambiance but at the same time offering a level of comfort and service comparable to luxury hotels. The nostalgia for familiar surroundings has also been exploited by upmarket tourism development elsewhere. For example, the new resort at the Pedu Dam has also used the theme *Balik Kampung* in its promotion, besides using ornamentals such as trishaws and vernacular architecture to complement the luxury chalets.

Another trend observed at many of the small-scale tourism development areas, especially along the mainland beaches, for example, Cherating, Batu Rakit and Penarik, was the increase in the number of tourists travelling in large groups. These groups of between 50 and 200 people were usually from the same workplace in the cities, staying at these places to participate in office activities such as in-house seminars, sports carnivals, staff family outings, and staff reward schemes. They were mainly factory workers, lower category government servants, and university/college students who spent two to three nights at the destination engaged in daytime activities on the beach (games, water sports) and attending seminars, discussions, karaoke sessions, camp fires in the evenings. Needless to say, most of the establishments that catered for such groups were fully booked during the school or public holidays. Most of the activities that are carried out by these groups could have been done at a community hall in the city, or within the office premises or compound, so the main reason for travelling to another place to carry out such activities, in combination with other leisure pursuits, can again be attributed to the need 'to escape'.

The main attraction for domestic tourists at Cherating and Langkawi was the beach. This corresponds with the findings of the Frank Small and Associates (1986a,b) study, which stated that the most popular holiday destinations among domestic tourists were beach resorts. At Langkawi, the beach (57%) was an even more important attraction than the lure of duty-free shopping (27%), despite the fact that the latter was the main 'tourist gaze' (Urry, 1990) the government had intended for Langkawi. 'Village life' was not mentioned at all but the 'natural environment' and 'natural beauty', which are attributes usually associated with the *kampung* environment, were the other significant 'pull factors' mentioned.

There could be three reasons why domestic tourists did not consider 'village life' as an important 'pull' factor. Firstly, it may be due to the fact that the image of 'village life' at Cherating was totally different from the individual's own authentic *kampung*. Secondly, it is also possible that the 'longing for a familiar setting' was actually hidden in the subconscious (Krippendorf, 1987) and not revealed as a reason for going to Cherating/Langkawi. Finally, it might be that, when one travels in a large group for only a short period of time, the overall holiday experience, especially the camaraderie among friends, was more important than the physical attractions (Dearden and Harron, 1994).

Tourist subgroups

Based on the survey findings, it was possible to classify the tourists who were attracted to small-scale tourism development areas into five sub-groups: three foreign and two domestic. The classification is not meant to be a static or absolute blueprint, but more as a framework for understanding the changing demand for small-scale tourism development in Malaysia.

Foreign tourist I – the youth travellers The youth travellers are mainly students with limited financial resources, but with plenty of time to see the world before they settle down to a routine job upon the completion of their studies. Socio-cultural images of the destination are essential to them, even if they seldom appear interested in getting to know more about the local people or the local way of life. A clean and quiet beach is their most important 'pull' factor and they will always choose the establishment that offers the best value for money. Outlets that offer do-it-yourself batik paintings are also valued by them as they can paint their own T-shirts and take back souvenirs that are different from those purchased by the mass tourists. Otherwise, they prefer little addition to the tourist facilities currently available. This sub-group is likely to be attracted to the 'non-touristic' destinations within the budget travellers' network and they are not likely to revisit the same place again as 'there are too many other places in the world to visit'.

Foreign tourist II – the professionals   The professionals can well afford to stay at the conventional hotels but prefer the informality and authenticity of the *kampung* tourism areas. Comfort and cleanliness are very important to them but so is the tourist image of the place. They have been to many parts of the world and know that there is no such thing as a truly authentic holiday destination. Thus, they will appreciate the staging of traditional performances and the reproduction of traditional architecture and handicrafts as long as they are 'professionally' done. In the same light, they welcome the construction of new attractions that promote the cultural attributes of the area. As 'green tourists', however, they object to environmental degradation and will revisit the area, again and again, if their perceived image of village life is maintained and environmental degradation minimized. They are also likely to avoid the shabby establishments preferred by the youth travellers.

Foreign tourists III – the claustrophobic islanders   Although not included in the survey, they form the biggest tourist sub-group attracted to the small-scale tourism development areas along the mainland and on the off-shore islands of Johor. This group refers to the Singaporeans who take frequent short breaks to these areas to recuperate from the stresses of modern city life. They seek mainly diversions in the form of fishing, snorkelling and scuba diving, and avoid the up-market establishments for the more spartan accommodation facilities. The tourist image of 'village life' is immaterial to them and satisfaction is gained by 'gazing' at the various attractions in the marine environment.

Domestic tourists I – the emerging middle class   The middle class requires clean, comfortable, air-conditioned and up-market tourist accommodation, fitted with facilities/amenities comparable with those offered by luxury hotels. It is also essential that the buildings simulate a *kampung* image in their design and materials used. In addition, they not only welcome new and 'contrived' cultural attractions but also urban-based facilities, such as water scooters and other active forms of water sports. Privacy is another important requirement and they normally avoid staying close to large groups, restaurant and bars. They also detest the environmental degradation caused by tourism development.

Domestic tourists II – the down-market convention groups
This particular group is not concerned about the images of the destination area as long as it can accommodate and serve everyone. Rooms should be plentiful, cheap and able to accommodate as many people as possible. Privacy is not a prime consideration but the provision of ample public areas is paramount. Some of the facilities required by this group are a large multi-purpose hall, prayer room, and a location close to the beach. The demand from this group, however, is seasonal, being extremely high during weekends and public holidays and extremely low during weekdays.

The Supply of Small-scale Tourism Development

The supply of small-scale tourism development has reflected the changing tourist motivation and demand. At the new destination areas, the type of accommodation provided was mainly spartan frame huts, chalets and longhouses with communal bathrooms and toilets. They were mostly constructed by the locals and operated by members of their family. Although the construction of this cheap tourist accommodation involved minimal site clearing or grading, wastewater treatment was very basic. Overall, they exuded a seedy atmosphere and were popular with the youth travellers. Nonetheless, interviews with the chalet operators revealed that most of them were planning to upgrade their chalets once financial resources are available. According to them, their present chalets were not economically viable and the construction of this basic accommodation was just a stepping stone.

At the mature destination areas, smarter chalets have been built, many with air-conditioning and relatively luxurious furnishing. Major site clearing was carried out to prepare the building platforms as well as to facilitate the construction of paved footpaths. Most of the development was carried out as a joint venture between the local landowner and entrepreneurs from outside. The design of the chalets was based on the vernacular architecture and septic tanks were commonly used to treat wastewater. Overall, these chalets offered comfort, cleanliness and good service and were popular with the foreign professional and the local middle class.

Barrack-style development or tourist factories could be found at destinations which have reached the 'saturation' stage of development, such as Cherating, Pulau Langkawi and Pulau Tioman. Site clearing, backfilling and grading were carried out on a large scale, removing or killing many of the mature trees within the site. The majority of this form of development was an extension of the original development carried out by the local entrepreneurs with backing from the commercial banks. The design of the chalets was plain and repetitive, and the buildings were arranged to appear like a holiday camp. At Pulau Tioman, however, some of the second generation chalets were being tastefully planned and constructed to 'stage' a clean, comfortable and aesthetically pleasing *kampung* environment. Using the design expertise available at a local university, the redevelopment project was carried out by the local community, using standard chalets provided by a semi-government agency, Majlis Amanah Rakyat. These chalets were bought through a loan scheme designated for small-scale chalet operators and the construction of the chalets was closely supervised by the owner and design and planning consultant so as to minimize unnecessary site clearing and tree felling. Given the deterioration of the tourist image and physical appearance of destination areas such as Cherating, the popularity and success of the second generation chalets at Pulau Tioman will set a good example for the other local communities involved in tourism.

Implications of Uncontrolled Development

Small-scale tourism development areas have been allowed to grow without any form of development control, deteriorating into grotesque enclaves at locations such as Cherating, Pantai Chenang and Pantai Tengah (Langkawi), Kampung Tekek (Pulau Tioman), and Pulau Besar. The reasons for the lack of planning and development control are: (i) most small-scale tourism development areas are located outside the boundary and jurisdiction of local planning authorities; (ii) the district offices in charge of development within these areas lack technical knowledge and personnel to plan, manage and enforce development guidelines; and (iii) local authorities and district offices adopt a patronizing attitude towards such development in the hope that unrestrained development will bring economic benefits to the local population. Therefore, a number of repercussions of uncontrolled development occur.

Depletion of marine habitats and ecosystem

Uncontrolled development has mainly caused the depletion of corals and the deterioration in water quality, and, to a lesser extent, the depletion of mangroves. In 1984, more than half of the corals at Pulau Tioman were found to be damaged by boat anchors (Ridzwan, 1994). In 1995, between 20% and 40% dead corals were found in the waters fronting the popular tourist spots on the island, mainly due to sedimentation (Worldwide Fund for Nature, Malaysia, 1995). Also, the *E. coli* content in the coastal water exceeded the prescribed standard by 92 times (Voon, 1994).

Economic leakage

On most of the off-shore islands, building materials, fuel, and food supply, including fish, have to be transported from the mainland due to insufficient local production. This has resulted in high economic leakages, comparable to those of mass tourism. At Pulau Tioman for instance, 47 sen out of every RM 1.00 (47%) spent by tourists on the island had leaked to other areas on the mainland (Ho, 1993).

Displacement of the local community

The sudden rise in the land value had urged many of the locals to sell or lease their land to outsiders. The local youths, who had left these areas even before the advent of tourism, do not want to return as they will only be employed as waiters, cooks and so forth. Instead, imported workers have been brought in from as far as Thailand and the Philippines. In addition, beach boys from the mainland towns have set up outlets selling and making handicrafts and other souvenirs, which has altered the social fabric and cohesion of these former fishing villages.

Breakdown of traditional values

The breakdown of traditional values may occur as the number of tourists outnumber the hosts. At Pulau Tioman, tourists outnumbered the hosts by 58 times in 1990 and it was projected to increase to 110 times by the year 2000 (Voon, 1994). However, the fishing communities surveyed were found to be very resilient. At Pulau Perhentian Kechil, drinking, petty drug-taking and topless bathing took place within the 'tourist space', watched by the locals while they were at work. The 'alien' culture was tolerated by the locals as 'an occupational hazard' as long as tourists did not encroach and pry on their daily lives at the village, which was separated from the tourist area. It was the beach boys from the mainland towns that were affected by the so-called 'demonstration effect'.

Conflict over limited resources

Conflict over the limited water supply, grazing land and public right of way are common. Water shortage is acute in areas such as Pulau Perhentian and Pulau Sibu. Submarine pipes have been used at Pulau Redang to transport water from the mainland, but such technology is not affordable to small-scale tourism development.

Recommended Proposals Arising from the Study

The sustainable growth of small-scale tourism development can only be achieved if the development does not degrade the fragile physical environment and the development is economically viable from the perspective of the local community. The following proposals are put forward to serve as a starting point for future discussions on the strategies and policies required to achieve this aim.

Greater emphasis on tourism planning and management in local plans

The statutory local plans should highlight, promote and provide clear guidance for the development of tourism areas, with greater emphasis on the future role of small-scale tourism development.

Increasing the small-scale tourism development product base

Small-scale tourism development should not be solely concerned with the provision of accommodation but also the development of the other important aspects of this tourist product, for example, the use of local materials to produce handicrafts and commercial farming to supply fresh vegetables and fruits to the restaurants. This will help reduce out-migration and economic leakage and this concept of community-based tourism should be encouraged by the authorities. Ironically, rural development agencies have recently embarked on the construction and operation of tourist accommodation. Instead of helping the local community produce food and handicrafts for the tourism industry, they are now competing with them.

Regular dialogue with local community

Government officials involved with tourism planning/management should hold regular dialogues with the local community so that their aspirations and ideas are taken into consideration in the decision-making process. As Din (1993, p. 335) found out, 'the local residents are more knowledgeable than they are commonly presumed to be'. This can be done through the public participation process of the statutory structure (local) plan system but past practices have not fully exploited this opportunity. The District Development Plan (DDP) may be a better alternative. Although it is non-statutory, it uses a bottom-up approach that includes soliciting the views of the local community (Kechik *et al.*, 1991).

Improving the local community's stake in future development

Sooner or later, the local operators will require financial assistance to upgrade their initial venture, for example the upgrading of chalets and providing an effective wastewater treatment system. Instead of allowing them to be exploited, the local authority and other relevant agencies should identify opportunities in areas where outside entrepreneurs can invest in tourism development projects in joint ventures with the local community. They should form a body to represent the host community to ensure that the locals receive an equitable share of the profits from the investment.

Improving collaboration between government agencies and local cooperatives

Government agencies involved in rural development have been recently directed to allocate 10% of their annual budget for small-scale tourism development. Due to the lack of qualified personnel, their efforts have largely been ad hoc and in isolation. The emergence of financially stable and knowledgeable local cooperatives has been observed, especially at Pulau Tioman and Pulau Langkawi and close collaboration between these two parties is essential in the future.

Utilizing expertise at local universities

Since tourism planning/management is becoming more and more inter-disciplinary, local universities with their broad base of expertise will be able to help tourism managers in their forward planning tasks. In addition, local universities can conduct training courses to improve the management and public relations skills of the local entrepreneurs as well as disseminate knowledge on the importance of the principles of sustainable tourism.

Formulation of planning manual

A comprehensive planning manual will help local authorities plan/manage small-scale tourism development in a more knowledgeable and sensitive manner. Lack of knowledge on technical matters is common among planners, e.g. knowledge on coastal geomorphology to guide the siting of buildings/structures, in the light of the serious coastal erosion threat along the east coast (Wong, 1986, 1988).

Conclusions

It should be accepted that small-scale tourist development areas in Malaysia, which started as 'drifter enclaves', have now been penetrated by mass tourism. Consequently, 'tourist spaces' have been created at all these places, separating the tourists from the ordinary flow of *kampung* life. Nonetheless, foreign tourists have accepted this fact and are willing to play the 'tourist game', and enjoy the 'staging' of the authenticity of 'village life' as long as it is well performed. It is only when this fragile image of 'village life' is altered that tourism demand will be significantly affected. In the process, the host community has fully embraced tourism and should be regarded not as 'actors' in a tourist setting, but as proper entrepreneurs. In the early days of *kampung* tourism, most of the local entrepreneurs lacked capital and business experience (Din, 1988). While this problem persists in most of the small-scale destination areas, lately a section of the local community at Pulau Tioman has shown the ability to overcome this handicap. In the operation, promotion, upgrading and expansion of their chalet development, a few local entrepreneurs have displayed the business acumen of the highest standard. This was demonstrated in their planning, budgeting and liaison with authorities as well as professionals. More importantly, they have managed to do so on their own through the formation of a local cooperative. Despite continuous pressure to sell their land to outsiders, they have managed to resist this temptation and continue to depend on family members to operate their business. For such upgrading schemes to succeed elsewhere, local entrepreneurs should be given the financial assistance, technical advice and training to enable them to create a sustainable and uniquely Malaysian tourism product, complementing mass tourism.

References

Cohen, E. (1972) Towards a sociology of international tourism. *Social Research* 39, 164–182.

Cohen, E. (1973) Nomads from affluence: Notes on the phenomenon of drifter-tourism. *International Journal of Comparative Sociology* 14 (1–2), 89–103.

Cohen, E. (1982) Marginal paradises: bungalow tourism on the islands of southern Thailand. *Annals of Tourism Research* 9, 189–228.

Cohen, E. (1987) Alternative tourism – a critique. *Tourism Recreation Research* 12(2), 13–18.

Dearden, P. and Harron, S. (1994) Alternative tourism and adaptive change. *Annals of Tourism Research* 21, 81–102.

Din, K.H. (1988) *Keusahawanan Bumiputra dalam industri pelancongan di Cherat-ing dan Pulau Tioman*. Kertas Kadangkala Bil.1, Jabatan Geografi, Universiti Kebangsaan Malaysia, Bangi, Malaysia.

Din, K.H. (1993) Dialogue with the hosts: An educational strategy towards sustainable tourism. In: Hitchcock, M., King, V.T. and Parnell, M.J.G. (eds) *Tourism in South-East Asia*. Routledge, London, pp. 327–336.

Frank Small and Associates (1986a) *The Study on Opinions and Attitudes of Malaysian Tourists Visiting Local Tourist Spots and Resorts*. Market Study Commissioned by the Tourist Development Corporation Malaysia, TDCM, Kuala Lumpur.

Frank Small and Associates (1986b) *The Study on the Travelling Behaviour and Attitudes of Local Malaysians Interviewed in their Homes*. Market Study Commissioned by the Tourist Development Corporation Malaysia, TDCM, Kuala Lumpur.

Government of Malaysia (1991) *Sixth Malaysia Plan 1991–1995*. Government of Malaysia Printers, Kuala Lumpur.

Graburn, N.H. (ed.) (1976) *Ethnic and Tourist Arts: Cultural Expressions from the Fourth World*. University of California Press, Berkeley.

Hamzah, A., Kechik, A.T. and Ibrahim, M. (1991) Trend of resort development in offshore islands of Johore. Paper presented at *Regional Conference on Resort Development*, 26–28 July 1991, Desaru, Malaysia, pp. 1–22.

Ho, M.Y.J. (1993) Kesan pengganda sektor pelancongan ke atas penduduk tempatan, kajian kes: Kg.Tekek, Pulau Tioman. Unpublished BA dissertation, Universiti Teknologi Malaysia, Johor Bahru, Malaysia.

Hong, E. (1978) Tourism: its environmental impacts in Malaysia. Paper presented at *The Symposium on the Malaysian Environment*, RESCAM Complex, Penang, 16–20 September 1978, pp. 297–305.

Jabatan Kerajaan Tempatan (1993) *Cadangan kehendak-kehendak minimum bangunan dan pembangunan infrastruktur di pulau destinasi pelancongan*. Local government circular, Kuala Lumpur.

Jabatan Perancangan Bandar and Wilayah (1994) *Kajian keupayaan tampungan dan pelan pengurusan pelancongan/rekreasi Pulau Perhentian, Trengganu*. Unit Pengurusan Alam Sekitar, JPBW, Johor Bahru, Malaysia.

Kechik, A.T., Hamzah, A. and Ibrahim, M. (1991) Planning for community and sustainable tourism. Paper presented at *The International Conference on Tourism: Development, Trends and Prospects in the 90s*, 16–18 September 1991, Kuala Lumpur, Malaysia, Paper 11, pp. 1–17.

Krippendorf, J. (1987) *The Holiday-makers: Understanding the Impact of Ttravel and Tourism*. Butterworth-Heinemann, Oxford.

MacCannell, D. (1989) *The Tourist*, 2nd edn. Routledge, London.

Majlis Perbandaran Kuantan (1994)Tourism data compiled by the Town Planning Section, Kuantan Municipal Council, Kuantan, Malaysia.

Malaysia Tourism Promotion Board (1991) *Annual Tourism Statistical Report*. MTPB, Kuala Lumpur.

Malaysia Tourism Promotion Board (1992) *Annual Tourism Statistical Report*. MTPB, Kuala Lumpur.

Malaysia Tourism Promotion Board (1993) *Annual Tourism Statistical Report*. MTPB,Kuala Lumpur.

Malaysia Tourism Promotion Board (1994) *Annual Tourism Statistical Report*. MTPB, Kuala Lumpur.

Mansfield, Y. (1990) Spatial patterns of international tourist flows: towards a theoretical framework. *Progress in Human Geography*, 4, 372–390.

Masri, B.H. (1992) The growth and prospects of domestic tourism. Paper presented at *The International Conference on Tourism: Development, Trends and Prospects in the 90s,* 16–19 September 1991, Kuala Lumpur, Malaysia, Keynote Address 2, pp. 1–10.

Ministry of Culture, Art and Tourism, Malaysia (undated) *Malaysia National Tourism Policy Study.* Policy Document, MoCAT, Kuala Lumpur.

Murphy, P. (1985) *Tourism: A Community Approach.* Routledge, London.

Pearce, D.G. (1995) *Tourism Today: a Geographical Analysis,* 2nd edn. Longman, London.

Ridzwan, A.R. (1994) Status of coral reefs in Malaysia. In: Wilkinson, R., Sudara, S. and Chou, L.M. (eds) *Proceedings of the Third Asean–Australia Symposium on Living Coastal Resources.* Chuialongkorn University, Bangkok, pp. 49–56.

Riley, P.J. (1988) Road culture of international long-term budget travellers. *Annals of Tourism Research.* 15, 313–328.

Tourist Development Corporation Malaysia/Institut Teknologi Mara (1988) *Domestic Market Study.* TDCM, Kuala Lumpur.

Tourism Development Corporation Malaysia/Singapore Tourist Promotion Board (1983) *Report by Joint Task Force on Tourism in Johore.* TDCM Southern Section, Kuala Lumpur.

Unit Perancang Ekonomi Negeri Pahang (1995) *Kajian Semula Pelan Induk Pulau Tioman.* Deraf Laporan Akhir.

Urry, J. (1990) *The Tourist Gaze.* Sage Publications, London.

Voon, P.K. (1994) Land use and sustainable development: The case of Pulau Tioman. Paper presented at *The Third International Conference on Geography of the ASEAN Region,* 25–29 October, 1994, Kuala Lumpur, pp. 1–22.

Wong, P.P. (1986) Tourism development and resorts on the East of Peninsular Malaysia. *Singapore Journal of Tropical Geography* 7, 152–162.

Wong, P.P. (1988) Beach resort sites on the east coast of Peninsular Malaysia. *Singapore Journal of Tropical Geography* 9, 72–85.

Worldwide Fund for Nature, Malaysia (1995) *The Concept and Analysis of Carrying Capacity: a Management Tool for Effective Planning, Part III, Case Study of P. Tioman.* WWF Malaysia, Kuala Lumpur.

Anthropologists, Local Communities and Sustainable Tourism Development

15

Stroma Cole

Buckinghamshire College, Wellesbourne Campus, Kingshill Road, High Wycombe, Bucks HP13 5BB, UK

Introduction

The importance of saving the world's ecosystem cannot be overestimated but tourism can only be used as a tool to this end when the culture of the people has the potential to be developed. Not all people are naturally hospitable, not all want to serve. These are vital factors in successful tourism and so careful examination of cultures prior to planning projects in new areas is essential for the success of sustainable tourism development.

The impact of tourism on culture has been an important issue since the first edition in 1978 of Smith's (1989) book on the anthropology of tourism, but little reference is made to the importance of anthropologists' contribution in works on sustainable tourist development. Likewise, host perceptions and how they change with the development of tourism, first put by Doxey (1975), are widely known but scant attention has been given to the suitability of people's culture to receive tourists.

This chapter examines the suitability of two groups living in neighbouring areas of the island of Flores, Indonesia. It is based on the author's work as an anthropologist working first as a small specialist tour operator and then alongside consultants for the Asian Development Bank. As seen in the two parallel case studies, the importance of the anthropologist's in depth knowledge and resulting understanding of cultures can make or break the success of sustainable tourist development. In all but specifically nature-based tourism, the role of the host must be put as central to sustainable tourist development. In order to give them this central role it is crucial to have a clear understanding of their culture, including internal politics before development can be considered.

The majority of works on sustainable tourism put the physical environment as the hub of the issue and people as part of the periphery. Studies that do have an important human aspect either make reference to the impact on societies already involved in tourism, or lay emphasis on community involvement without getting to grips with the practice of how this involvement will be carried through. Anthropologists' knowledge of the people is a vital link to understanding local communities and therefore how to successfully get them involved. It looks at the ways community involvement can evolve causing antagonist relations in once harmonious communities, and offers the idea of community projects as an important solution.

The Study Area

Two regions, Ngada and Manggarai, are compared in terms of their suitability for tourist development. Physically the two areas are very similar, lying adjacent to one another on a mountainous volcanic island (Fig. 15.1). Both have a minimum of infrastructure, one main road that runs east–west and connects the market towns. Beyond this road, minor unmetalled roads and tracts connect the villages. The area of Manggarai lying to the west receives more rainfall. Socio-economically, the areas are also very similar; most of the inhabitants are peasants living close to or below the breadline.

Culturally, however, the areas are quite different. Table 15.1 outlines the main differences in terms of material culture, the outward signs of culture, which would be of interest for tourism (Graburn, 1976). Whereas in the Ngada region there are several villages with their traditional architecture intact, in Manggarai the only traditional houses are those that have been rebuilt with the hope of attracting tourists. The Ikat weavings from the Eastern Isles are renowned for their intricacy and beauty. In Manggarai modern brightly-coloured threads are used whereas the Bajawan weaving relies on the traditional blues and blacks with white and gold motifs.

All the features of the two areas shown in the table and discussed above could be identified by almost anyone educated in tourism as pluses and minuses in terms of the areas' attributes for potential tourism. However, there is a deeper level of understanding required to ascertain the suitability of the areas for tourism development: the people themselves must be understood. With an anthropologist's training in ethnography, getting an insider understanding, an emic perspective, it is possible to fully appreciate the differences that exist between the two groups. Historical knowledge also aids this understanding.

For centuries the Manggarai have been raided for slaves so they are fearful of outsiders and aggressive towards them. When an uninvited guest

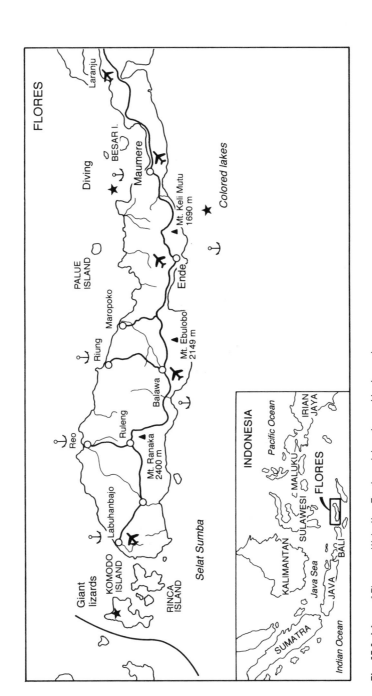

Fig. 15.1. Map of Flores within the Eastern Islands and Indonesia.

Table 15.1. A comparison of Ngada and Manggarai attractions for tourism.

Attraction	Ngada	Manggarai
Weaving	Often still produced traditionally, black with fine white or gold patterns	Navy blue or black with bright yellow or mixed coloured flowers
Music and dance	Unique bamboo instruments, varied dancing forms, some warrior type, some all female group dances	*Caci*, of great interest but dances inflict considerable injuries on opponents, more a fight than a dance
Other handicrafts	Small box bags used for betel nut chewing paraphernalia, some bamboo household objects	Brightly coloured hats
Village architecture	Still traditional, of outstanding interest to tourists, also village totems and megaliths in the villages	Previously large round communal structures. The only one that stands is a reconstruction to attract tourists
Walks	Moderately hard, very varied and interesting	Very hard, steep and arduous

walks into a Ngada village, it is likely that a villager will come out, greet the stranger, offer a drink or a banana, or a meal if one has just been prepared. On the other hand, when a stranger walks into a Manggarai village, most people will scatter and hide. If the visitor remains, the villagers will reappear slowly, hands on weapons (large sword), eyes piercing. Even when a stranger is properly introduced the people are cold, reserved and exceedingly wary.

To a trained eye, the dances of the two areas illustrate important differences between the groups and their suitability for receiving tourists. Many Ngada dances are based around war dances: feet stamping, shrieking and swords held high in the air are common features. However, dances are also performed by men and women or women alone, in circles. The circle expands as dancers join in, the movements welcome additional people to join the circle. In contrast there is only one traditional 'dance' still performed in the Manggarai region. *Caci*, often translated as whip dancing, is a competition of pride between two men, one man holding a whip, the other a shield. It is more of a fight than a dance and often inflicts considerable injuries.

The Manggarai have very high self esteem *(harga diri yang sangat tinggi)* or self respect, which is played out through the *caci* dance. They have no history of willingly serving others and their self respect precludes them from considering this a suitable occupation. They are not welcoming and hence lack one of the essential qualities necessary for a host population. The Ngada people, on the other hand, are very friendly and welcoming of outsiders and happy to perform the hosting function (Van Harssel, 1994).

From an anthropologist's in-depth understanding it is possible to ascertain the suitability or not of groups to be assisted in their development of tourism. In the next part of this chapter the further roles of anthropologists in sustainable tourist development are shown by contrasting case studies in these two regions.

Small-scale Tourist Development in Ngada

The first case study of small-scale tourism development underlines the prerequisites for successful development and how anthropologists are most able to contribute to it. The study is based on the work of a travel company run by the author and a fellow anthropologist, setting up cultural/ethnic tourism in a small village a few miles from the market town of Bajawa. Many authors (Wood, 1984; Smith, 1989; Harron and Weiler, 1992) distinguish between ethnic and cultural tourism, where the latter is seen to have a more generalized focus. However, the author believes that such a dichotomy would be better seen as a continuum, because, for example, a destination can change from the ethnic pole to a cultural one as development takes place and the less specialist tourist moves in. Such a transformation has been also discussed by Dearden and Harron (1994) in relation to the changing motivations of trekkers to northern Thailand. Within a single tour group some of the clients can take a more specialist interest and consider themselves partaking in ethnic tourism while others can enjoy the experience in a more general way and thus be partaking in cultural tourism.

Preparation and consultation

A physically suitable village, in terms of there being intact traditional architecture and *ngadu* and *bhaga*, variously translated as totems, shrines, and sacrificial posts, in place, was found, and a visit took place to discuss with the villagers the idea of bringing tourists to stay in the village. From discussions it was learnt that a few backpackers had passed through the village. The villagers were not, however, euphoric, as Doxey (1975) would have suggested, at this first contact with Western tourists. The villagers had been perplexed and somewhat upset by their initial contact with tourists for the following reasons:

1. female tourists bathed in the (deeper) male part of the villagers' bathing place, a natural volcanic pool.
2. post-adolescent, pre-childbearing women tourists had bathed without wearing a sarong. In the village females bathe topless if they are pre-adolescent and post childbearing; between these life stages women cover their breasts with a sarong at the pool.
3. amorous relations were made in public; hand holding, hugging and kissing

are all considered acceptable only in private.

4. too few clothes had been worn, particularly on Sundays. As a Catholic village, the villagers are more conservative in their dress code on the day of worship.

5. the villagers' small attempts to speak English, for example 'What your name?' or 'Where you from?' were largely ignored.

In addition the villagers felt ashamed they had so little to offer. From these initial discussions much was learnt about the traditional power structure in the village, which is largely matriarchical, such as who were the important people to be respected, and who should be consulted on what matters and in what order.

The tour organizers were also able to help the villagers feel more adequate and improve their confidence by teaching them some basics about hosting. For example, the villagers always drink their beverages with large amounts of sugar and are offended when guests do not drink beverages offered. It was explained that Westerners rarely take sugar and this averted embarrassing misunderstandings. Thus following the preparatory stay, the villagers were happy to receive its clients as their guests and trusted that the responsibility of explaining the correct social behaviour lay with the tour leaders.

Ensure guests behave

The first visit was arranged for 12 tourists and they behaved as requested by their tour leaders. The tour leaders had many roles (see Cohen, 1985, for a full discussion) but, as pointed out by Weiler and Davis (1993) in writing about nature-based tourism and its effects on the environment, the importance of the tour leader in monitoring visitors' behaviour to minimize the adverse effects of tourism is very great. Here the role was to minimize the negative socio-cultural impacts. As anthropologists it was possible to explain the reasoning behind the dos and don'ts. Rules with explanations are always more readily followed. The team could explain how to act with whom, when and why, which not only enhanced the experience for the tourists but also aided them to behave in an acceptable fashion, and hence enhanced the experience for the hosts too.

Two-way communications

As neither guest nor host could speak each other's language the researchers were put in an extremely powerful position. This point was made by Cohen (1985) but, from both the guests' and hosts' point of view, this was a very valuable part of the experience. In previous encounters with tourists the hosts had been unable to communicate, which had added to their feelings of inadequacy and vulnerability. They used the experience of the study to ask questions of Westerners on matters they had always wanted answers to. The information they gathered was very important to them.

Substantial permanent benefits for the hosts

Half way through the second day of the stay, the guests asked how much it was appropriate to tip the hosts. As it was felt it was the village as a whole (the musicians, the dancers, the cooks, the water carriers, the vegetable pickers, the palm wine collectors, etc.) that had contributed to the success of their stay, it was considered inappropriate to tip any particular individuals. Consultation was again very important, so it was necessary to gather together the appropriate members of the community to discuss the issue of tipping.

It was noticed that many children suffered from skin infections, which were explained by the villagers to be due to the short supply of water for bathing. However, it transpired that there was ample water but it was approximately three kilometres, uphill, from the village to the bathing places fed by a spring. There was an impromptu meeting and the idea of the water being brought down the hill in bamboo pipes to a tank behind the village was suggested, a project of benefit to everyone. This was a simple project that required the minimum of facilities, bamboo for the pipes and concrete for the container to store the water. The guests agreed with the idea, accepting that, while no individual would benefit, the host community would collectively benefit. The amount the guests contributed was voluntary. The older village women pointed out that it was important that the money should not be given to any individual because, however honest, the temptation to use the funds in a case of financial hardship might lead to them being misappropriated. So the money was donated openly at a public farewell ceremony.

A community project is an excellent way to show appreciation for the work put in by all members of the community. In this case it was recognized that members who were perhaps not directly involved and did not receive any of the direct economic benefits arising from tourism still had their lives interrupted and were generally inconvenienced by the presence of the guests. The community project benefited them as well as the villagers who took a proactive tourism role. This type of project meant that the fund could be added to in large or small amounts at any time. Thus there was no antagonism amongst the villagers about the money received through tourism.

In addition to this extremely important community project, the village experienced a second substantial permanent benefit from the visits. During the first stay the villagers entertained the guests with music and dance, and it was immediately realized that it was of an exceptionally high standard. They were the first outsiders to have experienced them. A tape of their music was made and played to potential interested parties and photographs of the musical instruments used were shown to them. Since then the villagers have played at regional, inter-regional and national competitions, eventually participating in the international bamboo music festival held in Bali. Thus, as a result of a musical performance locally to outsiders, these villagers now take great pride in their traditional music.

In summary, tourism on a very small scale with a minimal budget had four benefits for this community:

- economic benefits from the sale of weaving, small bags, bamboo household items and instruments and coffee.
- enhancement of their pride in their unique musical traditions, which had fostered and encouraged the attainment of international standards.
- the convenience of a handy water supply and a noticeable reduction in skin problems in children, which was evident once the water project was up and running.
- a chance to communicate with Westerners and have many of their questions about foreigners answered.

In contrast to this successful case study, there are some less successful, perhaps more common examples, where some or all of the above pre-requisites are not followed are now given.

Other Examples from Ngada

There are several villages in the Ngada region being visited by operators running seemingly similar tours to the one described above. The majority of visistors, however, do not stay overnight in the villages. The frequently observed pattern is for the tour bus to drive up and offload a group of tourists. Members of the group sit on the verandahs of the villagers' homes, often with their legs wide apart and their shorts or skirt pulled up high. They take photographs of the *bhaga* and *Ngadu* without having them interpreted and attempt to see the more private aspects of the villagers' lives by peeping through half-open doors or even cracks in the woodwork.

Villagers receive no advance notice of the tourists' arrival, although the tour companies know many months in advance exactly which date and at approximately what time groups will arrive. As Bates (1991) noted, Trans Nuigini Tours always booked in advance before passing through any village. It is not only a matter of courtesy that responsible tour operators do this: if a visit is known in advance the villagers have a chance to prepare and sell a few weavings, betel nut bags, coffee, or local palm juice, and therefore at least gain some economic benefit from these otherwise uninvited guests.

Less responsible tour operators thus fail on the grounds of inadequate preparation. It could be considered an unreasonable expense to visit remote villages to discuss in advance the idea of a visit but a letter would serve at least to inform the villagers of the date and approximate time of arrival. On a visit, the tour leaders are either ignorant themselves, or unwilling or unable to communicate the basics of the hosts' culture and thus improve the behaviour of the tourists. Thus, not only are there no substantial permanent benefits for the hosts, but even the opportunity for minimal economic return is denied as the villagers are unprepared for the tourists' visit.

A Big Bank Project in Manggarai

As already discussed in the first part of this chapter, the author does not consider that the Manggarai people are a suitable target for tourism development. To substantiate this, the different approaches to tourism development will be examined and contrasted, largely underlining the point made by many writers (De Kadt, 1992; Burns and Holden, 1995), that the scale of development is exceedingly important to the question of sustainability. This section also looks further at the question of community participation.

The region of Manggarai was one area chosen by the Asian Development Bank for its biodiversity reserve project because of its unique upland moist forest, which is threatened by severe deforestation. The regional town of Ruteng lies at the centre of Manggarai and close to the project area. As its name suggests, biodiversity preservation was the aim of the project and ecotourism was seen as one important route to achieving that preservation.

A discussion of the arguments for and against ecotourism development in general, or specifically in this area, would be inappropriate here. However, it should be pointed out that there are some rather more fundamental flaws in the project. Its ecotourism part was investigated and set up without the use of an anthropologist. Either no attempt was made to understand the local people or they were misunderstood. As pointed out at the beginning of this chapter, not all groups of people are naturally hospitable or willing to serve. In such cases it is hard to see how they can become appropriate hosts, an essential prerequisite for successful tourism development. If the Asian Development Bank had employed an anthropologist at the initial stages of this project they would have been alerted to the unsustainability of successful tourism development in this community.

Having lacked the correct personnel for the initial phase of this project, it continued in the same vein. Of the nine international consultants who worked on the project during the three months the author was there, only one, the team leader (who resigned), spoke any Indonesian. Most of the Manggarai, although amongst themselves they converse in the Manggarai language, speak good Indonesian as they have a long history of contact with other groups. Although there were local consultants, many of whom spoke reasonable English, their position was one of considerable hardship (one died in a local hospital), and, in general, working relations with internationals were ineffective.

Not only could none of the international consultants converse with the locals, they were not given any background about their culture that might help them with understanding or communications. Woodley (1993) discusses the problems that can arise because of cultural barriers between planners and residents. Many of those involved in the Manggarai project (including the community development worker) spent most of their time in Jakarta, the Indonesian capital, over one day's travelling time away.

Community participation is heralded as one of the great principles of sustainable tourism, as outlined by Pigram (1994) and Murphy (1985). Many writers have suggested the importance of community control and involvement at the planning stage (Burns and Holden, 1995; Hunter and Green, 1995). As pointed out by the ecotourism consultant in this project, there was very little community consultation, let alone participation in planning. If Tourism Concern's (1992) advice, that local people should determine their own development, had been heeded, or Cronin's (1990) suggestion that no project should proceed without the local population's approval, then the ecotourism component of the Asian Development Bank project would have been halted.

Conclusions

Community participation

Community participation is considered a basic principle of sustainable tourism development but, besides the question of equity, it is the least successfully put into practice. In communities that have hitherto had minimal education and little previous experience of tourism, it is unlikely that the goals of community planning and control of tourism can be fulfilled. The island of Flores is being rapidly developed for tourism, by outsiders. It is the responsibility of those outsiders to ensure the communities have a positive experience of tourism.

The communities need educating, beginning at the basic level of understanding the hosting function, a vital function in tourism, as pointed out by Van Harssel (1994). Beyond the needs of awareness, education and training, the communities need to feel involved; an initial positive experience of tourism will give the communities the confidence and desire to plan and control tourism fruitfully in the future.

Although a Ngada village water project is far from being community planning and control, it was the villagers' idea. It was a project that benefited the entire community and thus prevented antagonism between members. It gave the villagers, along with the other factors discussed, a positive experience of tourism and will thus enhance the chance to achieve sustainability of the industry at a local level in the future.

Anthropologists and sustainable tourist development

The importance of anthropologists for the development of sustainable tourism in less developed countries should not be underestimated. In all but specifically nature-based tourism, the role of the hosts must be put as central to the sustainable tourism development process. To give them this central role it is important to have a clear understanding of their culture.

Anthropologists should be involved from the outset when areas are being explored for the possibility of tourism development so that communities that are not naturally hospitable, are unwilling to serve or are likely to be hostile to the tourism development process can be identified. Anthropologists also have a central role to play to facilitate understanding between local communities and international consultants to prevent cultural differences between planners and residents becoming barriers to the development process.

When used as tour leaders, anthropologists contribute to minimizing the adverse effects of tourism on host communities. Indeed, it is anthropologists that have been at the forefront of the Western ethnocentrism and romanticism, whose bias is typified in the assumption, pointed out by Sharpley (1994), that it is better to preserve cultures as traditional rather than allowing them to modernize. Accepting that the process of modernization will occur with or without tourism, the anthropologist as tour leader can help the contact with Westerners to be inoffensive and as smooth as possible. Also, anthropologists as tour leaders will enhance the experience of the tourists.

References

Bates, B. (1991) Impacts of tourism on the tribal cultures and natural environment in Papua New Guinea. In: *Ecotourism 1991, Conference Papers.* University of Queensland, Australia.

Burns, P. and Holden, A. (1995) *Tourism: A New Perspective.* Prentice Hall, Hemel Hempstead.

Cohen, E. (1985) The tourist guide: origins, structures, and dynamics of a role. *Annals of Tourism Research* 12, 5–29.

Cronin, L. (1990) A statergy for tourism and sustainable developments. *World Leisure and Recreation* 32(3), 12–18.

Dearden, P. and Harron, S. (1994) Alternative tourism and adaptive change. *Annals of Tourism Research* 21, 81–102.

De Kadt, E. (1992) Making the alternative sustainable: Lessons from development for tourism. In: Smith, V. and Eadington, W. (eds) *Tourism Alternatives*. Wiley, Chichester, pp. 47–75.

Doxey, G. (1975) A causation theory of visitor resident irritants: methodology and research inferences. In: *Proceedings of the Travel Research Association 6th Annual Conference*, California.

Graburn, N. (ed.) (1976) *Ethnic and Tourist Arts: Cultural Expressions from the Fourth World*. University of California Press, Berkeley.

Harron, S. and Weiler, B. (1992) Review: ethnic tourism. In: Weiler, B. and Hall, C.M. (eds) *Special Interest Tourism*. Belhaven, London, pp. 83–94.

Hunter, C. and Green, H. (1995) *Tourism and the Environment: A Sustainable Relationship?* Routledge, London.

Murphy, P. (1985) *Tourism: a Community Approach*. Methuen, New York.

Pigram, J. (1994) Alternative tourism and sustainable resourse management. In: Smith, V. and Eadington, W. (eds) *Tourism Alternatives*. Wiley, Chichester.

Sharpley, R. (1994) *Tourism, Tourists and Society*. Elm Publications, Huntingdon.

Smith, V. (ed.) (1989) *Guests and Hosts: the Anthropology of Tourism*, 2nd edn. University of Pennsylvania Press, Philadelphia.

Tourism Concern (1992) *Beyond the Green Horizon*.World Wildlife Fund, Godalming, Surrey.

Van Harssel, J. (1994) *Tourism and Exploration*, 3rd edn. Prentice Hall, Englewood Cliffs, New Jersey.

Weiler, B. and Davis, B. (1993) An exploratory investigation into the roles of the nature based tour leader. *Tourism Management* 14, 91–98.

Wood, R.E. (1984) Ethnic tourism, the state of cultural change in Southeast Asia. *Annals of Tourism Research* 11, 353–374.

Woodley, A. (1993) Tourism and sustainable development: the community persective. In: Nelson, J.G., Butler, R. and Wall, G. *Tourism and Sustainable Development: Monitoring, Planning, Managing*. Department of Geography Publication No. 37, University of Waterloo, Waterloo, Ontario, Canada.

Sustainability and the Consumption of Tourism

16

R. Sharpley[1] and J. Sharpley[2]

[1]Faculty of Business, University of Luton, Park Square, Luton LU1 3JU, UK; [2]Department of Business and Finance, University of Hertfordshire Business School, Mangrove Road, Hertford SG13 8QF, UK

Introduction

Since attention was first drawn to the potentially harmful impacts of mass tourism some 25 years ago (Mishan, 1969; Young, 1973; Turner and Ash, 1975), there has been an increasingly vigorous debate about how best to achieve a more balanced approach to the development of tourism. This debate initially focused upon the relationship between tourism and the physical environment, a relationship frequently considered to be one of conflict rather than harmony (Budowski, 1976); the prime concern, therefore, was to find ways of integrating tourism with its natural environment. However, by the 1980s, it had been recognized that tourism also impacts both positively and negatively on host societies and cultures and that the development of tourism should not be 'prejudicial to the social and economic interests of the population in tourist areas, to the environment or, above all, to natural resources, which are the fundamental attraction of tourism, and historical and cultural sites' (World Tourism Organization, 1980). As a result, alternative forms of tourism, variously described as responsible, appropriate, soft or green tourism, were proposed as potential solutions to the tourism 'problem'.

More recently, the concept of sustainable tourism development has been embraced by tourism academics and practitioners alike. Building upon the three general principles of sustainability, namely, the conservation of natural

© CAB INTERNATIONAL 1997. *Tourism and Sustainability*
(edited by M.J. Stabler)

resources, long-term planning, and a more equitable global share of resources and opportunities, sustainable tourism has been promoted as the way forward to achieving a symbiotic relationship between tourism and the broader physical, social and cultural environment upon which it depends. However, despite widespread agreement that such a relationship is both desirable and necessary, there are still widespread disagreement and confusion about the viability of sustainable tourism. Indeed, beyond the publication of codes of practice (Mason and Mowforth, 1995) or development, the promotion of micro-solutions such as ecotourism, and limited attempts to introduce 'good practice', little practical progress has been made.

A variety of reasons have been put forward to explain this lack of progress towards achieving more sustainable forms of tourism development. Some have suggested that many of the proposed solutions are either impractical or, in the longer term, likely to be worse than the original problems (Butler, 1991), whilst others have identified an unwillingness on the part of the tourism industry as a whole to accept environmental responsibility (Robinson and Towner, 1992). It has also been argued that the promotion of sustainable tourism is little more than a marketing ploy, an attempt by the tourism industry to jump on the 'green bandwagon' without adopting a more positive approach towards sustainability (Wheeller, 1991; Wight, 1993). Certainly, relatively few tour companies in the UK have actively pursued environmental policies (Elkington and Hailes, 1992), whilst, more recently, financial pressures within the industry have forced many operators to cut back on expenditure supporting environmental programmes (Josephides, 1996).

Overall, however, it is evident that the majority of sustainable tourism development policies have been somewhat myopic in their approach. That is, their primary focus has been on the protection of the physical and socio-cultural environment in destination areas, which, although a fundamental requirement, is just one ingredient of any sustainable development policy. It is also necessary to take into account, and to balance, the requirements of the other factors of what has been described as tourism's 'magic pentagon' (Müller, 1994). These factors are:

- economic health
- subjective well-being of locals
- unspoilt nature, protection of resources
- healthy culture
- optimum satisfaction of guest requirements

It is suggested here that, by and large, the 'optimum satisfaction of guest requirements', or, more generally, the recognition and satisfaction of the needs of tourists, has been subordinated to the broader requirements of the environment and communities in destination areas. In other words, most sustainable tourism policies have failed to accept and to respond to a number of tourism 'truths' (McKercher, 1993) in general, and the fact that 'tourists are consumers, not anthropologists' in particular. This chapter argues, therefore, that it is essential to understand and exploit the nature and characteristics of the consumption of tourism and, through a case study of a tourism development in The Gambia, it demonstrates that it is in fact possible to develop forms of tourism which both benefit the host environment and satisfy the broader needs of tourists.

The Consumption of Tourism

Tourism is, undoubtedly, a major social phenomenon of the twentieth century and, arguably, one of the most influential factors in global social and economic development (Vellas and Bécherel, 1995). Relatively few, if any, parts of the world are not now tourist destinations and, for many countries, tourism represents an increasingly important element of broader socio-economic development policies. Therefore, it is, perhaps, inevitable that most attention has been focused on the growth in demand for tourism, its economic benefits and, of course, the more recent concerns about the impacts of tourism.

Conversely, although researchers have long been concerned with tourist motivation, it is only recently that efforts have been made to analyse and understand the consumption of tourism from a social point of view. That is, little attention has been paid to tourists as consumers of the tourism product (Taylor, 1994), to the significance and meaning of tourism to the consumer and, other than general observations, to changes in the nature of the demand for tourism. In particular, although 'explaining the consumption of tourist-services cannot be separated off from the social relations within which they are embedded' (Urry, 1990a, p. 23), there has been relatively little exploration of the link between cultural transformations in tourism generating countries and the resulting changes in tourism demand.

This is, arguably, one of the main factors that has restricted the progress towards the development and acceptance of more sustainable forms of tourism. Although the role of the tourist has certainly not been overlooked, nevertheless many of the proposed solutions to the problems of mass tourism have been based either upon the need to modify tourists' behaviour or upon inaccurate assumptions about perceived changes in tourists' (consumer) behaviour. In other words, most proposed forms of alternative or sustainable tourism have been dependent on adapting the tourist to the product (or assuming that tourists' demands are changing) rather than designing the product to suit the existing needs of tourists. This is confirmed by Butler (1991), who has identified changing tourist types, curbing tourist numbers and educating tourists as three principal categories of alternative tourism proposals.

More generally, the perceived emergence of the environmentally-aware, green or 'good' (Wood and House, 1991) tourist has often been used as the justification for the promotion and potential viability of alternative, sustainable forms of tourism. There is little doubt that, over the last decade, there has been increased public demand for green, environmentally friendly products in general; surveys of consumer behaviour have found that almost 40% of adults in the UK always buy, or try as far as possible to buy, environmentally friendly products (Mintel, 1991). However, this level of environmental concern has yet to become apparent within the purchase and consumption of tourism. For example, a recent evaluation of 21 sustainable rural tourism projects in England found that their 'green message' appealed only to a minority of tourists (Countryside Commission, 1995), whilst it has also been pointed out that one reason for the 'difficulty of achieving sustainable tourism ... [is the fact that] ... the trend towards indulging in pleasure and enjoyment and living life to the full continues virtually undiminished' (Müller, 1994, p. 134).

There is no doubt that, since the emergence of mass tourism, the consumer behaviour of tourists has been dynamic, responding both to changes and developments in the supply of tourism and to other factors, such as the availability of leisure time and levels of income. At the same time, styles of travel have transformed as tourism has matured into a widespread social activity. Increasing demands for self-catering accommodation, fly–drive holidays and special interest tourism, combined with the recent falls in the demand for traditional summer package holidays, are evidence of the greater confidence and independence of what has been described as the 'new tourist' (Poon, 1993); the new tourist is flexible and independent, with values and lifestyles different from those of the more traditional mass tourist. Such a process is, perhaps, inevitable as tourism comes of age. However, of greater importance in the context of sustainable tourism development is the relationship between the consumption of tourism and cultural change or, more specifically, the role and meaning of tourism as a form of consumption in modern society.

Tourism in Modern Society

Mass tourism has been described as 'one of the quintessential features of modern life' (Urry, 1988, p. 35). It is an accepted and expected mass leisure activity and, whilst technological and socio-economic advances within modern society have facilitated the growth of tourism, the characteristics of life in modern society have also created the need or motivation for tourism (Krippendorf, 1986; Sharpley, 1994). However, identifying the relationship between modern society and tourist motivation represents only the first step in understanding the consumption of tourism. That is, it is not only the *need* to consume tourism, but also the *style* of that consumption that may be linked to developments and transformations in society. For example, it has been suggested that structured, mass tourism to British seaside resorts in the 1950s reflected the nature of modern, industrial society at that time, and that more recent changes towards post-industrial or post-modern society have directly influenced both the style and the role of tourism in society (Urry, 1988, 1990b).

Whether or not society is indeed becoming post-industrial or developing cultural traits that may be described as 'post-modern' is a matter of debate beyond the scope of this chapter. Nevertheless, the production–consumption relationship within tourism has undergone a fundamental transformation in recent years (for example, the emergence of specialist tour operators, the promotion of niche markets, or the development of more individual, flexible forms of tourism) and this, in turn, may be linked to changes in consumer culture which result from broader cultural transformations. The questions to be addressed, therefore, are how have these changes in consumer culture influenced the consumption of tourism and how may this relate to the development of more sustainable forms of tourism?

Consumer culture and tourism

Consumer culture, for the purposes of this chapter, may be defined as the character, significance and role of the consumption of commodities, services and experiences within modern society. Three different approaches to the examination and explanation of consumer culture have been identified (Featherstone, 1990, 1991). Firstly, consumer culture may be seen as resulting from what is described as the *production* of consumption. That is, consumer culture is directly related to the mass production of goods and services for purchase and consumption, with the producers being able to dictate styles, taste and fashion. As a result of this culture of mass production and consumption, and the need for all commodities to appeal to the widest possible market, high and low culture become merged with the cultural value of commodities tending towards the lowest common denominator. Thus, from this perspective on consumption, the dominant role of the producer leads, perhaps inevitably, to a diluted and homogeneous cultural value of goods and services. Within the context of tourism, this approach is of most relevance to the analysis of earlier forms of mass, package tourism when the producers of tourism (tour operators) were, to a great extent, able to control the development and style of the mass consumption of tourism and (cultural) product quality was sacrificed to price.

In contrast, the second approach focuses on the *mode* of consumption, highlighting the culture of consumption rather than simply viewing consumption as the inevitable result of production. It is based upon the notion that, within post-industrial society, traditional social groupings are being replaced by a new and expanding middle or 'service' class. Within this new social class, there are two identifiable groups, namely, those who possess both economic and cultural capital in significant quantities (the new bourgeoisie), and those who possess cultural capital but less economic capital – the new petit bourgeosie (Bordieu, 1984). The latter, larger group are seen as the new taste-makers; having less financial resources, they seek social differentiation and status through different styles, rather than values, of consumption. Thus, for the new middle classes as a whole, the mode of consumption of goods, services and experiences, including tourism, has become one the primary means of demarcating social status and relationships. It has been argued that it is the emergence of this new service class and associated culture of consumption that has led to the development of specialist, niche market tourism products; 'while travel has remained an expression of taste since the eighteenth century, it has never been so widely used as at present' (Munt, 1994, p. 108–109). As a result, it has also necessitated a more responsive approach on the part of the producers of tourism to the changing demands of tourists.

Of particular relevance to the consumption of tourism, the third perspective on consumer culture concentrates on 'the emotional pleasures of consumption, the dreams and desires which become celebrated in consumer cultural imagery' (Featherstone, 1990, p. 5). In this case, consumption neither flows logically from production, nor does it play a role in the determination of social status. Rather, consumption is viewed as the 'fulfilment of dreams', as a search for pleasurable experiences, as a means of escaping from the rigidity and structure of day-to-day culture and society. Reference is frequently made to the traditional role of fairs and carnivals in the pre-industrial era; they were both local markets and places to indulge in pleasure, to experience unusual or exotic images. It is not surprising, therefore, that tourism is seen by some as a continuation of the carnivalesque tradition into modern consumer culture; the spectacle of mass tourism at the seaside resorts in the late nineteenth and early twentieth centuries and, more recently, the emergence of theme parks are both seen as evidence of this trend (Urry, 1990b). More generally, of course, the desire to escape from the ordinary and mundane, to consume the dreams and fantasy of travel, is also considered to be a major tourism motivating factor (Dann, 1981).

If these perspectives on consumer culture are applied to tourism, it is evident that, in recent years, the nature of the consumption of tourism has developed from a producer-led to a consumer-led form of consumption. That is, early forms of mass tourism, in particular mass package tourism, were symptomatic of the dominant role of tour operators in shaping holiday styles and tastes, whereas, more recently, the industry has had to become increasingly responsive to the changing demands of the consumer. Underlying this shift in the dominance of consumer culture is the alleged cultural transformation of society from industrial to post-industrial, resulting in the emergence of a broad new middle class which constantly seeks new and different styles of consumption as a means of establishing and maintaining social differentiation. 'Tourism has emerged as a key commodity' (Munt, 1994) in this process and it is this new consumer culture that has challenged the traditional mass production/consumption character of mass tourism. New styles of tourism, albeit within the traditional framework of the 'package holiday', have been developed to satisfy the changing needs of the consumer and it has been suggested that many alternative forms of tourism, including green tourism or ecotourism, have enjoyed increasing popularity more as a result of people's need to maintain their 'cultural capital' and social differentiation rather than from a genuine concern for the environment.

In short, tourism, as a form of consumption in modern society, has developed as a marker of social status, a cultural signifier of taste. The implication, therefore, is that tourists will increasingly wish to consume styles of tourism that deviate from collective or mass forms of tourism. Thus, the producers of tourism, in order to satisfy this underlying purpose, or culture, of tourism consumption, will have to respond by developing tourism products which possess the aura of positional goods (Walter, 1982) but which are, at the same time, affordable to the so-called 'new petit bourgeosie'. For example, the recent entry of two major tour operators into the cruising market has brought what is normally perceived to be an exclusive form of tourism within reach of a broader market. Therefore, the challenge facing the tourism industry is to continue to move away from the traditional, lowest common denominator approach (the production of consumption perspective) to development based on quality, for it is quality that goes some way to fulfilling the broader requirements of tourists within the mode of consumption perspective. It is also this concept of quality that provides the link between the consumption of tourism and sustainable tourism development.

The consumption of tourism and sustainable tourism development

As stated earlier, one of the main factors contributing towards the lack of progress towards the development of more sustainable forms of tourism development has been the widely held, yet inaccurate, assumption that tourists are increasingly becoming more environmentally aware. Undoubtedly a proportion of tourists fall into this category, such as those who participate in active conservation holidays, but viewing the consumption of tourism from a consumer culture perspective suggests that non-environmental considerations, such as the pursuit of status and emotional pleasure, are the more influential factors in tourism purchasing decisions.

Thus, the consumption of low impact, individualistic, green or alternative (to mass) forms of tourism arguably signifies taste and style decisions, rather than environmental concern.

At the same time, many attempts to develop sustainable forms of tourism have been characterized by small scale projects which, by concentrating on minimizing environmental impacts, have overlooked the comfort, quality and status needs of tourists. In some instances, this has proved to be a successful and appropriate approach, such as in the case of the village tourism projects in Casamance, Senegal, but developing sustainable tourism 'requires consideration of much more than counting the negative effects of mass or conventional tourism' (Butler, 1992). All five sectors of Müller's (1994) 'magic pentagon' require equal attention, implying the need for compromise rather than allowing environmental concerns to dominate. Thus, from a consumer culture perspective, tourism developments should seek to satisfy the consumption-orientated needs of tourists, as well as optimizing the benefits to destination environments and local communities, by injecting quality, taste and comfort into the product. As the following case study of a tourism development in The Gambia demonstrates, such an approach can fulfil the broader (though not purist) requirements of sustainable tourism development whilst appealing to a wider sector of the market.

Tourism Development: an Illustrative Example in The Gambia

The Gambia is not a country that is normally associated with the development of sustainable tourism. Lying within six hours' flying time from northern Europe and blessed with virtually uninterrupted sunshine during the winter months, The Gambia's potential as a winter sun destination has been increasingly exploited since the first tourists arrived in 1965. During the 1965/66 season, just 300 tourists, mostly from Scandinavia, arrived in The Gambia; by 1993/94, that figure had risen to an estimated 130,000, almost 70% arriving on charter flights originating in northern European countries (Sharpley *et al.*, 1996). This rapid growth is mirrored within the development of the accommodation sector. In 1967, there were just two tourist hotels with a total of 52 beds; there are now 17 large hotels, a number of guest houses and one 'club-share' complex, together offering almost 5800 beds. Tourism has also represented an increasingly important sector of an economy that still depends largely on the production and export of groundnuts. Although the country's annual net foreign exchange earnings from tourism of about $25 million are relatively insignificant in global tourism terms, nevertheless the tourism sector represents almost 11% of The Gambia's gross domestic product. Some 7000 jobs are also directly or indirectly dependent on tourism, although many of these are low grade and seasonal. For example, over 40% of employment in hotels is seasonal.

The development of tourism in The Gambia is considered in detail elsewhere (Economist Intelligence Unit, 1990; Dieke, 1993, 1994; Thomson *et al.*, 1995). However, the characteristics of tourism in The Gambia, for the most part sun–sea–sand package holidays dominated by the British market, indicate the inherent lack of sustainability of the tourism sector. This has long been recognized by the Gambian authorities, who have been trying to diversify both the tourism product and their markets to spread the economic benefits of tourism at the same time as reducing their dependency on the British market. Unfortunately, the events of 1994, following the military *coup d'état*, were a sad demonstration of the lack of sustainability of the Gambian tourism industry (Sharpley *et al.*, 1996), whilst the one local tourist development adhering rigidly to the concept of alternative, ecotourism, the Kololi Inn, has not proved to be successful. In contrast, however, one tourism site located away from the main tourist areas has shown that, by striking the correct balance between the needs of tourists and environmental integration, it is in fact possible to develop a sustainable product with a more mass market appeal.

The Kemoto Hotel

The Kemoto Hotel, also known as the African Club, is a relatively new hotel development some 150 kilometres up-river, or inland, from the main coastal resorts and from Banjul, the capital of The Gambia. Taking its name from the nearby village of Kemoto and occupying a dramatic position overlooking the River Gambia at Mootah Point, the hotel is accessible either by river (although there is no regular transport service along the River Gambia) or by road. The drive from Banjul takes approximately four hours; the route follows the main road eastwards before turning off north along a tortuous dirt track for the last thirty kilometres.

Although in one of the more remote locations in the country, the hotel is remarkably luxurious. It comprises 50 rooms, most of which are in circular huts set amongst a banana plantation, with a total of 124 bed spaces. The accommodation is relatively simple, yet modern, clean and comfortable with *en suite* shower and toilet facilities in each room. Central facilities include a well-stocked bar and lounge area within a traditional, African style building, a restaurant serving both Western and local dishes and, perhaps most surprisingly for a hotel in the middle of the African bush, a large, modern swimming pool. The hotel has running water, which is pumped from a nearby well, and electricity supplied by its own generator. In short, the Kemoto Hotel provides the highest level of comfort and facilities within the constraints of its location.

The visitor experience is further enhanced by the activities offered. The hotel arranges day safaris to local villages and schools, dancing and wrestling are frequently provided as entertainment in Kemoto village and bird-watching excursions along the river and its tributaries are a particular highlight. Visitors are also welcome to wander through the village to meet local people, gaining, as far as is realistically possible, an insight into Gambian village life, and the local primary school is also within walking distance of the hotel.

However, despite the apparent concentration on Western-style comfort and quality, there is a surprising degree of integration with the village, optimizing the benefits of tourism to the local community. Although the hotel is not locally owned, 80% of its 35 employees come from the village, albeit mostly occupying the more menial positions. Local labour and crafts are also utilized in the continuing development of the complex and in the hotel's 'kitchen garden', which supplies all the fresh produce required by the restaurant. The local villagers are paid by the hotel for dancing and wrestling displays, virtually all the fish caught by local fishermen is bought by the hotel and it is also the local fishermen who provide, and earn important additional income from, the river bird-watching trips. An interesting arrangement also exists between the hotel and the local baker. In return for the use of ovens on the hotel site, the baker pays a daily 'rent' of 20 loaves of bread to the hotel, ensuring a regular local supply of bread to the restaurant; the remainder he sells to local people. However, perhaps most importantly, fresh water is pumped from the hotel's own supply to a standpipe in the village and, having its own electricity generators, the hotel also provides free street lighting in the village.

Thus, although from the perspective of green or ecotourism develop-ment the quality and range of facilities at the Kemoto Hotel might be considered to be inappropriate to its setting, the hotel has gone a long way towards achieving, albeit almost by accident, a sustainable form of tourism. Furthermore, it is a form of tourism which is not dependent on an up-market, special-interest or niche market. Most of its customers, prior to the collapse of tourism to The Gambia in the 1994/95 season, came on thrice-weekly organized up-river excursions from the resort hotels on the coast, with Airtours providing the majority of visitors. At a relatively cheap price (in 1994, the hotel charged £24 per person for dinner, bed and breakfast plus entertainment), tourists are able to enjoy a reasonably authentic taste of Gambian culture but with a level of comfort and quality commensurate with the overall standard of their holiday. At the same time, the interdependence between the hotel and the local community has optimized the benefits to visitors, local people and the hotel itself.

Therefore, although the Kemoto Hotel faces a number of problems, such as poor accessibility, a lack of entertainment or activities for longer-staying guests, and its current dependence on large tour groups 'supplied' on excursions, all of which represent a threat to its longer-term sustainability, it nevertheless provides an alternative, and potentially more successful, model for sustainable tourism development, which may be applied to a variety of tourism scenarios.

Conclusions

This chapter has identified two major themes in the current approach to the development of sustainable tourism, namely, the need to protect the physical and socio-cultural environment in tourist destination areas and the reliance on the ability to change tourists' needs and behaviour. For example, the laudable attempts by a number of large tourist organizations to adopt 'sustainable' policies have been centred upon mainly environmental concerns (Middleton and Hawkins, 1993). At the same time, the entire sustainable tourism debate has been based upon the notion that green, ecotourism is 'good' and that traditional mass tourism is 'bad'. As a result, the majority of sustainable tourism developments have been small-scale, local projects, whilst tourism consumption habits have been largely overlooked. In particular, little or no attention has been paid to the way in which transformations in consumer culture could be exploited in order to encourage demand for more sustainable tourism.

The analysis of consumer culture and the case study of the Kemoto Hotel in The Gambia presented here together suggest that sustainable tourism development policies do not need to adhere to 'text-book' principles. When the shift from production-led consumption to a consumer culture based upon both status and hedonistic objectives is taken into account, it becomes clear that the needs of the emerging service class or petit bourgeoisie represent an opportunity to expand sustainable tourism to a broader audience. If these social tourist objectives are met through the provision of a quality experience with the needs of the local environment and community taken into account, there is no specific need for a 'sustainable' tourism product to be labelled and sold as such. In other words, a sustainable product can be successfully sold to tourists not specifically seeking it. Whilst this might require a 're-definition' of the form and objectives of sustainable tourism, there is no doubt that its future lies in a broad approach, the fundamental basis of which must be the satisfaction of consumer needs. Without the consumer there can be no sustainable tourism.

References

Bordieu, P. (1984) *Distinction: A Critique of the Judgement of Taste*. Routledge, London.

Budowski, G. (1976) Tourism and environmental conservation: conflict, co-existence or symbiosis, *Environmental Conservation* 3(1), 27–31.

Butler, R. (1991) Tourism, environment and sustainable development. *Environmental Conservation* 18(3), 201–209.

Butler, R. (1992) Alternative tourism: the thin end of the wedge. In: Smith, V. and Eadington, W. (eds) *Tourism Alternatives*. University of Pennsylvania Press, Philadelphia.

Countryside Commission (1995) *Sustainable Rural Tourism: Opportunities for Local Action*. CCP 483, Countryside Commission, Cheltenham.

Dann, G. (1981) Tourist motivation: an appraisal. *Annals of Tourism Research* 8, 187–219.

Dieke, P. (1993) Tourism and development policy in The Gambia. *Annals of Tourism Research* 20, 423–449.

Dieke, P. (1994) The political economy of tourism in The Gambia. *Review of African Political Economy* 62, 611–627.

Economist Intelligence Unit (1990) *Senegal and The Gambia*. International Tourism Reports No. 3, EIU, London.

Elkington, J. and Hailes, J. (1992) *Holidays That Don't Cost the Earth*. Victor Gollancz, London.

Featherstone, M. (1990) Perspectives on consumer culture. *Sociology* 24(1), 5–22.

Featherstone, M. (1991) *Consumer Culture and Postmodernism*. Sage Publications, London.

Josephides, N. (1996) A black time for 'green' issues. *Travel Weekly*, 21 February, p. 11.

Krippendorf, J. (1986) Tourism in the system of industrial society. *Annals of Tourism Research* 13, 517–532.

McKercher, R. (1993) Some fundamental truths about tourism: understanding tourism's social and environmental impacts. *Journal of Sustainable Tourism* 1(1), 6–16.

Mason, P. and Mowforth, M. (1995) *Codes of Conduct in Tourism*. Occasional Papers in Geography No. 1, Department of Geographical Sciences, University of Plymouth.

Middleton, V. and Hawkins, R. (1993) Practical environmental policies in travel and tourism. *Travel and Tourism Analyst* No. 6, 63–76.

Mintel (1991) *The Green Consumer*. Market Report, Mintel, London.

Mishan, E. (1969) *The Costs of Economic Growth*. Penguin, Harmondsworth.

Müller, H. (1994) The thorny path to sustainable tourism development. *Journal of Sustainable Tourism* 2(3), 131–136.

Munt, I. (1994) The 'other' postmodern tourism: culture, travel and the new middle classes. *Theory, Culture and Society* 11, 101–123.

Poon, A. (1993) *Tourism, Technology and Competitive Strategies*. CAB International, Wallingford.

Robinson, M. and Towner, J. (1992) *Beyond Beauty: Towards Sustainable Tourism*.

Business Education Publishers, Houghton-le-Spring.

Sharpley, R. (1994) *Tourism, Tourists and Society*. Elm Publications, Huntingdon.

Sharpley, R., Sharpley, J. and Adams, J. (1996) Travel advice or trade embargo? The impacts and implications of official travel advice. *Tourism Management* 17, 1–7.

Taylor, G. (1994) Styles of travel. In: Theobald, W. (ed) *Global Tourism: The Next Decade*. Butterworth Heinemann, Oxford.

Thomson, C., O'Hare, G. and Evans, K. (1995) Tourism in The Gambia: problems and proposals. *Tourism Management* 16, 571–581.

Turner, L. and Ash, J. (1975) *The Golden Hordes: International Tourism and the Pleasure Periphery*. Constable, London.

Urry, J. (1988) Cultural change and contemporary holiday-making. *Theory, Culture and Society* 5, 35–55.

Urry, J. (1990a) The consumption of tourism. *Sociology* 24(1), 23–35.

Urry, J. (1990b) *The Tourist Gaze*. Sage Publications, London.

Vellas, F. and Bécherel, L. (1995) *International Tourism*. Macmillan, Basingstoke.

Walter, J. (1982) Social limits to tourism. *Leisure Studies* 1, 295–304.

Wheeller, B. (1991) Tourism's troubled times. *Tourism Management* 12, 91–96.

Wight, P. (1993) Eco-tourism: ethics or eco-sell? *Journal of Travel Research* 33(3), 3–9.

Wood, K. and House, S. (1991) *The Good Tourist: A Worldwide Guide for the Green Traveller*. Mandarin, London.

World Tourism Organization (1980) *Manila Declaration on World Tourism*. WTO World Tourism Conference, Manila.

Young, G. (1973) *Tourism: Blessing or Blight?* Penguin, Harmondsworth.

Tourism and Social Responsibility: A Philosophical Dream or Achievable Reality?

17

M. Ireland

College of St Mark and St John, Derriford Road, Plymouth PL6 8BH, UK

Introduction

Social responsibility, like sustainability, has a number of meanings which are derived in part from the context in which the concept is used. In Britain, the lay person will be most familiar with that used in connection with the Church, which has 'boards of social responsibility'. In the United States, it is used when referring to business practice. Irrespective of context, the idea of social responsibility implies that an institution is asking its members to act in a certain manner towards the community in which it is located. This chapter takes as its context the tourist industry in the Celtic periphery of the south west peninsula of Britain, and looks for evidence of socially responsible tourism practice. A central theme of this chapter is to ascertain to what extent socially responsible management strategies can secure a sustainable long term future for tourism in Cornwall. Cornwall has been chosen because of its importance as a destination for domestic tourists and a small but significant European market.

Part of the county's attraction is its relative isolation, which helps maintain the cultural identity of its people (Griffin, 1993). It is estimated that Cornwall attracts 3.4 million visitors each year, who spend a total of £613 million (1991). However, the economic benefits of tourism are not obvious to many who live and work in Cornwall. Cornwall County Council reports that 'In 1991 average gross weekly earnings for men of £246.30 were the second lowest in Great Britain, 23% below the national average of £318.90' (Griffin, 1993). Set against this picture of relative poverty is a vision of opportunity to develop a socially responsible tourism industry to benefit both the environment and the community. Devon and Cornwall Training and Enterprise Council (TEC) have described this objective as a challenge.

> The challenge is to bring about fundamental change: to move this vital industry from one that is fragmented and reactive to one that is cohesive and can pro-actively offer a tourist product that benefits both the customer and the host communities.
>
> (Bell, 1995b)

This chapter examines the recent history of tourism development in Cornwall (1985–1995) in an attempt to explain why a tourism product which is socially responsible is increasingly becoming a shared objective among the business community, local authority and local people.

The Concept of Social Responsibility

Following the line of argument put forward in the introduction, social responsibility will be defined in relation to its meaning within Western capitalism. Because capitalism is translated from economic and political theory into practice through the medium of the firm or corporation, this would seem an appropriate starting point to develop an understanding of what is meant by corporate and individual social responsibility. At its most fundamental, as capitalism itself has raised ethical questions, so do the firms which operate within that system.

Two competing schools of thought can be identified in the literature regarding what is understood by social responsibility in business. The first is within the doctrine of a free market, represented here by the work of Friedman (1993) in his essay 'The social responsibility of business to increase its profits'. The second argument, which can be broadly termed collectivist, is the one taken by Donaldson (1993), Rackman *et al.* (1990) and Miller and Ahrens (1993), who introduce concepts central to any theory of social responsibility, those of obligation, justice and economic democracy.

The aim is to arrive at a working definition of social responsibility which can be applied to the tourism industry. To achieve this we begin with the work of Friedman (1993), who identifies the central problem surrounding any discussions of social responsibility in business: they are known for their analytical looseness and lack of rigour. This methodological weakness arises because of the failure to distinguish between the corporate and individual social responsibility. Friedman (1993) takes the view that it is people not businesses who have social responsibility. It is not always easy to differentiate between the social responsibilities of the firm and those of the individual managers and employees within the organization. For example, some companies in the tourism industry employ people on a seasonal basis who would otherwise be unemployed. This policy could be seen as a socially responsible action by the managers to help the long-term unemployed.

However, such a policy may not be in accord with the shareholders or customers, who may hold prejudiced views about the long-term unemployed, seeing them as scroungers, lazy or dishonest. If this was an employment policy initiated by the manager without reference to the shareholders, this, Friedman (1993) argues, is the manager acting on the grounds of political principle. There is for Friedman only one social responsibility of business, to make a profit. Notwithstanding this argument, it would be perfectly possible for managers to exercise individual social responsibility behaviour and act in an economically rational manner to maximize profits for the shareholder. In the scenario outlined, the unemployed are likely to work for low wages and incur tourism business minimum cost in national insurance and pension contributions. This view is confirmed by research carried out by Cornwall County Council, who describe the tourism industry as 'a seasonal employer and which provides relatively low rates of pay' (Griffin, 1993).

This poses an interesting research question. How do employers publicly justify their employment policy in the tourism industry? This is a grey area which needs further research when trying to establish whether socially responsible behaviour has been exercised towards a local community heavily dependent on the tourism industry for employment. However, in reality, a high proportion of the tourism industry is made up of sole traders, who are unlikely to have shareholders in a formal sense. This leads to an overlap between the examination of individual and corporate social responsibility. Arguments about the firm's responsibility in the community have not been taken into account by Friedman. The firm does not operate in a vacuum: it is important to take into account the influence of the economic and social environment on the firm's activities. Expressed simply, the degree to which a firm's business practice is socially responsible is a function of local and national markets, conditions in which it operates, profits made and other variables. It is the strength and influence of these other variables which are the determining factors. Donaldson (1993) points out that the firm whose activities affect the social and natural environment enters into a 'contract' with the society. This contract results in obligation being fulfilled by the firm to members of society or vice versa. The question remains, however, what guarantee is there that obligations will be fulfilled? Into this equation it is necessary to add another variable, regulation. Regulation of firms in the environment is usually undertaken by local or national agents of the state. In the context of this chapter, regulation of the tourism industry has become the responsibility of tourist boards, conservation and heritage agencies and local government. This chapter will examine the effectiveness of the policies of such agencies, using Cornwall as its context. Set against the need to regulate the tourism industry in order to ensure social responsibility is the demand for individual freedom by tourists. The way in which individuals exercise freedom on holiday will depend on their 'world view'. Attempts have been made in America to bring the interest of the individual as a consumer and the firm together under Federal legislation. The Corporate Democracy Act (USA 1990) proposed the setting up of a public policy committee, which will be responsible for policies concerning community relations, consumer protection and environmental protection (Miller and Ahrens, 1993). However, there is evidence to show such attempts at legislation have limited success. Patterson (1992) describes tourism as an out of control global industry that has no understanding of limits or responsibility or concern for the host people of a land.

If one is to accept this viewpoint, then regulation appears to have had limited success in fostering socially responsible behaviour in the tourism industry. An alternative method of bringing about change may be through codes of conduct, which in effect establish a social contract between the various interest groups in the tourism industry. Such codes need to educate participants in tourism to accept limits to the growth of this industry. There is among those promoting tourism a belief that the problem of production has been solved (Schumacher, 1973). The tourism industry, like other industries, has failed to distinguish between income and capital. In this context, capital is the irreplaceable culture and environment which comprise the host destinations. If we accept the view that the capital of the tourism industry is finite, then demand management systems which are going to help maintain destinations must be put in place before any further growth in tourism takes place. Demand comes from at least three sources, the tourist as a consumer, the tourism industry and the host community. Wherever competing demand arises, social responsibility should be exercised by all parties. Miller and Ahrens (1993) have put forward three useful criteria when looking for evidence of social responsibility. These are:

- the principle of respect for individual rights.
- the principle of responsible recommendations (which in the context of the tourism industry recognizes the central place of profitability).
- the principle of moral consideration (which is of particular importance when considering the impact of tourism on host cultures).

The next section of the chapter looks for evidence of forms of management, taking Cornwall as a case study, which provide evidence for socially responsible tourism development.

Socially Responsible Tourism Management: Evidence from the Field

The central problem which inhibits the development of a responsible approach to tourism in Cornwall is similar to that elsewhere in the world. Tourism is a diverse activity with many small units in competition. Tourism entrepreneurs have traditionally planned for the short term. What need to be taken into account by business are the long run trends in visitor demand. The following case material has purposely taken this perspective, with a chronological account of popular concerns about tourism development in Cornwall (*circa* 1985–1995).

It is important to note that this period predates international concerns about the global effects of tourism and other industries in the environment, as expressed at the United Nations Conference on Environment and Development (1992). Evidence for the emergence of anything which can be described as socially responsible tourism in Cornwall is likely to be implicit, rather than explicit. Anthropologists working in the county and nearby in North Devon (Bouquet, 1985; Gilligan, 1987; Ireland, 1987, 1993) all studied the impact of tourism on local people, but made no specific reference to socially responsible tourism. More recent work by Payton (1995) and Shaw and Williams (1993) tends to consist of accounts of the history of Cornish culture as seen by the historian, visitor surveys and economic geography studies. Social responsibility is a concept which has been imported into the vocabulary of the discipline that traditionally studies tourism from theology and business ethics (Mahoney and Vallance, 1992).

It has, therefore, been necessary to try and identify the emergence of popular concerns about tourism in Cornwall which demonstrate some element of social responsibility. Such concerns are unlikely to manifest themselves in academic writing, but first in the local press. In the period covered by this review of the popular press, recurrent themes emerge: the nature of the host/guest relationship; economic development tempered by demand management; and arguments in favour of a sustainable tourism product. These themes reflect problems which have come to be associated with the steady growth in the numbers of tourists coming to Cornwall from the 1960s to the late 1970s, and which peaked in 1978. Planning legislation as a reaction to visitor numbers was incorporated into the 1981 structure plan for Cornwall. The plan introduced the concept of tourism pressure areas (TPAs). These TPAs (see Fig. 17.1), as they became known, were coastal areas which were considered especially sensitive from an environmental point of view, or severely congested or both (Griffin, 1993). Public awareness of TPAs is graphically expressed in a report which highlights the lack of individual social responsibility among visitors. The diverse pressures reported include tourists failing to control dogs and the inability of local services to cope with needs of the growing retired population. The lack of individual social responsibility which existed is encapsulated in the following quotation: 'Has nobody ever really seen the muck, the filth, the litter and the detritus of permissiveness created by the holidaymakers ... ?' The question is asked 'who pays to clean up all this human mess?' (O'Connor, 1984). The answer to this question and one of the first mentions of a strategy to encourage social and environmental responsibility among visitors come in a report on the

Fig. 17.1. Location of tourism pressure areas (TPAs) in Cornwall.

work of the North Cornwall Heritage Coast Service (Earl, 1990).

The emphasis made to the visitor by the North Cornwall Heritage Coast Service is on quiet informal enjoyment of the area. Value is placed on solitude and a touristic experience without the car. The aim of the Heritage Coast Management Plan is to: (i) deal with problems and conflicts of interest; (ii) promote cooperation between those who live in and use the coastal area; and (iii) increase public understanding and enjoyment of the coast (North Cornwall Heritage Coast Management Plan). This initiative is complemented on the South Cornwall coast by the efforts of small businesses to develop socially and environmentally responsible tourism. Neillands (1990) describes the attempt by a small tourism company (Adventure-time), based on the Lizard peninsula and offering walking holidays, to bring tourists close to nature. One of the proprietors, Martin Hunt, explained that the company offers walking holidays with a difference. 'The difference is that we try to put our clients back in touch with nature, the various birds, mammals, flowers and rock strata to show why Cornwall is the way it is' (Neillands, 1990).

These brief examples illustrate the efforts of the public and private sectors in Cornwall to educate tourists to be appreciative and responsible towards the host environment. They have also identified the product Cornwall has to offer the tourist – a rich landscape. This landscape, which serves as a commodity for the tourism industry, also has cultures which need to be sustained. Citing a study commissioned by the West Country Tourist Board, Macaulay (1990) points to the fact that Cornwall's farmers are slowly acknowledging the importance of tourism to support the farm business, with one in ten farmers not involved in tourism realizing its potential and considering developing tourist accommodation. What is interesting is the kind of tourism product being developed in the countryside, i.e. self catering accommodation which emphasizes seclusion and rural character rather than large camp sites or chalets. What this suggests is the development of a product and tourist infrastructure sympathetic to responsible tourism. The call for closer links between the tourism industry and farmers in Cornwall comes in a report which claims 'Tourism lives on the back of farmers' (Gregory, 1991). The point is made that farmers and landowners provide the resources for tourist to enjoy by maintaining the landscape. Therefore, in order to sustain these farming communities, the tourism industry should purchase more local food. In other words the tourism industry is being asked to make a net investment in the cultural and environmental capital it exploits to achieve growth. Evidence of a new philosophy toward the management and investment in the cultural and environmental capital of Cornwall by the tourism industry began to emerge in the early 1990s. Cairns Boston (Managing Director of the Land's End and John o'Groats Company) appears to be advocating a responsible approach by tourism entrepreneurs to facilitate environmental conservation and the need for a genuine experience for the visitor (*Western Morning News*, 18th March 1991). In the east of the county, Caradon District Council has put these sentiments into practice with a green tourism scheme, Project Explore (Taylor, 1991). The aim was to create a partnership between environmental improvement and tourism marketing to benefit both the local economy and the visitor. The report specifically identified the areas of beauty, culture, history and wildlife resources to attract the 'discerning tourist' (Taylor, 1991). The first clearly identifiable socially responsible tourism policy in Cornwall came from Penwith District Council in their green tourism strategy for West Cornwall (Williams, 1991). Penwith aimed to achieve a balance between local needs and tourism. These examples point to a welcome trend emerging in Cornwall in the early 1990s, a demand management strategy for tourism which is both socially and environmentally responsible. The origins of this philosophy can be traced to the work of the Cornwall Tourism Development Action Programme (1988–1991). The prime objective of the programme was 'to achieve a balance between profitable tourism development and conservation of natural resources and amenities' (Bryce, 1991).

However, the development of this new philosophy into policy and practice has not been without its problems, largely due to competing demands for tourism growth which have been most successful in periods of economic recession. In a report titled 'Holiday crisis for West resorts' (*Western Morning News*, 6th July 1992, p. 3) Paul Tyler, Liberal Democrat MP for North Cornwall, argues that it is irresponsible not to have a healthy tourism industry in Cornwall. The indicators of economic recession in North Cornwall which call for tourism as the solution were an unemployment rate seven times higher and small businesses failing 20 times faster than 15 years ago. Housing repossessions were 800% higher than in 1977. In the short term, growth at any price would seem to be welcome to a local community besieged by economic blight. However, short term prosperity has to be set against the need to sustain resources of the tourism industry for the future. Notwithstanding the effects of economic cycles, it has been suggested that some of the problems experienced by the tourism industry in Cornwall are due to the Cornish people themselves. A change in attitudes to produce a sustainable tourism product which will result in employment being created has also been argued. To achieve this type of sustainable growth, the Cornish tourism industry was, and continues to be, calling for not only a new product but a new market. In this context Emma Nicholson, MP for Torridge and West Devon, was quoted as making a plea for quality and less of the 'bucket and spade brigade'. The hidden agenda in a move toward socially responsible tourism is to market the product to those tourists who are likely to behave in just such a way! It could be argued that socially responsible tourism also involves social engineering. What follows is an examination of how this goal of improved quality in the tourism product will be achieved through partnerships between local authorities and the business community in Cornwall.

It has been argued that the economic recession is antipathetic to the development of tourism which is socially and economically responsible. This is evident from the demands being made in the wake of the recession by the Cornish Tourist Board and the West Country Tourist Board. The Cornish Tourist Board members were asking for a meeting with the then tourism minister, Viscount Ullswater, to petition for grant aid to help small- and medium-sized businesses. In this context, responsibility took on a different meaning. With the representative for Restormel protesting that 'It is the Board's responsibility to look after the industry, particularly after such a lousy season' (*Western Morning News*, 12th September 1991, p. 3), the responsibility referred to is economic, not social, to ensure growth in visitor spending and employment. This was quantified by Ron Morrison-Smith, Director of the West Country Tourist Board, who predicted the creation of 20,000 new jobs and additional revenue from tourism of £460 million annually by 1996 (Caruth, 1991).

Limits to growth in the tourism industry are not solely as a consequence of supply and demand factors, but as a result of factionalism among local interest groups. The unintended consequence of a failure to agree a tourism policy can be to curb growth. The evidence of factionalism and fragmentation are apparent in a report by Trevail (1989), who cites criticism of the Cornwall Tourist Development Action Programme (CTDAP), by councillors in Restormel. The criticism made by the then chairperson of Restormel Tourism Committee was that: 'The CTDAP was going to solve all the problems of our industry in Cornwall. It has a £500,000 budget over three years, but I think it is a fragmented organisation' (Trevail, 1989).

This belief that the tourism industry is fragmented continues today (Bell, 1995a). The exact causes of this fragmentation in the tourism industry in Cornwall remain outside the scope of this chapter, and a question for future research. However, what is clear is the damaging way this disagreement is reported in the press. For example, there have been conflicting headlines, such as 'Cornwall "super tourist board" bid rubbished' (Jobson, 1992a) and 'Tourist board is backed' (Jobson, 1992b). What lies behind the headlines is a disagreement between local council representatives and a trade association about the recommendation to enlarge the Cornish Tourist Board. These recommendations came in a report prepared by Professor Anthony Travis for the Cornwall Tourism Development Action Programme in 1991. The report's objective was to prepare a tourism strategy for the 1990s. The cost to Cornwall of such fragmentation is the county's inability to present a convincing case to government for financial aid to develop tourism. This point was made clear by Cairns Boston as former Chair of the Cornwall Tourism Development Action Programme, currently Chair of Devon and Cornwall Training and Enterprise Council. He said, 'For the future, particularly as Britain becomes more integrated into Europe, projecting Cornwall and its single clear identity is going to be more important that ever' (Boston, 1991).

Thus, this section of the chapter has sought to demonstrate that the failure to resolve tensions between representatives who comprise the different bodies which promote tourism in Cornwall has led to failure in marketing the county as something other that a 'bucket and spade holiday' destination. There have been successes such as the Lizard Peninsula, Project Explore in the Caradon district (*Cornish Times*, 11th February 1994) and the Heritage Coast in North Cornwall. Much of this success has been due to the policies put forward by the county and district councils within Cornwall. The penultimate section of the chapter attempts to bring together the policies of the councils and other agencies within Cornwall which could be described as in some way socially responsible.

Tourism Policy in Cornwall

Cornwall is divided into six districts for the administration of local government, with the second tier authority being Cornwall County Council located in Truro. The six districts are Penwith, Kerrier, Carrick, Restormel, North Cornwall and Caradon. The tourism policy for each of these districts takes into account the policies set out in the Cornwall Structure Plan (1993). There are a total of 17 policies governing tourism and recreation in the county, which cover all forms of development from serviced and self-serviced accommodation to tourist attractions and facility provision and the environment. What is important is to identify the conceptual basis for these policies and how they correspond to creating a tourism industry which is responsible and has a sustainable product. The key concepts underlying all county and district policies is that of tourism pressure areas (TPAs). Tourism pressure areas have the advantage of being defined conceptually and spatially. They usually consist of areas of particular landscape, scientific or historic value where growth in certain types of tourist accommodation is unacceptable (Griffin, 1993). The geographical location of the TPAs can be seen in Fig. 17.1, which shows their close proximity to sensitive landscapes. Cornwall County Council is not only concerned with what might be broadly termed green issues; the deposit draft copy of the structure plan of 1995 stresses that one of the foundations of the holiday industry is Cornwall's distinctive and unique culture. Culture as an attraction is interpreted in the plan as arts and crafts and not the way of life of the ordinary people, in contrast to the view expressed by Richie and Zins (1978). Although protection and enhancement of the social fabric of Cornish communities is not mentioned specifically, tourism policies do make reference to new developments being sensitive to the locality, rural activities and culture and the quality of life of local people.

At the county level, policies have to be general enough to be interpreted for local needs. As such, the council, in its policies, implies rather than makes explicit a responsible approach to the future development of tourism. In contrast with some of the calls from the private sector reported in the press, the council appears to advocate a low growth strategy for tourism in Cornwall. They make the point in their 1993 strategy document that there is little to suggest any substantial increase in tourist numbers in the future (Griffin, 1995). Instead the council predicts a change in the type of tourism, to attract the tourist seeking the 'away from it all' holiday. The trend is in accord with the council's wish to promote green and heritage tourism in Cornwall.

The County Council has recently been joined by a new player in the field of tourism development, Devon and Cornwall TEC, who operate at a sub-regional level. In a discussion paper the Devon and Cornwall TEC set out their vision for the future of tourism in Devon and Cornwall, to the turn of the century. The theme of the development plan is to promote the benefits of tourism to the sub-region and spell out the cost of doing nothing, which would lead to further economic and social decline. The key objective is growth through improved quality of the tourism product and an increase in the number of visitors. To achieve growth the following targets need to be met: (i) to increase turnover by £6 billion on the 1995 figures; (ii) to create employment for a further 10,000 people; and (iii) to attract an additional one million visitors to the region (Bell, 1995a). The rhetoric used in the document to support such a policy is powerful: 'people will leave the dole queue when more visitors come' (Bell, 1995a). The questions have to be asked about the type of employment and how long people will be employed and for what levels of pay.

There is, however, one major barrier to growth according to the plan, the host culture. As a consequence, one key recommendation it contains is the need to promote the benefits of tourism to the local communities, which will in turn improve visitor experience (Bell, 1995a). Nevertheless, a fundamental weakness of the development plan is the failure to give an explicit statement about the need to make the tourism product sustainable, the emphasis being on better marketing and the generation of increased profits. The only statement in the plan which can be deemed to be a socially responsible approach to tourism development is embedded in a section promoting benefits. The Devon and Cornwall TEC acknowledge that a balance needs to be struck between the attraction, the host and the visitor, i.e. that the volume of visitors does not either alienate the host community and/or detract from the actual attraction by sheer volume of numbers (Bell, 1995a).

Considering the two approaches to tourism development together, it is possible to find common ground between Cornwall County Council and the Devon and Cornwall TEC. Both acknowledge the importance of uniqueness, which needs to be maintained and promoted (Bell, 1995b; Griffin, 1995). There is also a shared recognition that a balance must be maintained between the competing demands of host and guest. Where the two plans differ is in the way in which these goals are to be achieved alongside the need to sustain economic prosperity for Cornwall. The council advocate a low growth strategy based on their predicted trends in visitors numbers. In contrast, Devon and Cornwall TEC base much of their vision of a rejuvenated tourism industry on increasing visitor numbers by one-third of their present levels. This calls into question the wisdom of the Devon and Cornwall TEC plan. It is a solution to economic problems which is likely to place severe strain on the local communities and threaten cultures, as local communities are being asked to improve the quality of the tourism product by enriching the experience of the tourists. The Devon and Cornwall TEC must not make the mistake, all too common in tourism development plans initiated nationally (Greenwood, 1977), alienating local people from their cultural heritage.

The County Council and the Devon and Cornwall TEC have taken some of the responsibility for providing an overall view of development of the tourism industry in Cornwall. However, there are other initiatives taking place in the county which lie outside the remit of this chapter.

What becomes clear from Table 17.1 is that there is likely to be overlap in the remit of these organizations, which further adds to the view that tourism policy in Cornwall is confusing and fragmented. This is most likely to be apparent at the point at which policy is put into practice, the local authority district level within Cornwall. Each of the policies of the six district councils on tourism has been reviewed for evidence of responsible tourism practice.

Penwith, in the far south west of Cornwall, advances three themes in its 1994 local plan (Giddens, 1994): caring for the environment, the local community and the visitor. The council puts forward two concepts which can be used to classify forms of tourist attraction. Attractions are divided into primary and secondary. Primary attractions include natural physical features, heritage sites and culture. Secondary attractions include theme parks and amusement parks which are commercially orientated. Penwith District Council make this distinction in order to provide justification for their tourism policies 'that the identity of Penwith is its main resource, which should not be devalued by the proliferation of secondary attractions' (Giddens, 1994). In summary Penwith District Council policies on tourism development appear to exhibit a high degree of social responsibility. This is clearly expressed in a statement from the council's 1991 Environmental Tourism Strategy, which goes beyond a traditional marketing approach. According to Giddens (1994), the strategy aims to achieve a balance between maximizing the economic benefits of tourism whilst ensuring the maintenance of a high quality environment and helping to sustain the local community and its culture.

In sharp contrast to Penwith, the adjoining district council of Kerrier, which includes the tin mining areas around Camborne and Redruth, is not a traditional tourism destination. It is acknowledged in the Environmental Strategy for the Camborne–Redruth Area that it is 'not a tourist destination but the objective is to integrate heritage and countryside attractions, to attract visitors and locals'. Kerrier has chosen to merge its efforts to develop tourism with neighbouring Carrick Council and promote the districts under the generic title, 'Falmouth and South West Cornwall'. In doing so, Kerrier and Carrick have produced what they term a unique partnership of the private and public sector (Carrick and Kerrier District Councils, 1995). The brochure they have produced retains the usual directory of accommodation and short features to promote the major resorts. In addition there is a section devoted to Cornish heritage and language, both in the Cornish language and in English. Visitors are requested to adopt a responsible approach to this Celtic land. The joint approach of the district councils toward marketing tourism in a responsible way must be welcomed. In marketing Falmouth and

Table 17.1. A summary of the activities of some organizations.

Organization	Activity
European Union	A Green Paper on tourism which may have major implications when the Maastricht Treaty is re-examined in 1996 (The Role of the Union in the Field of Tourism, April 1995)
English Tourist Board	1. Action programme for the ETB and 11 Regional Boards (Tourism – Competing with the Best, March 1995) 2. Review of marketing Britain. In future the West Country may feature as part of a new brand – the South of England
West Country Tourist Board	A major re-evaluation of its role and organization by a consultancy at a cost of £30k.
West Country Development Corporation and Devon and Cornwall Training Enterprise Council	A discussion paper (Cornwall and Devon Tourism Development Plan 1995–1999, May 1995)
Devon and Cornwall Training Enterprise Council	1. Three pilot tourism projects (The Lizard, Newquay and Bude). 2. Welcome Host training programme. 3. Multimedia Interactive System for Tourism Information (MISTI) in collaboration with the European LEADER Project, Rural Development Commission, Cornwall County Council, Penwith District Council and Isles of Scilly Council. Phase 1 £92k.
Rural Development Commission	1. Business support and advice. £30–40k per annum 2. Collaboration with several tourism-related projects. Funding of £500k per annum available.
Local Government Review Audit Commission	Recommendations for reducing duplication between county and district councils (Review of Tourism, January 1995)
Joint County and District Working Group	Working group to study Audit Commission recommendations. A separate panel has been set up to examine the economic development and tourism aspects.
Consortium of Tourism Officers	The Consortium are also involved in a study of the Audit Commission recommendations and are examining ways in which to combine the districts into new marketing areas.
County Council Policy and Resource Committee	An in-house review of the role and organization of the CTB leading to the recent restructuring of the CTB. Agreement on an alternative means for the private sector has yet to be reached.

south west Cornwall, they overcome one of the criticisms levelled at district councils, that they promote the area as a political and administrative unit which has no meaning to people outside the county. Independently, Carrick District Council, which includes the City of Truro, does have a policy to promote the sustainable enjoyment of its cultural assets, with the stated objective of seeking government and European Union funding to maintain and protect these assets (Carrick District Council, 1995).

Restormel Borough Council covers part of mid Cornwall and the china clay district around St Austell. The resort of Newquay is in the north west of the borough while to the south west lies Fowey. Restormel is an interesting case inasmuch as the resort of Newquay epitomizes the 'bucket and spade' holiday image which the county wishes to escape. In response to pressures for a quality tourism, Restormel's Tourism and Leisure Strategy (1995) put forward one of the clearest statements of any district for socially responsible tourism policy. This policy is spelled out in the council's Tourism and Leisure Strategy objective 3B. For example, strategic objective 3B.7 aims to facilitate 'green tourism and the right to achieve a balance between recreational and environmental interest' (Restormel Borough Council, 1995).

Notwithstanding this responsible approach to tourism development expressed in the Restormel policy statement, there is cause for concern over the way in which the policies and strategy objectives have been written. There is an excessive use of inverted commas around phrases, which imply ambiguity of meaning. For example, 'quality tourism', 'rural tourism', 'tourism diversification', 'green tourism' and 'broad base tourism' (Restormel Borough Council, 1995).

North Cornwall District Council has no such ambiguity in its policies toward tourism management and development. The council has its own heritage coast service jointly funded by the district council and the Countryside Commission. The 1995 North Cornwall Heritage Coast Management Plan aims to provide 'support for tourism with careful environmental management' (North Cornwall District Council, 1995). The district council's philosophy on tourism stated in the local plan

> is to strike the right balance between supporting a flourishing tourism industry, protecting the natural and historical assets of the area which attract the tourists and avoiding the adverse impact on the communities of local people and the functioning of other local businesses.
> (North Cornwall District Council, 1995)

Caradon District Council in south east Cornwall has a similar philosophy toward tourism development to their neighbour, North Cornwall. In the Caradon Coastal Rural Areas Local Plan (1991) the council sets out six principles which advocate a sustainable tourism policy. However, it is worth pointing out that the council's ability to sustain rural communities through retaining housing stock of permanent residence is limited. Although policy

R4 states that 'Chalet and holiday let sites will not normally be permitted' (Caradon District Council, 1991), the council is powerless under existing planning legislation 'to stop a permanent residence becoming a second home or holiday let' (Caradon District Council, 1991, p. 75). However, it is doubtful if changes in the law will do much to curb the blight which has occurred in Caradon's premier coastal resort of Looe. Recent fieldwork in the resort (in March 1996) has identified a small number of properties in a derelict state, in prime locations in the town overlooking the harbour. It was discovered that these properties had once been holiday accommodation of various forms.

From the policy documents reviewed for the six councils it is evident that each has acknowledged the need to implement a tourism development strategy which is sustainable. Penwith and North Cornwall appear to be the leaders in a policy which can be described as socially responsible. Each council recognizes the value of local culture and expresses this in its policies. In this respect the district councils display a surprising degree of uniformity. This is in accord with the recommendations of the recent Local Government Audit Commission Report (Audit Commission, 1994), which looked for ways in which duplication between county and district councils could be reduced (see Table 17.1). The problem of achieving coordination in Cornwall's tourism industry is much wider than this. There is a need also to include the private sector, which is comprised mainly of small family businesses, in any effort to unify the Cornish tourism industry. The latest proposal to overcome fragmentation is the creation of a Tourism Forum for Cornwall, to include both public and private sectors. The setting up of this forum was one of the recommendations of the Cornwall Tourism Business Plan Working Group (Tregoning, 1995). This forum is now in existence and would seem the best way to achieve a sustainable future for the tourism industry in Cornwall.

Conclusions

This section has two purposes, to summarize the arguments presented and set out an agenda for socially responsible tourism. The evidence presented from Cornwall indicates that socially responsible tourism is an achievable reality under certain conditions of development. There needs to be a careful balance between regulation of the tourism industry via planning legislation and efforts to effect change through voluntary codes of conduct (Mason and Mowforth, 1995). Any codes of conduct infer the obligation to behave in an agreed manner, between host, guest and the tourism industry, by following principles of social responsibility (see the discussion by Miller and Ahrens, 1993). In Cornwall this level of agreement is only now being realized through the setting up of the tourism forum.

The chapter has shown, using Cornwall as a case study, the barriers to achieving socially responsible tourism which exist. The three main barriers to socially responsible tourism can be expressed conceptually with a view to applying them to other tourist regions in the UK. Economic cycles of growth and depression are the most obvious barrier to achieving socially responsible tourism in any region. Tourism as a form of economic development usually takes place in areas which have previously suffered decline. In development terms, tourism is often planned by local entrepreneurs on a short term basis; the British tourism industry continues to be dominated by small firms. This concentration of small tourism businesses results in local market structures and strategies which are not conducive to socially responsible tourism, simply because of the high degree of competition. In these market conditions, businesses will compete rather than cooperate to improve the tourism product. Cultural barriers to socially responsible tourism are the most difficult to overcome, because of the difficulty of changing local attitudes and convincing people that their culture will benefit.

An agenda is normally considered as a list of items for discussion; accordingly, an agenda for socially responsible tourism need not be an extensive treatise. Thus, this short agenda is offered as a trigger for facilitating further discussion. First of all, those concerned in the development of tourism need to be familiar with the principles of social responsibility set out in this chapter (Miller and Ahrens, 1993). Limits to the desirable growth of tourism in any region should be discussed with the local community and its views respected, especially if they are at odds with promotional policies of national and regional tourism bodies. The tourism industry should plan for the long term and invest in the cultural capital, which adds uniqueness to the tourism product, by developing one which is sensitive and accepted by local people, and attractive to potential visitors. The tourism industry must take the responsibility to educate the tourists, their host and its members. Finally, socially responsible tourism will only be understood and achieved when the industry is not feared or resented by host cultures and can be seen as of real benefit to them.

References

Audit Commission (1994) *Time for Change? A Consultation Paper on Work to Support the Implementation of Local Government Reorganisation.* Audit Commission for Local Authorities, HMSO, London.

Bell, M. (1995a) *Cornwall and Devon Tourism Development Plan 1995–99: Discussion Paper.* Devon and Cornwall Training and Enterprise Council, Plymouth.

Bell, M. (1995b) *Regional Challenge: A Vision for Tourism in Devon and Cornwall.* Devon and Cornwall Training and Enterprise Council, Plymouth.

Boston, C. (1991) In: Williams, D. Pull as one, says tourism chief. *Western Morning News*, 9th August.

Bouquet, M. (1985) *Family Servants and Visitors: The Farm Household in Nineteenth and Twentieth Century Devon.* Geo Books, Norwich.

Bryce, A. (1991) The future of Cornish tourism. In: *From Pilgrimage to Package Tour.* Cornwall Tourism Development Action Programme.

Caradon District Council (1991) Leisure, recreation and tourism. In: *Caradon Coastal Rural Areas Local Plan.* Caradon District Council, Liskeard.

Carrick and Kerrier District Councils (1995) *Falmouth and South West Cornwall Holiday Guide.* Carrick and Kerrier District Councils, Redruth.

Carrick District Council (1995) *Economic Development Strategy.* Carrick District Council, Redruth.

Caruth, R. (1991) Spending creates jobs in tourism. *Western Morning News*, 29th January.

Cornish Times (1994) Caradon's new tourism initiative launched in Looe. *Cornish Times*, 11th February.

Donaldson, T. (1993) Constructing a social contract for business. In: White, T. (ed.) *Business Ethics: A Philosophical Reader.* Macmillan, London.

Earl, B. (1990) The hidden coast. *Western Morning News*, 30th March.

Friedman, M. (1993) The social responsibility of business is to increase its profits. In: White, T. (ed.) *Business Ethics: A Philosophical Reader.* Macmillan, London.

Giddens, K. (1994) *Penwith Local Plan. Draft for Consultation.* Penwith District Council, Penzance.

Gilligan, H. (1987) Visitors, tourists, and outsiders in a Cornish town. In: Bouguet, M. and Winter, M. (eds) *Who From Their Labours Rest? Conflict and Practice in Rural Tourism.* Avebury, Aldershot.

Greenwood, D. (1977) Culture by the pound: an anthropological perspective on tourism as cultural commoditisation. In: Smith, V.L. (ed.) *Host and Guest: The Anthropology of Tourism.* Blackwell, Oxford.

Gregory, C. (1991) Tourism lives off the back of farmers. *Western Morning News*, 9th January.

Griffin, C.G. (1993) *Cornwall Structure Plan: Explanatory Memorandum.* Cornwall County Council, Truro.

Griffin, C.G. (1995) *Cornwall Structure Plan. Deposit Draft.* Cornwall County Council, Truro.

Ireland, M. (1987) Planning policy and holiday homes in Cornwall. In: Bouquet, M. and Winter, M. (eds) *Who From Their Labours Rest? Conflict and Practice in Rural Tourism.* Avebury, Aldershot.

Ireland, M. (1993) Gender and class relations in domestic employment in tourism: a case study of Sennen, West Cornwall. *Annals of Tourism Research* 20(4), 666–680.

Jobson, R. (1992a) Cornwall 'super tourist board' bid is rubbished. *Western Morning News*, 4th January.

Jobson, R. (1992b) Tourist board is backed. *Western Morning News*, 3rd February.

Macaulay, D. (1990) Tourists bring cash to save farms. *Western Morning News*, 26th November.

Mahoney, J. and Vallance, E. (1992) *Business Ethics in a New Europe.* Kluwer Academic, Dordrecht.

Mason, P. and Mowforth, M. (1995) *Codes of Conduct in Tourism.* University of Plymouth.

Miller, D.F. and Ahrens, J. (1993) The social responsibility of corporations. In: White, T. (ed.) *Business Ethics*. Macmillan, London.

Neillands, R. (1990) On the trail of the Lizard. *Weekend Telegraph*, 14th July.

North Cornwall District Council (1995) *North Cornwall District Council Local Plan. Deposit Draft*. NCDC, Bodmin.

O'Connor, M. (1984) Who pays to clean up the holiday mess? *Western Morning News*, 4th September, p. 3.

Patterson, K. (1992) Aloha for sale. *Focus* No. 4. Tourism Concern, London.

Payton, P. (ed.) (1995) *Cornish Studies* (second series) 3. Exeter University Press, Exeter.

Rackman, D., Mescon, M.H., Bovee, C.L. and Thill, J.V. (1990) Ethical and social responsibilities of business. In: *Business Today*, 6th edn. McGraw-Hill, New York.

Restormel Borough Council (1995) *Tourism and Leisure Strategy for Restormel: Appraisal and Policy Statement*. Restormel BC Tourism and Leisure Department, St Austell.

Richie, J.R.B. and Zins, M. (1978) Culture as determinant of the attractiveness of a tourism region. *Annals of Tourism Research* 5(2), 252–267.

Schumacher, E.F. (1973) *Small is Beautiful: A Study of Economics as if People Mattered*. Blond & Brigg, London.

Shaw, G. and Williams, A. (1993) *Cornwall Tourist Visitor Survey 1992*. Exeter University Press, Exeter.

Taylor, M. (1991) Eco-tourism guide to sell nature and history. *Western Morning News*, 22nd July, p. 3.

Tregoning, A. (1995) *The Organisation of Cornwall's Tourism Industry. Discussion Paper*. Cornwall Tourism Business Plan Working Group.

Trevail, M. (1989) New tourism body replies to critics. *Cornish Guardian*, 26th January.

Williams, D. (1991) Royal backing for green tourism plan. *Western Morning News*, 20th August, p. 3.

Western Morning News (1991a) Land's End lessons in leisure blueprint. *Western Morning News*, 18th March, p. 3.

Western Morning News (1991b) Holiday area needs help. *Western Morning News*, 12th September, p. 3.

Western Morning News (1992) Holiday crisis for West resorts. *Western Morning News*, 6th July, p. 3.

Practical Approaches to Sustainability: A Spanish Perspective

18

P.A. Hunter-Jones,[1] H.L. Hughes,[2] I.W. Eastwood[1] and A.A. Morrison[1]

[1]Department of Environmental and Leisure Studies, Crewe and Alsager Faculty, Manchester Metropolitan University, Crewe Green Road, Crewe, CW1 5DU, UK; [2]Department of Hotel, Catering and Tourism Management, Manchester Metropolitan University, Hollings Faculty, Old Hall Lane, Manchester M14 6HR, UK

Introduction

The aim of this chapter is to review the concept of sustainability by considering the practical application of this concept to the Spanish tourism industry against the background of the separate literature on sustainable development and sustainable tourism. The Spanish tourism industry's policies and practices will be delineated and contributory factors identified.

Spain is an established popular tourism destination exhibiting many features of mass or volume tourism. Traditionally, three countries (UK, Germany and France) collectively have accounted for over half of Spain's international market. Contemporary research, however, notes fluctuations in this demand, with the international image of Spain cited as a contributory factor. Quality tourism has been proposed as a solution to declining markets. Wight (1993), for example, infers quality tourism to be an interchangeable term for sustainable tourism. The literature suggests that the application of any formal sustainable tourism practices is relatively recent. Much of the work has tended to focus upon specific projects, for example, work by the International Federation of Tour Operators (IFTO), and the same destinations, for example, Mallorca and the Costa del Sol. This chapter extends this review by considering additional areas of the mainland and the Canary Islands, notably Tenerife.

Sustainable Development and the Sustainability Ethos

The literature on sustainable development has focused primarily on the economic motives driving policy and practice. As a result of this, as indicated in Chapter 1, the significance of sustainable development in contemporary development studies is acknowledged. The evolution of sustainable development has included reference to environmental issues and strategy development. Furthermore, the components of sustainability appear central to the concept of sustainable tourism. Sustainable tourism has been extensively studied, although a lack of one common definition has prompted a degree of ambiguity. Generally speaking, there is no shortage of rhetoric in support of sustainable tourism as a desirable and essential policy option. Much of the literature, however, has been couched in policy terms and there is little evidence of any one definitive policy applicable across a range of situations. The environment is recognized as an integral part of sustainable tourism development but has been considered as part of the economic environment, with comparatively little documentation about the physical and socio-cultural tourism environment. The same is apparent for the effects of tourism. The relationship between tourism and the environment is presented as being dependent upon many conditional factors. These factors vary in their coverage and emphasis and are seldom considered collectively. The industry's response to sustainable tourism has been varied and has included the development of strategies, initiatives and codes of ethics. The value of such responses has been frequently questioned, with codes of ethics singled out as potentially harmful to the sustainable tourism cause (Wheeller, 1994), whilst initiatives and strategies are often small in scale, failing to take account of the dynamic nature of tourism (as exemplified by Butler, 1980) and the effects of mass tourism. One mass tourism destination attempting to respond to the challenge of sustainability is Spain.

The Case Study: Spain

Spain occupies approximately 85% of the Iberian peninsula and consists of 17 autonomous regions, the most popular for tourism being Andalucia, Catalonia, the Balearic Islands and the Canary Islands. The climate is both warm and Mediterranean, making summers on the mainland and the Balearic Islands ideal for beach tourism, whilst winter for the Canary Islands provides sub-tropical conditions, making them ideal for winter sun destinations.

Background

There is a considerable amount of literature reviewing the development of tourism in Spain, although the outlook presented pre-1989 tends to differ from that discussed post-1989. Valenzuela (1991) and Barke and France (1992) consider the evolution of Spanish tourism, the Organization for Economic Cooperation and Development (1980) and later Fayos Solá (1992) determine the place and significance of tourism to Spain, Albert-Piñole (1993) establishes the context within which the industry developed, whilst Williams and Shaw (1991) classify the leading role of tourism as a factor contributing to Spanish economic development post-World War II. The majority of research focuses upon the demand for Spain, supply of facilities and services and the role of government in tourism development.

Tourism demand has been reviewed to varying degrees, with the literature mainly considering the volume, form and timing of tourism activity. Valenzuela (1991) determines contemporary demand to be on a mass scale, with beach tourism the most prevalent form. Figures for 1993 confirm that 71% of visitors to Spain chose beach-based holidays (Secretaría General de Turismo, 1993), with Spain accounting for 46% of all beach holidays marketed by Thomson that season (Travel Trade Gazette, 1993). Furthermore, analysts note (see, for example, Middleton, 1991; Ryan, 1991) that such holidays are bought as an inclusive or package holiday by a price-sensitive consumer, particularly in the UK. Implicit throughout most studies is the regional and seasonal concentration of tourism activity and the heterogeneous nature of the market.

Commentaries on the tourism facilities and services in contemporary Spain are notably selective, with certain regions and locations, such as Andalucia (Torremolinos) and Mallorca being extensively reviewed, whilst others, for example, Navarra and Extremadura, receive only a brief mention. This clearly reflects the significance of certain areas in terms of the numbers of tourists, again demonstrating the regional concentration of activity. The problems of determining a typology of the Spanish product are further compounded by the fragmented nature of the tourism industry. Two recent reports by the Economist Intelligence Unit (1994a,b) go some way to redressing this problem by providing a general outline of the inter-sectoral linkages in Spain, although largely from a Spanish domestic perspective. In these reports, airlines, airports, travel agents and tour operators, railways and hotels are reviewed and statistics presented, but little reference is made to attractions, whether natural or man-made. Studies of a more specific nature, which can be categorized into four main areas, have been conducted:

- transport developments, notably improvements in infrastructure, air and ferry services (see Hillier and Raffael, 1992; Travel Trade Gazette, 1993).
- travel agents and tour operators, with reference to traditional and new

forms of tourism activity to be encouraged (Tourism Marketplace, 1993).

● accommodation provision, noting the concentration and limited availability of quality accommodation (Travel Trade Gazette, 1993).

● attractions, determining the significance of the various autonomies and differentiating the natural and man-made attributes of the regions.

Government strategy in Spain, in general terms, is supportive of tourism, with investments in promotional campaigns and infrastructural improvements frequently reported. Part of the tourism role of government is to develop a national tourism policy. Conflicting approaches to policy development are noted by varying political parties, although overall policy development appears to favour tourism as a panacea for economic problems. For example, to support the development of mass tourism in the 1960s, planning controls were relaxed and incentives offered to pump-prime the tourism industry (Organization for Economic Cooperation and Development, 1980), although later, to counteract falling visitor numbers, land use planning policies were reintroduced (Fayos Solá, 1992). The development of the Mallorcan tourism industry in particular exemplifies such an approach when, during the 1960s, to support tourism development, building restrictions were relaxed and many British, German and Spanish tourism operators began to construct accommodation *en masse*, many close to the beaches of Illetas, Magaluf and Palma Nova (Barke *et al.*, 1996). This *laissez-faire* approach to development has today been replaced by a structured and controlled quality tourism policy, which includes the reintroduction of land use policies. Irrespective of the approach, there appears to be an inherent assumption that tourism is viewed by the Spanish government as worthy of support, justifying government intervention.

Current issues

In 1989, Spain experienced an overall decline in visitor arrivals. Davidson (1992) summarizes the problem, noting a 0.2% decline compared to 1988 figures, reflected throughout the traditional main international markets, i.e. Germany (-1.7%), France (-0.8%) and the UK (-3.9%). Post-1989, a return to former levels of visitor numbers is observed, albeit at a slower rate of increase. Research indicates a variety of contributory factors which explain the decline in arrivals, including:

● market maturity, e.g. turning away from beach holidays and a greater diversity in demand/supply.

● image impediment, e.g. lager louts.

● temporary problems, e.g. unpredictable currency fluctuations and terrorist threats in certain resorts.

- structural problems, e.g. regional and seasonal concentration of the tourism industry, increasing competition, shortfalls in accommodation provision.
- environmental problems, e.g. pressures on both the natural and built environment, reduction in the quality of life.

Such problems are illustrated by Morgan (1991), who, reviewing tourism in Mallorca, defines five specific problems facing the destination:

- overdependence on the UK and German market.
- overdependence on the UK inclusive tourism market offering a high volume but low yield.
- excess supply (60%) of accommodation in the one- to three-star categories resulting from a relaxation in building restrictions to facilitate the original growth of the tourism industry.
- environmental deterioration, particularly as a result of hotel construction, with many developments permitted close to the beach.
- image.

Overall, fluctuations in visitor numbers have received varying coverage and analysis, with the importance of environmental issues in the debate being particularly questioned.

The seminal paper produced by the Organization for Economic Cooperation and Development (1980), although in some respects dated, offers a directed discussion, reviewing environmental issues and Spanish tourism. The various impacts of tourism are presented and two critical problems facing the future of the tourism industry identified, namely a 'divergence between tourism and environmental objectives' and 'the impact of environmental degradation induced by tourism' (pp. 21–22). A later study by Williams and Shaw (1991) develops this theme by inferring that the environment, in terms of the physical features, plays an integral role in encouraging (or discouraging) tourists to frequent a destination. Sofield (1991) extends this view to include the social and cultural attributes of a destination. The contribution of environmental factors to the debate is perhaps best demonstrated by Pollard and Rodriguez (1993), who, in a survey of visitors to Torremolinos, identified the environment as the main factor prompting the decline of the resort. Problems cited include contamination, architectural pollution and poor standards of behaviour by tourists.

The importance of environmental issues in explaining decline is challenged by Ryan (1991), who, in a general study of significant factors in destination choice, found environmental factors to be seldom mentioned by visitors. How specific such problems are to Spain is questionable. Williams and Shaw (1991), reviewing temporary and structural problems, emphasize that the wavering popularity of Spain is merely a reflection of the falling demand experienced by Mediterranean resorts in general due to the increasing availability of alternative, more 'exotic' destinations. Ryan (1991) and Middleton (1991), in examining the nature of contemporary demand, concur with this viewpoint proposing that the problem lies firmly with the increasing expectations of the consumer, who now demands a more complex, improved holiday product. It is, therefore, arguable that the problems experienced by Spain are a microcosm of a global problem.

The application of quality tourism

Mainland Spain

Studies published in recent years, notably post 1992, have suggested a turn around in Spanish tourism fortunes (see, for example, Fayos Solá, 1992), identifying both the reorganization of the tourism role of the Spanish government and the introduction of a number of development initiatives as contributory factors. One initiative of note is the FUTURES initiative, which was signed in 1992, following a consultative period between regional and local authorities and the Spanish institute of the promotion of tourism, Turespaña. The aim of the initiative is to provide a common focus and philosophy for tourism development throughout Spain covering the period 1992 to 1995. The primary objective of the initiative is to increase competitiveness, promote new technologies and ensure balanced and integrated quality tourism developments (Albert-Piñole, 1993). Apparent throughout is an emphasis upon a quality tourism industry. Quality developments are taken to include:

- environmental improvements through accommodation upgrading and the introduction of planning controls.
- product diversification by encouraging inland travel and focusing upon art and cultural heritage.
- national and local marketing plans.
- updating business structures.

Quality tourism has been the subject of various studies. A general consensus is noted in terms of the contributory components of quality tourism in Spain. Such components are summarized by Valenzuela (1991) and Williams and Shaw (1991) as:

- diversification of tourism supply.
- developments in the interior and the spatial extension of tourism.
- return to traditional quality of service.
- upgrading of coastal tourism.
- encouragement of domestic tourism.
- targeting of pensioners.

These components are reviewed to varying degrees, with the first two most often being considered. The diversification of tourism supply generally includes two areas, namely improvements to infrastructure and accommodation. Hillier and Raffael (1992) note that infrastructural improvements include the introduction (in 1992) of a high speed train link between Madrid and Seville, the completion of a new terminal at San Pablo airport (Seville) and the continuation of the road building programme initiated in preparation for the 1992 festivities. In addition, Travel Trade Gazette (1993), in a series of articles reviewing tourism policy development in Spain, highlights the extension of P & O's ferry operation to include Portsmouth–Bilbao and the small but steady development of charter flights to the northern area of Spain ('Green Spain'), traditionally served by Iberia, mainly through scheduled flights. Improvements to accommodation provision are considered in terms of investment in the country house hotel sector and *paradores* (Tourism Marketplace, 1993).

Policies to encourage the spatial extension of tourism include the introduction of 'new' locations in the tourist choice. To market such opportunities, a plethora of information sources, including radio programmes, for example, Breakaway Radio Four, television programmes, for example, The Travel Show, and travel writers for national newspapers, carry regular features on the 'new' improved Spain, albeit from a consumer and industry viewpoint. Evident throughout is the routine reporting of traditional and new forms of tourism activity to be encouraged, supplemented by various travel details. Whilst the quality of this latter information source is often variable, it is arguable that such information sources currently reflect the tourism industry's state.

Balearic Islands

The Balearic Islands, situated approximately 120 miles south of Barcelona, are frequently singled out as forerunners in the application of quality tourism initiatives. Of particular note in the development of quality initiatives was the agreement in 1991 between the International Federation of Tour Operators (IFTO) and the Balearic Islands Tourist Council (BITC) to work together on a private sector led in depth study of tourism in Mallorca. The intention of the study was to identify the maximum tourism flows Mallorca could absorb without damage to the island and in doing so to develop a model for sustainable tourism which could be applied to other destinations. The developments resulting from this study have been widely reported (see, for example, Morgan, 1991; Barke *et al.*, 1996) and are taken to include the following.

Accommodation provision

- The reduction of Mallorca's bedspace capacity through building licences and hotel inspections.
- Restrictions on the location and height of new hotel developments.

Infrastructural improvements
- Pavement widening.
- Traffic management.
- Vegetation improvement.
- Beach upgrading (including policies to bring sand to replenish existing resources).

New markets
- For example, diversification of consumer type and country of origin.

Conservation management   A rolling programme of projects administered through the Mallorcan Conservation Council and Conservation School, which include:
- dry stone walls.
- footpath management.
- litter management.
- signposting.
- water quality monitoring.
- beach management.
- noise and air monitoring.

Canary Islands
Tenerife ... A New Approach (Tenerife Tourist Development Bureau, 1994), provides an example of a quality tourism initiative operational within the Canary Islands. The approach embraces initiatives similar to the mainland and Balearic Islands with respect to infrastructural improvements, total quality programmes and conservation management techniques.

Infrastructural improvements, introduced primarily to enhance the image of the island, are designed to offer a broad range of leisure facilities, easier access around the island and quality accommodation. Such improvements include expansion plans for Reina Sofia airport, such as the construction of a second runway, new terminal buildings and improved passenger facilities. Furthermore, the regional authorities have passed legislation aimed at establishing a bed limit capacity for the island to provide protection from overdevelopment. Total quality programmes, aimed at investing in training programmes for staff, are designed to cover all aspects of quality management. An integral part of all quality development strategies is an emphasis on retaining the ecological richness of the island. To achieve this, 50% of the island's surface is now under a protection order, a clear network of footpaths has been introduced and managed, whilst particular areas are subject to zoning schemes. Mount Teide National Park (the second largest national park in Spain) exemplifies such a scheme. Under the park management master plan (Instituto Nacional Para La Conservacion de La Naturaleza, 1990), the park is divided into different areas of control and use. The scheme, originally established in 1985, is policed by park rangers, who, in recent years, have also received training under the total quality programme.

Reserve Area (Class I) These areas are closed to the public and free access is forbidden. Access for scientific purposes and environmental control is regulated by the park administration; for example, the summit of Mount Teide is classed under this category, with access closed off during the late 1980s.

Restricted Use Area (Class II) Public access is controlled, with visitors not allowed to leave the marked trails except in those cases expressly authorized by the park administration.

Moderate Use Area (Class III) These areas have free public access.

Area of Special Use (Class IV) These are areas set aside for buildings and services that are necessary for the management of the park. There is free public access.

Also of note is the role of international field study projects. Such projects encourage partnerships between local authorities and community associations. Their purpose is to counteract environmental decay and promote the aims of sustainable tourism. Proyecto Ambiental Tenerife is an example of a charitable organization involved in such initiatives, which include those under the auspices of the European Union:

- conservation (whale and dolphin projects).
- rural tourism initiatives.
- specific research projects.
- workshops, including traditional arts and crafts.
- traditional farming techniques.

Discussion

To what extent the quality tourism strategy introduced successfully addresses the environmental issues raised in the Organization for Economic Cooperation and Development (1980) report, i.e. 'a divergence between tourism and environmental objectives' and the 'impact of environmental degradation induced by tourism' (pp. 21–22), remains to be seen. Wight (1993) infers quality tourism to be an interchangeable term for sustainable tourism. Yet no common agreement on sustainable tourism appears to exist. From the outset, therefore, quality tourism lacks clarity and potentially will be, as sustainable tourism is, inhibited in progress as a result. Furthermore, what is actually meant by 'quality' remains ambiguous. If, as the majority of literature indicates, it is considered in terms of product and service quality, how then can it be determined? By whose standards is quality measured? With the exception of total quality management programmes, as evident in Tenerife, there appears little documented in the way of response to these questions. The application of, for example, scales such as SERVQUAL (examining service quality) is expected but seldom, if ever, considered.

Development in the interior and the spatial extension of tourism appear geared towards encouraging tourism to 'new' destinations – predominantly interior towns and cities, developing also a greater emphasis upon culture. How 'new' many of these destinations are is questionable, as many feature in the past work of a number of travel writers/poets, etc. Nonetheless, much of the debate has been developed in support of this approach, commending it, for example, for informing people about additional parts of Spain whilst at the same time reducing the regional concentration of tourism and enabling a fairer distribution of economic gains. Tourism to inland destinations (for example, Madrid) traditionally reflects a more constant level of demand all year around. By encouraging more travel inland this has the potential to reduce the problems of seasonality, as evident on the Mediterranean/Atlantic coasts, the islands and the north coast. Whether it will reduce or increase the environmental degradation induced by tourism remains to be seen, as little of the debate has considered the possible environmental impact resulting and the capacity and capability of these 'new' destinations to receive tourists. Furthermore, much of the commentary has reviewed tourism in isolation from other development issues and objectives. Seldom is the historical context within which contemporary mass tourism evolved considered. Instead there is a frustrating tendency for authors to criticize the transformation of Spain, as has occurred (for example, in Torremolinos) over the last 30 years, questioning why such developments were allowed. Politics has clearly played a central role. The problems of economic development, coupled with the Civil War of little more than 50 years ago and the subsequent dictatorship, provides a context for tourism development which is frequently neglected. History often has been shown to repeat itself, with the cyclical nature of resort destination development, as proposed by Butler (1980), one example of this. It is, therefore, arguable that much more can be learned from considering the original patterns of tourism development in Spain when developing solutions to present day problems.

Conclusions

In the literature, Spain has been identified as a leading tourism destination in post-war development. Fluctuations in current demand were attributed to various factors, particularly temporary, structural and environmental. Much of the commentary on environmental factors was of a general nature and tended to review issues in isolation. The introduction of policies to combat fluctuations in demand has been well documented. The main findings indicate that such policies include a diversification of the traditional tourism activity, with mainland Spain focusing upon the spatial extension of tourism to the interior and the cultural attributes of it, the Balearic Islands gearing policies towards environmental protection through planning and building controls whilst the Canary Islands have extended this to include proactive environmental zoning techniques. How suitable such policies will be in the long term remains questionable. In general, policies do not plan for mass tourism or differentiate between package or independent travel. Instead they promote the spatial extension of tourism to destinations previously 'untouched' by the main market. It is arguable that such an approach fundamentally conflicts with the aims of sustainable tourism by failing to protect resources for future generations. If Spain is a microcosm of a global problem as previously suggested, a period of reflection may serve to evaluate the success and portability of contemporary policies and initiatives.

References

Albert-Piñole, I. (1993) Tourism in Spain. In: Pompl, W. and Lavery, P. (eds) *Tourism in Europe. Structures and Developments*. CAB International, Wallingford (UK), pp. 242–261.

Barke, M. and France, L. (1992) *The Development of Torremolinos as an International Resort: Past, Present and Future*. Business Education Publishers Ltd, London, 37 pp.

Barke, M., Towner, J. and Newton, M.T. (eds) (1996) *Tourism in Spain. Critical Issues*. CAB International, Wallingford (UK), 432 pp.

Butler, R.W. (1980) The concept of the tourist area cycle of evolution: implications for managers of resources. *Canadian Geographer* 14, 5–12.

Davidson, R. (1992) *Tourism in Europe*. Pitman, London, pp. 150–154.

Economist Intelligence Unit (1994a) International tourism reports database. Spain. *EIU International Tourism Reports, London* 1, 92–97.

Economist Intelligence Unit (1994b) Destinations. *Travel Industry Monitor* July, 10–11.

Fayos Solá, E. (1992) A strategic outlook for regional tourism policy. The White Paper on Valencian tourism. *Tourism Management* 13, 45–49.

Hillier, C. and Raffael, M. (1992) Fare trade. *Caterer and Hotelkeeper* 16 April, pp. 44–46.

Instituto Nacional Para La Conservacion de La Naturaleza (1990) *Teide Parque Nacional*. ICONA, Spain.

Middleton, V.T.C. (1991) Whither the package tour? *Tourism Management* 12, 185–192.

Morgan, M. (1991) Dressing up to survive. Marketing Majorca anew. *Tourism Management* 12, 15–20.

Organization for Economic Cooperation and Development (1980) *The Impact of Tourism on the Environment. General Report.* OECD, Paris, pp. 21–47.

Pollard, J. and Rodriguez, R.D. (1993) Tourism and Torremolinos. Recession or reaction to environment? *Tourism Management* 14, 247–258.

Ryan, C. (1991) UK package holiday industry. *Tourism Management* 12, 76–77.

Secretaría General de Turismo (1993) *Tourism Statistics for Spain.* Madrid, Spain.

Sofield, T. (1991) Sustainable ethnic tourism in the South Pacific: some principles. *Journal of Tourism Studies* 2, 56–72.

Tenerife Tourist Development Bureau (1994) *Tenerife.....A New Approach.* TTDB, Tenerife.

Tourism Marketplace (1993) UK targeted in multi-million pound Spanish push. *Tourism Marketplace* 92, 10.

Travel Trade Gazette (1993) Spain and the Islands. *Travel Trade Gazette* no. 2077, 22 September, 54–55.

Valenzuela, M. (1991) Spain: the phenomenon of mass tourism. In: Williams, A.M. and Shaw, G. (eds.) *Tourism and Economic Development. Western European Experiences*, 2nd edn. Belhaven, London, pp. 40–60.

Wheeller, B. (1994) Egotourism, sustainable tourism and the environment – a symbiotic, symbolic or shambolic relationship. In: Seaton, A.V. *et al.* (eds.) *Tourism. The State of the Art.* Wiley, Chichester, England, pp. 647–654.

Wight, P. (1993) Ecotourism: ethics or eco-sell? *Journal of Travel Research* 31(3), 3–9.

Williams, A.M. and Shaw, G. (1991) *Tourism and Economic Development. Western European Experiences*, 2nd edn. Belhaven, London.

Policy Perspectives on Sustainable Tourism

Chapters 19 to 24 move the debate towards the implementation of sustainable tourism policies on both a 'top-down' and 'bottom-up' basis. The theme of these chapters is essentially that neither regulation from above nor self-regulation from below, in isolation, is sufficient to achieve sustainable tourism. International agencies, national governments and organizations, local authorities, the industry, voluntary bodies and communities need to cooperate. The general tenor of the contributions on the attainment of sustainability in the short to medium term is somewhat pessimistic unless there are far-reaching and fundamental changes in the attitudes and behaviour of tourists and the industry.

Goodall and Stabler in Chapter 19 give a broad review of the current position, concentrating on the policy implications of working from the principles of sustainable tourism, of which five are examined in terms of development and environmental interactions, with implementation of both policies and codes of practice to meet specific standards. In doing so, they give an economic interpretation which postulates a synthesis with ethics, natural science and politics in determining limits to tourism development.

The nature of environmental standards arising from the five principles is outlined, with a view to showing how the environmental capacity approach in relation to acceptable change must be applied. Using a number of examples, it is argued that environmental quality standards must square with environmental performance standards with which tourism entities must increasingly comply. Preventive rather than ameliorative instruments, reflecting the best available technology not involving excessive cost, are the most appropriate. It is argued that, for sustainable tourism to be feasible, minimum impact codes and self-regulation by the tourism industry need to be complemented by command and control and market-based instruments exercised by governments and specific environmental agencies.

MacLellan, in the following chapter, complements the case put by Goodall and Stabler but raises a question mark over policy objectives. The focus of the chapter is the extent to which sustainability is the most appropriate means of resolving tourism problems. The difficulties of defining and measuring it are illustrated by reference to policies introduced in Scotland. Two critical issues are identified and examined, the political and policy-making context and the monitoring and measuring of the effectiveness of policies.

MacLellan, in the context of his case study, while echoing the questions raised at the outset in this book concerning the unclear definitions and the confusions over and difficulties of translating the principles of sustainability into practice, suggests that confining attention to specific disciplines, sectors or location or issues is the only way forward, rather than despairing at the task of considering the problems holistically. Governments pay lipservice to sustainability but their actions are often misguided, tending to tinker with the margins, relying too readily on cooperation and local community involvement and being more concerned with organizational issues than environmental ones *per se*.

Government policy in Scotland is traced and assessed, it being concluded that it is difficult to draw inferences as to the effectiveness of initiatives because most are still in their infancy. It is, however, argued that understanding and implementation of policy are confused and possibly contradictory. In any event, the thrust has been blunted by public sector expenditure cuts, which have weakened conservation measures. MacLellan's final comment is that sustainable tourism policies have enabled the government to avoid recognition and establishment of national parks in Scotland, which might have been a more effective means of attaining desired objectives and resolving tourism problems. In any event, assessment of effectiveness depends crucially on establishing a hard database.

Sustainable tourism policies do not only relate to those emanating from the public sector, which Goodall and Stabler and MacLellan emphasize. Taylor and Davis, in Chapter 21, in examining the commonly held beliefs about residents' responsiveness to, and the community involvement in, tourism in their locality, raise policy issues which apply to both businesses and the public sector. They question the 'correctness' of community involvement in the sense that there is a supposition by businesses and policymakers that all residents not only endorse tourism and its development but participate as part of what they call 'the community show', that is as part of the product. They cite the Greek cultural characteristic of *'philoxenia'* (the welcoming of visitors in a warm, friendly way) as an illustration. It is argued that there is a conflict between those involved in tourism and those not involved who are affected by it, often adversely. The main issue, Taylor and Davis consider, is the conflict which arises over the planning of the growth and development of tourism where local participation is encouraged by public agencies but a vociferous minority, in favour or against, influence decisions, the silent majority remaining unheard, suggesting a passive but tacit acceptance. Residents as part of the product can also often feel that their traditions have been degraded for commercial ends.

Chapter 22 by Prinianaki-Tzorakoleftheraki to an extent develops the policy prescriptives implicit in Taylor and Davis's contribution but also relates to Burton's concerning the destination life cycle, which concludes the book (Chapter 24), in that reference is made to community involvement with the tourism industry and the role it can play.

The chapter, after giving the background to the state of tourism in Crete, shows that, like other Mediterranean destinations, Crete is suffering from an unbalanced geographical distribution of tourism, intense seasonality, overuse of natural resources and competition from the development of resorts in other countries. Furthermore, it draws on a narrow market from European countries, making it vulnerable to changes in tourist preferences and holiday destination choices.

In striving for both a viable and sustainable tourism, the author argues that the tourism industry is unlikely to maintain, let alone enhance, its natural and built heritage or culture. Although legislation is being enacted, its implementation de facto depends on more than simply the enforcement of the law. The author identifies responsible tourism management, which entails sound education and training, as well as commitment, as one key to achieving long-term viability, especially with regard to the response of those in the industry and Cretan communities, through what is termed 'philoxenia', 'physi' and 'philia'. Another important factor, also founded on education and training, is to diversify away from mass tourism to cultural, conference, eco- and rural tourism. The sectors of the tourism industry and technical educational institutes are seen as playing a significant role, particularly in broadening the scope of curricula to include environmental and sustain- ability issues. Educational and informational initiatives should extend beyond the industry to involve both local communities and tourists themselves, so that politicians, developers, the tourism workforce, local and tourist populations should then be in a position to show respect for the environment and act responsibly towards protecting it.

Parsler's contribution is apposite as a policy-orientated one because in two case studies he illustrates the tensions between the market and government regulation on two dimensions: central as opposed to local control and environmental conservation versus economic and tourism development. He points to the loss of biodiversity, soil degradation, water availability and sanitation problems as key environmental issues. While it is shown that tourism has the potential to engender conservation because both tourists wish to enjoy seeing the flora and fauna and the local population can earn a living from it, recognizing the commercial value of protecting natural environments, the implementation of policies and the financial resources devoted to conservation are inadequate. The policing of nature reserves is poor and control of tourism lax. For example, the tour guides system is badly organized and left almost entirely in the hands of individuals, who run it on a competitive basis, to the detriment of an effective long run environmental strategy. Parsler argues for local control in order that communities have a vested interest in safeguarding wildlife habitats, so endorsing the observations made by Taylor and Davis and Prinianaki-Tzorakoleftheraki.

The final chapter by Burton considers Butler's life-cycle model and examines the extent to which intervention is required to sustain open-access resources, but she gives it a remarkable and insightful twist by suggesting intervention is also needed for eco-tour operators who are committed to responsible ecotourism (in Ziffer's spectrum) to remain viable. She argues, from an examination of ecotourism in two North Australia areas (Wet Tropical – Queensland, and Top End Wetlands – Northern Territories) that the former is in the declining phase of the cycle as far as tour operators are concerned and that this suggests the same future for the Northern Territories. Therefore, intervention is necessary to restrict the number of tour operators and for national park authorities to define more clearly market conditions so that nature-based, and ostensibly more sustainable tourism orientated, tour operators remain viable. She, therefore, nicely considers the conjunction of sustainable development, sustainable tourism and business viability. It is ironic, however, that national park authorities, should the policies be introduced, would exercise a monopoly. Burton also touches on the case for quotas or licences as an instrument for controlling numbers of tour operators and licences.

Principles Influencing the Determination of Environmental Standards for Sustainable Tourism

19

B. Goodall and M.J. Stabler

Joint Directors, Tourism Research and Policy Unit, Faculty of Urban and Regional Studies, University of Reading, Whiteknights, PO Box 227, Reading RG6 2AB, UK

The Concept of Sustainable Tourism

In common with most other industries, tourism exerts a series of environmental impacts – requiring specific infrastructure, consuming scarce resources, and generating wastes. Also tourism activity interacts with natural environmental systems, from local to global scales, and has a capacity to initiate environmental change. Its dependence on the quality of natural resources, as well as the built and cultural environment, is greater than for many industries. Therefore, it might be expected that the tourism industry, national governments and destination public authorities, host communities and the tourists themselves share a common interest in maintaining the environmental quality of the resource base, by promoting the sustainability of tourism. However, tourism continues to be a growth industry and that growth has the potential to over-consume resources and to damage the natural environment in an irreversible way.

Evolution of the concept of sustainable tourism has paralleled that of the related wider concept of sustainable development (Krippendorf *et al.*, 1988; Inskeep, 1991; Bramwell and Lane, 1993; Goodall and Stabler, 1994; Murphy, 1994; Hunter and Green, 1995). Both concepts involve problems of definition and are not without their critics and sceptics. Although commonality of ideas between the two concepts is clearly discernible, sustainable development does not wholly subsume sustainable tourism. A view exists that sustainable tourism is interpreted as the viability, in terms of the continued long-term existence of business, of the tourism industry (Goodall and Stabler, 1994; Wheeller, 1994). Such a 'tourism-centric' approach does not conform to all the general concerns and requirements of sustainable development (Hunter, 1995).

Central to the concept of and debate on sustainability/sustainable development is the extent to which a holistic approach should be pursued which allows for development in such a way that productivity can be sustained over the long term for future generations whilst preserving essential natural systems and protecting human heritage and biodiversity. To this Brundtland added equity and opportunity between nations (World Commission on Environment and Development, 1987). The concept contains the two seemingly paradoxical ideas of development and preservation but the focus on stewardship of natural resources is undeniable, even though the debate has broadened to include economic, social and cultural issues alongside the environmental concern. Of two principal stances in the debate – the weak and the strong – which is adopted is very relevant to how sustainable tourism is pursued.

Sustainable tourism development has been defined as involving the 'management of all resources in such a way that economic, social and aesthetic needs are fulfilled while maintaining cultural integrity, essential ecological processes, biological diversity and life support systems' (Tourism Canada, 1990, p. 3). The currently dominant, tourism-centric paradigm of sustainable tourism interprets this as:

- meeting the needs and wants of the host community in terms of improved living standards and quality of life in both the short and long term.
- satisfying the demands of a growing number of tourists and of the tourism industry and continuing to attract them in order to fulfil the first aim.
- safeguarding the environmental resource base of tourism, encompassing natural, built and cultural components in order to achieve both of the preceding aims (Pigram, 1990; Cater and Goodall, 1992; Cater, 1993).
- maintaining or enhancing the competitiveness and viability of the tourism industry.

Essentially this entails modifying patterns of economic and tourism development and growth, through the adaptation of conventional mass tourism as well as the introduction of new forms of alternative tourism (Smith and Eadington, 1992). Implied from the environmental viewpoint is qualitative growth, whereby tourism development can be achieved with less use of non-renewable resources and less stress on the environment (Müller, 1994), depending on the relative importance attached to the weak and strong sustainability stances.

The destination emphasis of this tourism-centric view is understandable but, for truly sustainable tourism, impacts on tourist origin areas, transit regions, and the global environment must not be overlooked (Goodall and Stabler, 1994; Hunter, 1995). Acknowledging environmental or resource limits to tourism development/growth is one thing, determining where those limits are is another! However, it is also clear that, unless much existing tourism activity is adapted, or even curtailed, and, indeed, even if new tourism development conforms to current best environmental practice, further tourism growth will give rise to problems of accumulating incremental impacts on natural systems. Understanding of those problems is compounded because of the uncertainty surrounding the nature of the environmental changes due to the inherent variability of natural environments in the absence of anthropogenic disturbance and the current limits to scientific knowledge pertaining to many of the complex intermeshing relationships and the natural time lags involved.

There is evidence to demonstrate tourism's increasing commitment to sustainability/sustainable tourism. At the international scale the World Tourism Organization (1995) has considered the major themes of *Agenda 21* (United Nations Conference on Environment and Development, 1992) applicable to tourism and the World Travel and Tourism Council (1991) has established a set of environmental principles to help companies and governments with policy formulation, based on the International Chamber of Commerce's (1991) *Business Charter for Sustainable Development*. This is echoed in the many guidelines on principles produced at national and regional destination scales by government organizations and tourism industry associations (United Nations Environmental Programme: Industry and Environment, 1995). The emphasis is on environmental commitment and recognition of overall responsibility on the part of tourism for its environmental impacts. Whilst there is clear acknowledgement of the need to take environment into account in the planning and development of tourism, much less guidance is available to existing tourism businesses seeking practical advice on how to make best environmental practice operational. Whilst there are sector-specific initiatives, for example the International Hotels Environment Initiative (1993), and anecdotal case material of environmental improvements by individual tourism businesses (Goodall, 1994; Wight, 1994a) there still exists a tremendous gap between theory/principles and practice. How do/should principles translate into practice?

What principles are identifiable as underpinning sustainable tourism? To what extent do these principles reflect a multi-dimensional and interdisciplinary viewpoint, combining the economic, ethical, natural science and political approaches to the determination of limits to tourism development? What are the environmental standards or targets which follow from these principles and which the tourism industry should be seeking to achieve? What measures are necessary to achieve these targets and what indicators of improved and improving environmental performance on the part of the tourism industry might be apposite? The discussion that follows highlights the difficulties in answering these questions unequivocally.

Principles for Sustainable Tourism

The principles underpinning sustainable tourism are not unique to the tourism case but apply to sustainable development in general. Inherent in the concept of sustainability is some form of environmental constraint. The ultimate aim of all economic activity should be to maximize social welfare or the real net benefits to society. This overarching principle, from an economic standpoint, needs to be considered if tourism development or activity is to be truly sustainable. There are a number of interrelated facets or subsidiary principles which can be identified and should govern tourism activity. In addition to the overriding principle of net benefit maximization, four are discussed below in terms of their relevance to tourism and natural environments interactions, recognizing that they are not mutually exclusive. These are to:

- treat the environment as natural capital.
- act with caution in the absence of conclusive scientific evidence of the impacts of any activity (exercise the precautionary principle).
- use resources so that environmental quality is not discernibly changed, or at least confined within some critically acceptable limit.
- correct any environmental damage on the 'polluter must pay' principle.

It is acknowledged that there are other subsidiary principles which should be examined, such as those concerning the moral issues or ethics of tourism, for example, intra- and intergenerational justice and the right of any species to exist as part of the biodiversity principle. These are only touched on in the discussion below, as these wider implications of tourism are reviewed in the introductory chapter to this text. Another matter not explicitly pursued here is the time dimension associated with the principles. For instance, that appertaining to the limits of acceptable change relates to proposed tourism development while the 'polluter must pay' principle concerns past or existing activity. Each of these principles identified above is considered in turn.

Justification for tourism development: real net benefits to society

Economic efficiency or optimality, defined in general terms as maximizing social welfare and in narrow terms as minimizing the cost of using resources, should be a fundamental criterion of decision-making concerning the environment and its use. It implies that the benefits of using scarce resources should exceed the costs of using them in each case. Each environmental input to tourism has an opportunity cost and should contribute to tourism output at least as much as it could produce in the next best alternative (non-tourist) use. This applies equally to tourism's primary use of resources, that is, the natural environment, such as landscape and wildlife, as a tourist attraction, and to its secondary use of resources, for example land and materials for the construction and operation of, for instance, hotels and transport systems. In addition to the direct use values involved in such tourism activity there are also indirect use values, one such value being the vicarious tourism referred to by Winpenny (1991, p. 25), which encompasses television programmes about wildlife and high value landscapes. Others are those posited by economics relating to option, existence and bequest demand (Stabler, 1995), considered below. Providing the real benefits and costs of tourism development, inclusive of the environmental impacts, are fully accounted for, only 'good' tourism projects will be developed. Moreover, there would be no need for any separate environmental criterion if environmental impacts were, in all cases, effectively valued in these economic terms.

A critique of this approach is called for because markets do not always achieve the most efficient allocation of resources. Due to the malfunctioning, distortion and even total absence of markets, prices do not always reflect the real social benefits and costs of resource use. In the market, net benefit from tourism development or use would be maximized at a level of provision or activity where the marginal benefit from the last unit of development/visitor equals the marginal cost of provision for that unit/visitor, as indicated in Fig. 19.1 at point P, where marginal revenue (MR) equals marginal private cost (MPC). This will not bring about the optimum level of tourism development, where the balancing of benefits and costs considers only the private benefits received and the private costs incurred, that is, benefits and costs directly received/incurred by a hotel proprietor or operator of ski lifts/pistes. If the full social costs, inclusive of environmental impact, for example, resulting from pollution associated with the hotel's consumption of energy or disposal of wastes into an 'environmental sink' or the effect of chemicals used by the ski resort operator in preparing glaciers for summer skiing on nitrogen and phosphorus levels in drinking water, are not taken into account, there is over-consumption of environmental resources by the tourism activity in question. This is the problem of bioeconomic inefficiency (Steele, 1995) and such detrimental externalities should be taken into account in the economic equation. Conversely, where social benefits accrue, such as to residents not directly involved in tourism provision, for example services and facilities which would not otherwise be provided, the market would undervalue tourism resources and supply. These considerations are demonstrated in Fig. 19.1, where the marginal private cost (MPC) plus marginal social cost (MSC) and the marginal revenue (MR) plus the marginal social

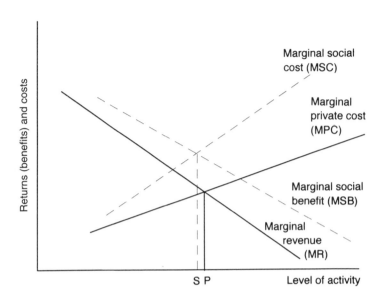

Fig. 19.1. Private and social optimal levels of tourism provision.

benefit (MSB) give rise to an optimum position at S, which suggests that the
marginal social costs outweigh the marginal social benefits, i.e. the optimal
level of provision is lower than P, the privately determined level of activity.

These problems are especially severe where markets are missing but the
environment is not 'free' even though a conventional market for its services
is absent. In such cases, resources, called public goods, or perhaps more
appropriately collective consumption goods in economics, are subject to
open access (sometimes even if privately owned) and therefore unpriced.
Because of non-exclusivity, individuals suffer no constraints on demand and
businesses have no incentive to reduce use of these environmental assets, even
less to invest in their preservation, and consequently they are overused and
are in danger of being degraded. In effect, overuse gives rise to a sub-optimal
situation. In Fig. 19.2, Q_p represents demand at price P_p for a resource which
is traded in the market. Where a resource is unpriced (P_0), given the demand
curve shown, the level of use is Q_0. This is likely to be incompatible with
protection of the resource, especially if optimal bioeconomic use lies to the
left of Q_p, that is, even if priced use would be sub-optimal. The problem is
further exacerbated where the environmental resource is depletable, as in the
case of sea sport fishing or irrigation water for golf courses in arid areas.

Further difficulties arise in applying the economic efficiency principle
because total economic value includes not only current use values but also
option, bequest and existence values. The problems of estimating future use
values (option values) and non-use values (bequest and existence) are fraught

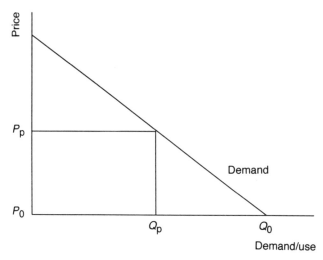

Fig. 19.2. The use of priced and unpriced resources.

with complications even though economic appraisal techniques are becoming more sophisticated (Pearce and Markandya, 1989; Barde and Pearce, 1991; Winpenny, 1991, 1995). Option value occurs where a potential tourist may have no intention of visiting a particular environmental attraction in the foreseeable future but may wish to retain an option to visit for herself or himself, or others, that asset at some future date. Whilst many such options may be foreclosed by non-tourist development, even new tourism facilities can foreclose options. In the case of golf course development on coastal sand dunes at Lough Gill, County Kerry, plant diversity has been reduced by almost half and species such as pyramidal orchids and the rare lady's tresses have been lost (Pearce, 1993). Existence value is derived from the pure pleasure in knowing that something exists, for example, an endangered species in a protected habitat, whilst bequest value, which some perceive as a form of option value (Pearce *et al.*, 1989), seeks to pass opportunities on to one's descendants. Both raise problems of valuation in so far as future generations may wish to satisfy their leisure needs in very different ways from at present. For example, if future generations rely more heavily on the technology of virtual reality to satisfy many of their leisure needs then tourist travel and site visits would diminish and the global and certain local environments benefit accordingly.

In addition to such intergenerational considerations there are also intragenerational ones which reflect issues of social justice or equity. Economic valuations are dependent upon the distribution of wealth and income within and between societies. Where this is very unequal, the interests of the poorest groups and nations will be the least considered. The issue is actually even more significant, as advanced industrial nations have used their economic wealth to influence the development of resources elsewhere, including for instance tourist enclaves in developing countries for their holidaymakers. Such tourists, in the main, expect advanced nation standards at their destination. Indeed, the World Tourism Organization has recently stipulated that hot and cold water should be available to tourists 24 hours a day (Georama, 1996). Local resources, especially energy and water, which tend to be in short supply in many developing countries, are diverted from local activities. Similarly, advanced industrial nations are critical of certain developing countries' resource use decisions, which, whilst benefiting the material standard of living of the local population, may be environmentally damaging in terms of reducing the value of environmental resources for tourism.

Lastly, the market-dependent economic approach can be criticized because economic optimality, as demonstrated in Fig. 19.3 (derived from Fig. 19.1), at S (which equates the private and social benefits of tourism's natural environment use with its private and social costs) does not coincide with the ecological and physical optimality at E proposed in the natural sciences, where any social costs generated just equal the assimilative capacity of the environment (EAC). This is the case because, as shown, the social cost, denoted by the area of triangle abc, is not necessarily totally eliminated and its effect may be cumulative. Again, the problem is most severe in the case of environmental resources which are open access public goods. For example repeated visits, even when restricted in number and party size and well-regulated, can still destroy/disturb fragile environments such as the Antarctic wilderness. Tourist visitation in the Antarctic summer coincides with the peak breeding periods of many species and even limited numbers of visitors

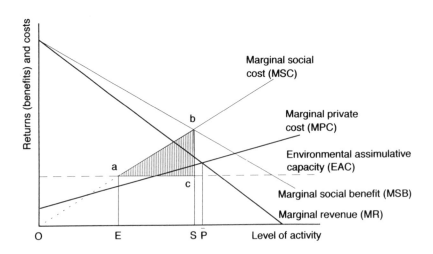

Fig. 19.3. Private, social (economic) and ecological optima.

can disturb wildlife breeding sites, which are a key factor of Antarctica's tourist attraction (Hall, 1992). An economic optimal position fails to meet ecological or physical desiderata because the recovery or assimilative capacity of the environment is impaired through persistent residuals or because the opportunity cost of eliminating all pollution or environmental damage is greater than the benefits to be derived. This highlights the anthropomorphic basis of any economic (or social) criterion and raises issues of the morality of human activities which lead to the extinction of other living species.

The environment as natural capital

The global environment, i.e. the conditions, circumstances and influences under which humans, flora and fauna exist, provides four fundamental functions or kinds of service for society.

1. It acts as general life support since the environment comprises components essential for life, health and human welfare, for example, the composition of the atmosphere, biodiversity and natural beauty.
2. It supplies energy and raw materials, which are the physical inputs to current production and consumption and which may be either finite or renewable.
3. It has a 'sink' function, which is the ability to absorb the waste products of social and economic activity through any of the environmental media (air, soil/land, water).
4. It has aesthetic and amenity value and enhances the quality of life.

Tourism is at least as dependent as any other industry on the environment in respect of the first of these functions as it is on the second for its consumption of resources in infrastructural and facility developments and especially of energy to support tourist travel and accommodation activities. Even though tourism may be regarded as less destructive to the environment in general than many industries in terms of generation of wastes, its very size and presence has created negative environmental impacts in many tourist destinations and its contributions to global environmental problems, such as stratospheric ozone depletion, cannot be ignored.

The environment can be viewed as a form of natural capital, similar in many ways to physical capital assets (Winpenny, 1995). Damaging the environment can then be likened to running down capital, which must, eventually, reduce the value of the recurrent functions or services it provides. Some of the life support functions are being lost or modified as a result of present directions of development, for example, the stratospheric ozone layer, and may suffer irreversible loss, for instance, biodiversity as a result of species' extinction. Finite resources, such as fossil fuels, will always be depleted when used, whereas renewable resources, such as water supply, may be depleted by excess use and insufficient maintenance. What level of environmental use is sustainable in the sense that it is consistent with preserving environmental capital? The literal view, of both natural scientists and many environmental economists, is of the environment as a capital stock that should not be diminished. Development should leave natural capital assets intact over time so that the same natural capital is bequeathed to each future generation (Winpenny, 1991).

The idea of a constant natural assets rule (Pearce, 1991) may be simple and appealing but interpreting and applying it is difficult. Environmental economics distinguishes three broad types of capital:

1. *man-made capital* (including tourism infrastructure such as hotels, airports, theme parks and holiday villages), which can be increased or decreased at society's discretion subject to opportunity cost, including environmental, considerations. Essentially the creation of man-made capital involves the transformation (in part) of natural environment or capital into built environment and cultural resources. The latter are just as much part of the total capital available to society and bequeathable to future generations as natural capital.

2. *critical natural capital*, being those natural assets (global climate, ozone layer, biodiversity, wilderness, etc.) essential to life that cannot be replaced or substituted by man-made capital and which, therefore, impose ultimate, absolute constraints on all activities. Loss of stratospheric ozone, for example, brings increased exposure to ultraviolet radiation, so increasing the risk of skin cancer in white-skinned people (amongst them Western tourists who expose themselves to the sun) and of eye cataracts (especially in Third World countries), as well as influencing the functioning of phytoplankton on the surface of the sea (Haynes, 1995).

3. *other (non-critical) natural capital*, which may be categorized as renewable natural resources, such as fish, forests, and agricultural products, which, with wise husbandry, can be sustained almost indefinitely, and non-renewable resources, for example those mineral resources that can be wholly or partly recovered (depending on the relationship between recovery costs and exploitation costs of primary resources) or substituted by man-made capital.

The idea that the next generation should inherit a capital stock no less than that inherited by the present generation has been criticized on the grounds that 'there is nothing sacrosanct about the stock levels ... inherited from the past' and that the size of the natural resource base to be passed on should be determined 'from considerations of population change, intergenerational well-being, technological possibilities, environmental regeneration rates and the existing resource base' (Dasgupta and Maler, 1990, p. 10). Moreover, does capital stock mean natural capital alone or natural capital together with man-made capital? To what extent can man-made capital be substituted for natural capital? There is no consensus: both 'soft' or 'weak' and 'hard' or 'strong' approaches are advocated (Daly, 1991). The former, whilst accepting that environmental considerations need to be taken into account in policy or decision-making, allows substitutability between man-made and natural environments and market-based instruments as opposed to more stringent regulation. It reflects the views of orthodox market economists and industry. Conversely, the latter approach, advocated by natural scientists, con-servationists and environmental economists, imposes the imperative of environmental absolutes and the constancy of natural capital. Rather than a dichotomous situation, the 'weak–strong' categorization is increasingly being viewed in environmental economics as a four group continuum, from very weak, through weak and strong, to very strong, which has policy and implementation implications (Turner *et al.*, 1994).

If natural capital is considered alone, is it a constant physical level of capital that is passed on or a level that preserves its value in economic terms (Pearce *et al.*, 1990; Winpenny, 1991)? The latter would allow the depletion of finite or non-renewable resources since, as these become scarce, so they become economically more valuable but where critical natural capital is damaged or depleted long-term problems of sustainability must ensue. By definition, the use of finite resources is non-sustainable in physical terms.

Even in the case of renewable resources, the idea that there is a level of environmental use which is sustainable and consistent with preservation of the stock of other natural capital is not without difficulty of application. The sustainable yield of a renewable resource is maximized at one and only one level of stock. It would appear obvious that a good policy for renewable resource management would be to aim for this particular stock size and thus for maximum sustainable yield (MSY). However, because the long-term population dynamics of many living resources is poorly understood, MSY cannot be established with any accuracy (Barkham, 1995). Moreover, MSY is basically a physical or biological optimum, not simply an economic optimum expressed in terms of the costs involved in collecting or harvesting the resource and the revenues generated in relation to the biological growth rate of the resource and prevailing interest or discount rates. The economic optimum may leave the stock in a subsequent period above, below or coincidentally at the MSY level depending on the cost structure relative to revenue. In Fig. 19.4 the total revenue curve, the shape of which is determined by the state of the stock, reproduction rate and harvest rate of the resource, shows that its highest point coincides with maximum sustainable yield (MSY). The profit maximum level of exploitation is at PM, which denotes the greatest vertical distance between total revenue and total cost. The point BE indicates the break-even position, where total revenue equals total cost. In economic terms, this is the shut-down point so further exploitation should cease. However, this is to the right of the MSY level and

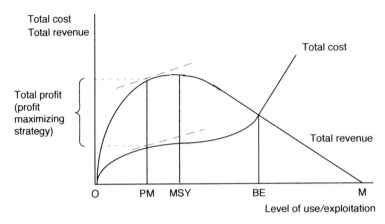

Fig. 19.4. A bioeconomic model of maximum sustainable yield applied to tourism use.

may indeed be a position where the resource cannot reproduce itself (or recover) even though it might be possible to go on gaining revenue to point M, where the resource is totally degraded or, in the case of biological species, total extinction occurs. The terms *open access resource*, the *commons*, *common access* or *common property resource* have often been used synonymously in the case of such renewable resources discussed above but a distinction should be made. True open access resources, such as the atmosphere and the oceans, are not owned so that there is no control over access, i.e. they are unregulated. The only constraint on open access resources is the cost of access, for example boats and labour for ocean sport fishing for marlin. However, independent decisions without regard to those made by others mean that the resource in aggregate is very likely to be over-exploited, so that total depletion may actually occur. Common access resources, for example, public rights of way in the UK, are regulated in that users have a legal right of access, subject to certain constraints. Likewise, those responsible for common property or common resources may be granted commoners' rights and often institutions are set up to oversee their use and exercise sanctions for misuse, as in the case of publicly owned resources, such as the New Forest in the UK. It is possible for common property resources to be overused and degraded where technical advances lead to changes in the nature and/or intensity of use, for instance, trail-bikes and off-road vehicles, and the controlling organization fails to adapt.

Caution in the absence of conclusive scientific evidence - the precautionary principle

In the face of uncertainty, irreproducibility of natural resources and the possible irreversibility of decisions, it should be assumed that a tourist activity or development might damage the environment. Unless there is clear scientific evidence to the contrary, decision-making should err on the side of caution where uncertainty exists as to the long-term consequences of current tourism resource use. There are three levels of uncertainty pertaining to present understanding of environmental systems: these relate to data shortages, model deficiencies and simply 'the unknowable' (O'Riordan, 1995a). No one knows, for example, what is the full value of stratospheric ozone: therefore the cost of depleting it by a third or more is unknown. Hence the significance of the precautionary principle, which, for example, imposes a responsibility on those tourism businesses and organizations wishing to develop and use resources to show that their actions will not cause potentially important and damaging environmental impacts.

Data shortages and model deficiencies can be reduced by investing in improved information services but it should be remembered such investment itself incurs an opportunity cost.

In its extreme form the precautionary principle is consistent with the very strong approach to sustainability, holding that no tourism development or activity should be undertaken if there is the remotest risk of substantial environmental damage. Thus critical natural capital must be safeguarded under the principle, as should renewable resources where MSY is not definable. In a more practical mode, opportunity cost is explicitly recognized, with the risk of substantial environmental damage being avoided provided the cost of doing so (including the opportunity cost of no action) is acceptable.

Applying the principle is again difficult since many tourism developments and activities can have double-edged impacts. Take nature-based tourism, which provides an incentive to conserve the natural assets on which it is based but where the very activity also threatens that wildlife since even small numbers, viewing in a controlled way, can disturb wildlife. In Venezuela, where low-impact ecotourism is being promoted, there is evidence that animals are no longer seen along the river below the Angel Falls between June and November, the main period when canoes are plying the river with tourists (Sykes, 1996).

Criticality: limiting resource use to no discernible environmental change

Environmental systems can be remarkably resilient and do have a capacity to withstand a degree of use. For example, the environment's 'sink function' is dependent upon its ability to assimilate and dissipate potential pollutants up to a certain level. This 'buffering' capacity of natural systems to protect the environment at large against stress or threat is part of critical natural capital and criticality lies in maintaining the functionality of environmental systems (O'Riordan, 1995b). Use of the environment's buffering potential, at least up to its capacity, is implied, as is the prevention of overuse (which would cause environmental damage). Essentially this is an environmental or ecological capacity issue but one which takes on economic dimensions as soon as the tolerance level is exceeded. The practical approach to limiting resource use in order to protect the environment tends to adopt a minimum impact stance. Some environmental change (damage) may be acceptable in opportunity cost terms but determining the 'limits of acceptable change' (Stankey *et al.*, 1984) brings a subjective component into the assessment.

Limiting resource use to bounds of acceptable change requires source reduction of certain inputs to tourism developments and operations, particularly so where resources are non-renewable and non-recyclable, for example, improved efficiency of extraction and use of finite resources, such as non-metalliferous minerals used in hotel and airport construction, and of fossil fuels to provide energy for travel. The minimization of waste and reduction of overconsumption during the production of tourism services, the substitution of environmentally benign inputs and equipment wherever possible, and the recovery and recycling of wastes or their safe disposal where unavoidable, all contribute to limiting resource use and therefore reducing or preventing environmental damage (Goodall and Stabler, 1994).

The possibility of defensive or preventive expenditure (Winpenny, 1995) should also be considered, whereby the functioning of environmental systems can be maintained or supplemented by providing buffering capacity by artificially equivalent means, for instance, sewage from a seaside resort being subject to full treatment facilities rather than being discharged untreated through outfall pipes into the sea. However, an opportunity cost is involved since resources to construct and operate the sewage treatment works are not available for other developments. With renewable resources, criticality requires the level of tourism use to be within the MSY (where this can be established). For example, big game hunting has been made environmentally acceptable in a number of African countries by exploiting the controlled culling of game as a tourist attraction (McNeely, 1988). Nevertheless, given that tourism is a growth industry, environmental change (damage) may follow from the cumulative effects of incremental tourism development even where those developments conform to current best practice and technology.

Correcting damage from current tourism activity - the 'polluter pays principle'

Existing forms of tourism activity, ranging from downhill skiing, development of coastal resorts, theme parks and golf courses, to increased visitor numbers to national parks and protected areas, have already brought about and continue to generate unmanaged environmental impacts. Such impacts include the loss of biodiversity and habitats, air and water pollution, noise and congestion, and increased competition for scarce resources (Tolba and El-Kholy, 1992; Stanners and Bourdeau, 1995). Where such environmental damage is directly attributable to tourism and the threshold of irreversibility has not been breached, the tourism firms and tourists should be held responsible and encouraged/required to alter their practices and/or behaviour, in effect to act in an environmentally more friendly manner. This is fully consistent with the 'polluter pays principle' (PPP), a long established tenet of general environmental policy applying to the producer or supplier (Opschoor and Vos, 1989) and its extension in the 'polluter (user) pays principle' [P(U)PP] (Tolba and El-Kholy, 1992; Opschoor *et al.*, 1994), where user means the tourist as consumer. The implication is that tourism firms and tourists who damage the environment should bear the cost of measures which public authorities have to implement to ensure an acceptable environmental quality. Whilst application at the destination scale would be most likely, tourism's impact on the global environment should not be ignored.

Basically PPP is a non-subsidization principle although it is not always applied in that form. It is therefore not unambiguous, especially since 'acceptable state' of the environment is subjective and transition periods of varying duration are commonplace to allow firms to adjust their practices. The 'polluter pays principle' has its origins in welfare economics, which argues that ideally the price of tourism products should reflect the full social costs, including environmental costs as related to pollution, resource exploitation and environmental degradation. Ideally, market prices should embody the full costs of producing and consuming goods and services (Weiszaecker, 1989). Making the polluter pay is a convenient way of internalizing these environmental externalities by ensuring that tourism firms and tourists actually and fully meet the cost of the environmental damage their practices and activities cause. With correct pricing to reflect this principle, tourism firms and tourists would adjust their practices and behaviour accordingly since the optimum use of the environment would occur where the marginal pollution reduction costs are equal to the marginal environmental damage costs. Where environmental goods and services are not marketable, such as open access resources, as in the case of many tourist destination environmental features and attractions, a second best approach is to equalize the marginal cost of environmental protection by imposing a price per unit on pollution discharged to give a least cost solution (Organization for Economic Cooperation and Development, 1991a).

It must be noted that investment in pollution abatement by tourism firms and adoption of alternative behaviours by tourists involve an opportunity cost. Investment will only be undertaken up to the point where the marginal benefit of reduction in environmental damage (or improvement in environmental quality) is equal to the marginal cost (or price per unit under the second best approach) of environmental protection. Whilst some reduction in environmental damage is certain, there is no guarantee that this will attain an economic optimum, let alone fall within the assimilative capacity and recuperative powers of the environment. Operationalization of the principle is, once again, not straightforward, as identifying the source and perpetrators, as well as the optimum and attaining it through policy instruments and business practices, is extremely difficult in real world situations.

From Principles to Practice

How do these principles translate into practice within the tourism industry? Clearly, influenced by the concepts of sustainable development and sustainable tourism, policy statements and commitments embodying the essence of the principles outlined above are increasingly commonplace, from the international level through continental and national scales to industry sector, destination, host community and site-specific responses. Voluntary codes of conduct have thus been developed by a wide variety of organizations, ranging from governments, including national tourism offices and tourist boards; non-government organizations, such as the Ecotourism Society, the Alliance Internationale de Tourisme/Fédération Internationale de l'Automobile and the World Wide Fund for Nature/Tourism Concern at the international level, as well as tourism industry and trade associations (United Nations Environmental Programme: Industry and Environment, 1995). Such codes may be targeted at tourism businesses, host communities or even the tourists themselves.

First and foremost, such initiatives seek to create an awareness within the tourism industry and governments of the importance of sound environmental policies and management practices. Not only is the environmental dimension to the fore but due consideration is given to socio-economic and cultural issues as well. Only when awareness has been achieved is it possible to encourage the tourism industry sectors, government agencies, destination planning authorities, host communities, non-government organizations and tourists to adopt business practices and behaviour which may facilitate the attainment of the goals implied by sustainable tourism. Thus the principles enunciated in the many codes concentrate on urging the tourism industry to:

- use resources sustainably, that is, to reduce the use of environmentally unfriendly inputs and products, to avoid the depletion of natural resources, and recognize the capacity limits of different resources/ locations.
- reduce its environmental impacts, for example, to control atmospheric emissions and disposal of sewage, encourage forms of tourism activity and transport which are environmentally benign.
- reduce waste and overconsumption, for example, by improved water management and recycling of glass, metals and paper.
- maintain natural and cultural diversity by showing sensitivity to the conservation of species and respect for heritage and local culture.
- integrate tourism development into land-use planning in order to assess the potential impact of new developments and prevent the accumulated impacts of incremental growth.
- adopt internal management strategies and systems, such as environmental auditing, in order to improve the environmental performance of current operations.
- support and involve local economies/host communities, for example, by using local products and suppliers wherever possible.
- market tourism responsibly.

There is no shortage of general advice available. The focus, however, is inevitably localized by concentrating on the tourist destination or host community so that the wider damaging interactions of tourism with the global environment, including tourist origin and transit regions as well as destinations, tend to be ignored. Moreover, diffusing awareness throughout the tourism industry has to be worked upon. Codes of conduct and environmental initiatives are not necessarily known or available to all the organizations and firms who might benefit (Stabler and Goodall, 1997, certainly not at the destination level. To date, most of the codes of conduct, because of the emphasis on principles, are of very limited usefulness to tourism businesses seeking practical advice on what is current best environmental practice and how this can be introduced into their business unit or site activity (Goodall, 1996).

Technical advisory codes or guidelines setting out how tourism businesses can develop and maintain environmentally friendly products, operations and practices are just beginning to appear. These tend to be industry-sector specific: which is not surprising in view of the obvious differences between running, for example, airline and hotel businesses. These manuals cover such matters as waste management, energy and water conservation, product and input purchases, noise, emissions to air and water and hazardous materials. Examples range from the internationally available Green Globe initiative (World Travel and Tourism Council, 1994) through trade associations, for example, the Thai Hotel Association (Goodno, 1993), to local destinations, for instance, the South Devon Green Tourism Initiative in the UK (Dingle, 1995), and individual companies, such as that by Canadian Pacific Hotels and Resorts (Troyer, 1992). Adoption and operationalization of such technical guidelines are fraught with difficulty since many tourism firms are complacent about their environmental performance or perceive any improvements they might make as having minimal environmental effects. Yet others perceive increased costs of changes as a disincentive and may genuinely be unable to raise capital necessary to fund the change. Furthermore, many small tourism firms may not be members of an industrial or trade association and therefore remain ignorant of current best environmental practice. In any event, environmental management practices evolved in large firms are not always appropriate for small and medium-sized enterprises (Stanners and Bourdeau, 1995), where procedures are more informal.

The result of all this activity, even though the formulation, diffusion and use of codes of conduct and environmental management systems are piecemeal, is that tourism's environmental performance is improving and will continue to improve. Current environmental performance, however, falls way short of truly sustainable tourism, especially where the global environment is concerned. The principles underpinning sustainability may be seductively appealing but there is a 'grey' area between acceptance of these principles and their translation into workable environmental objectives or standards. Indeed, the range of appropriate indicators for measuring improving environmental performance is not clear-cut.

Taking a lead from pollution control legislation, environmental quality objectives, defining some desired state of the environment that can be met through the attainment of specific targets, need to be defined, for example the preservation of beach amenity. This requires the establishment of appropriate ambient *environmental quality standards* (EQSs), which prescribe the level of pollution or nuisance that must not be exceeded. Such EQSs would normally be determined at the level of the tourist destination, although in theory they are applicable up to the global scale. Any EQS, such as the EC quality requirements for bathing waters based on 19 parameters (five microbiological, 14 physico-chemical), should allow for the capacity of the environment to assimilate and disperse pollutants and recuperate from damage. Application of EQSs demands extensive knowledge of, for instance, dose–response relationships. Critical loads or *threshold limit values* (formerly maximum allowable concentrations) which identify safe minimum standards are often difficult to determine in many instances because of limited scientific understanding of the environmental processes involved and/or an absence of data.

To what extent does or should an EQS allow for environmental change? All tourism activity has an impact on the environment but change is not necessarily damage. A value judgement is implied in the case of damage and the notion of *limits of acceptable change* has to be addressed (Wight, 1994b). Differing interpretations may be placed on this concept, ranging from the *no observed effect level* (NOEL), which suggests no change and certainly no damage; through the *no observable adverse effect level* (NOAEL), allowing change but no damage, and the *lowest observable adverse effect level* (LOAEL), involving minimal damage; to *as low as reasonably achievable* (ALARA), in which change and damage are balanced in terms of costs and benefits (United Nations Environment Programme, 1986).

What is the relationship between, for example, a destination level EQS and the environmental performance required of tourism firms in that destination? Acting in an environmentally responsible way has consequences for a tourism firm's costs and revenues, the incidence of which is variable through time (Goodall, 1992). Theoretically an EQS needs to be dis-aggregated into a set of *environmental performance standards* (EPSs) applicable to individual tourism firms: as long as firms comply with their EPSs then the EQS objective will be met. By definition, EPSs apply to point-form sources of potential environmental damage/pollution.

What is feasible from the viewpoint of the tourism industry and the individual tourism firm? The minimum expectation, again borrowing from pollution control approaches and consistent with the polluter pays principle, is an EPS based on 'best practicable' or 'best available' technology. This is now interpreted as *best practicable environmental option* (BPEO), in which all environmental media are considered and the option identified provides the most benefit or causes the least damage to the environment as a whole at acceptable cost in the long term as well as the short term (Royal Commission on Environmental Pollution, 1988). This largely corresponds to the inter-pretation in the European Commission Directive on industrial air pollution which gave rise to *best available technology not entailing excessive cost* (BATNEEC), the criterion in fact incorporated into the 1990 UK Environ-mental Protection Act (Department of the Environment, 1990). The applica-tion of BATNEEC allows the up-grading of required environmental performance standards from time to time in line with technological advance. However, *best available technology* (BAT) normally applies, although the presumption can be modified where it can be shown that the costs of applying the BAT would be excessive compared with the environmental protection achieved.

Such an approach, applied at the level of the individual tourism firm, is generally consistent with the concepts of economic efficiency and opportu-nity cost. Conformance to such standards will mitigate environmentally damaging consequences of tourism but will not necessarily eliminate them – critical natural capital, such as the stratospheric ozone layer, will still be eroded by tourism travel, non-renewable resources will still be consumed in tourism infrastructural developments, renewable resources, such as water, may be strained beyond their maximum sustainable yield (MSY) in certain destinations, and the costs of correcting current damaging activities, such as the disposal of untreated sewage, may be judged unacceptable in some destinations.

Where tourism is expanding at a particular destination, even though existing firms comply with EPSs and similar standards are imposed on new firms, the incremental growth of the tourism industry could result in the capacity of the destination environment to absorb potential pollutants, such as sewage effluent from hotels, being exceeded. Under BATNEEC, EPSs can only be tightened for existing (and new) firms where there has been technological advance in pollution control. Short-term, rapid growth of tourism in a destination can therefore be problematic from the environmental viewpoint. Moreover, EPSs based on BAT are usually imposed as uniform emission standards. All firms, say, large and small hotels in a destination, have to comply with the same standard. This ignores the fact that firms differ in their economic efficiency and hence in the efficiency (or resource cost) with which they are able to meet the uniform EPS. Also, it has to be acknowledged that environmental considerations may be compromised on political or social grounds since a BPEO solution may be overruled in favour of one with a less acceptable environmental outcome in order, for example, to safeguard or generate local employment (Silbertson, 1993).

Given that truly sustainable tourism is an unattainable goal in the foreseeable future, an alternative but nevertheless complementary approach to the setting of environmental standards is to concentrate on measuring improvement in environmental performance. Subject to data availability, this could be undertaken at any spatial scale and for any tourism firm. Pioneering work on *environmental indicators*, which has been undertaken by the Organization for Economic Cooperation and Development (1991b,c, 1993), advocates a *pressure–state–response* approach, involving an assessment of the state of the environment, the pressures on it and the range of societal responses to the situation. Taking biodiversity as an example, the development of land for tourism, where an increasing proportion of land in tourism equals a pressure indicator, would destroy certain wildlife habitats; therefore the number of endangered species threatened as a proportion of known species can act as a state of environment indicator, whilst the designation of protected areas, as a proportion of total area, would constitute a response indicator. Monitoring of the indicator values over time would reveal whether environmental performances and/or targets are being met or exceeded, for example, if fewer species are present they are not, whereas if a greater number are found the position has improved.

The development of environmental indicators is in its infancy, with the current emphasis being on pressure and state indicators (MacGillivray, 1995), but the need has been acknowledged by governments (for example, UK Government, 1994). Although there are gaps in the availability and reliability of the necessary data and debate about whether a particular indicator measures state or pressure, the indicators approach can make a useful contribution to decision-making in tourism. Where such indicators are available as a time series, improvement in environmental performance can be clearly identified. This possibility has been acknowledged by those tourism firms (mainly large transnational corporations) which have recognized environmentally responsible behaviour as integral to maintenance of their competitive advantage and market share. British Airways (1993), for example, has moved some way towards developing quantitative measures of its environmental performance across its main environmental interfaces of emissions, energy consumption, noise, waste, congestion and tourism. However, problems of interpretation can arise, for instance, whilst significant reductions in aircraft emissions such as carbon dioxide, nitrogen oxides per available seat kilometre can be demonstrated for airline operations, there could still be an increase in the total amount of emissions if total seat kilometres flown have increased proportionally more.

Finally, it has to be admitted that the setting of environmental standards and the derivation of environmental indicators are undertaken from an anthropomorphic viewpoint. Sustainability and sustainable tourism is concerned with management of the human use of the environment, as is clear from the World Conservation Strategy 2 (International Union for Conservation of Nature *et al.*, 1991). What of nature's rights? The European Union quality requirements for bathing waters have more to do with the 'holiday-makers' environment' than with marine ecosystems and should not be regarded as a proxy for measures of the state of the natural environment. Whilst disinfection of sewage would improve a coastal resort's bathing water performance it could damage the marine ecosystem by reducing seabed diversity (MacGillivray, 1995).

Conclusion: Reconciling the Tourist and the Environment

It is obvious that tourism, despite the efforts to date to improve its environmental performance, is still an industry which contributes more to the creation of environmental problems than it does to their solution. On an eco-performance continuum (Peattie, 1995) tourism is low on sustainability. Most tourism activity has a long way to go to improve its environmental performance to a level that will be increasingly demanded by consumers, who, whilst becoming greener, are themselves light years away from wishing to live in a conserver society. However, the greener tourists become the greater the pressures on tourist destinations and tourism firms to behave environmentally more responsibly and to take steps collectively and individually to improve environmental performance. So far much of the tourism industry's effort has been reactive. Further improvement in environmental performance is therefore to be expected as efforts become more proactive: codes of conduct and manuals of best practice will become widespread as the tourism industry encourages environmental self-regulation of its activities. Much improvement will result from eliminating/reducing problems of current pollution or environmental damage and will be largely conditioned by consideration of economic efficiency in the case of individual tourism firms and net social benefit for tourist destinations.

Whilst tourists may certainly exert influence on the way tourism firms operate, it may well be destination or host communities which, ultimately, will object to the costs of tourism and take action, both against increases in the number of tourists and the tourism industry, whether it is indigenous or origin-based and controlled. Thus, at the destination level, greater significance will be attached to the assimilative and carrying capacities of the environment and operating within the MSY of its renewable resources. In destinations where such limits have already been exceeded, doubts must be expressed as to whether there will be any curtailment in tourist activity.

It is unlikely that the precautionary principle will have much credence in tourism, especially as the motivation for many, so-called ecotourists is to experience fragile natural environments before they are destroyed. Moreover, as long as growth in tourism continues to take place, man-made capital will continue to be substituted for natural capital and critical natural capital, such as the atmosphere, will be at risk. In this respect tourism is no different from many other economic activities and it may be that the prospect of imminent catastrophe will be the trigger to fundamental changes to consumption and production patterns.

What is certain is that it requires more than voluntary, industry-based action to translate principles into practice and bring about the necessary 'revolution' to transform tourism into a globally, environmentally responsible industry (Goodall, 1996). Parallel and coordinated action by the international community, the state, local government and local/regional agencies, as well as the host community, is required. A holistic or global perspective embracing complex and wide-ranging policies and measures across both public and private sectors is needed. This will involve not only 'command-and-control' approaches, including land-use planning, but also, increasingly, economic instruments of a kind which give both incentives to take action (such as 'pump-priming' loans/grants/subsidies, tax breaks such as investment allowances or lower rates of corporation tax and effluent charges) and deterrents to continued damaging activities (for example, by imposing pollution taxes, user charges and tradeable permits). The application of such approaches to tourism is discussed elsewhere (Goodall and Stabler, 1994; Stabler and Goodall, 1996).

However, the inevitable conclusion is that the needs of the tourist and of the environment cannot be wholly and fully reconciled, not least because of the nature of tourism's current travel mode dependence, and that non-renewable resources will be consumed and critical natural capital put at risk. Educating the would-be tourist to 'holiday at home' is a mammoth task!

References

Barde, J.-P. and Pearce, D.W. (eds) (1991) *Valuing the Environment*. Earthscan, London.

Barkham, J. (1995) Ecosystem management and environmental ethics. In: O'Riordan, T. (ed.) *Environmental Science for Environmental Management*. Longman, London, pp. 80–104.

Bramwell, B. and Lane, B. (1993) Sustainable tourism: an evolving global approach. *Journal of Sustainable Tourism* 1(1), 1–5.

British Airways (1993) *Annual Environmental Report*. Environment Branch, British Airways plc, London.

Cater, E. (1993) Ecotourism in the Third World: problems for sustainable development. *Tourism Management* 14(2), 85–90.

Cater, E. and Goodall, B. (1992) Must tourism destroy its resource base? In: Mannion, A.M. and Bowlby, S.R. (eds) *Environmental Issues in the 1990s*. John Wiley, Chichester.

Daly, H.E. (1991) Sustainable development: from conceptual theory to operational principles. In: Davis, K. and Bernstam, M.S. (eds) *Resources, Environment and Population: Present Knowledge, Future Options*. Oxford University Press, Oxford, pp. 25–43.

Dasgupta, P. and Maler, K.-G. (1990) The environment and emerging development issues. Paper produced for the World Bank Annual Conference on Development Economics, Washington DC.

Department of the Environment (1990) *A Guide to the Environmental Protection Act, 1990*. HMSO, London.

Dingle, P.A.J.M. (1995) Practical green business. *Insights* March, C35–C45.

Georama (1996) Water for all. *The Geographical Magazine* 68(3), 16.

Goodall, B. (1992) Environmental auditing for tourism. In: Cooper, C.P. and Lockwood, A. (eds) *Progress in Tourism, Recreation and Hospitality Management*, Vol. 4. Belhaven, London, pp. 60–74.

Goodall, B. (1994) Environmental auditing: current best practice (with special reference to British tourism firms). In: Seaton, A.V. *et al.* (eds) *Tourism: The State of the Art*. John Wiley, Chichester, pp. 655–664.

Goodall, B. (1996) The limitations of environmental self-regulation: the case of the Guernsey hospitality industry. Paper presented at the *International Conference on Integrating Economic and Environmental Planning in Islands and Small States*, organized by the Islands and Small States Institute, Foundation for International Studies, University of Malta in collaboration with the Directorate of the Planning Authority, Valletta, Malta, 14–16 March.

Goodall, B. and Stabler, M. (1994) Tourism–environment issues and approaches to their solution. In: Voogd, H. (ed.) *Issues in Environmental Planning, European Research in Regional Science*, Vol. 4. Pion, London, pp. 78–99.

Goodno, J.B. (1993) Leaves rate Thai hotels on ecology. *Hotel and Motel Management* 208(8), 52.

Hall, C.M. (1992) Tourism in Antarctica: activities, impacts, and management. *Journal of Travel Research* 30(4), 2–9.

Haynes, R. (1995) Preventing disease. In: O'Riordan, T. (ed.) *Environmental Science for Environmental Management*. Longman, London, pp. 335–346.

Hunter, C.J. (1995) On the need to re-conceptualise sustainable tourism. *Journal of Sustainable Tourism* 3(3), 155–164.

Hunter, C. and Green, H. (1995) *Tourism and the Environment: A Sustainable Relationship?* Routledge, London.

Inskeep, E. (1991) *Tourism Planning: An Integrated and Sustainable Development Approach*. Van Nostrand Reinhold, New York, Appendix A.

International Chamber of Commerce (1991) *Business Charter for Sustainable Development*. ICC, Paris.

International Hotels Environment Initiative (1993) *Environmental Management for Hotels: The Industry Guide to Best Practice*. Butterworth-Heinemann, Oxford.

International Union for Conservation of Nature/United Nations Environment Programme/World Wide Fund for Nature (1991) *Caring for the Earth: A Strategy for Sustainable Living*. IUCN, Gland.

Krippendorf, J., Zimmer, P. and Glauber, H. (1988) *Fuer einen audern Tourismus* [Towards an alternative tourism]. Fischer Taschenbuch Verlag, Frankfurt.

MacGillivray, A. (ed.) (1995) *Environmental Measures: Indicators for the UK Environment*. Prepared for the Environment Challenge Group, World Wide Fund for Nature – UK, The New Economics Foundation and the Royal Society for the Protection of Birds, London.

McNeely, J. (1988) *Economics and Biological Diversity: Developing and Using Economic Incentives to Conserve Biological Resources*. International Union for Conservation of Nature, Gland.

Müller, H. (1994) The thorny path to sustainable tourism development. *Journal of Sustainable Tourism* 2(3), 131–136.

Murphy, P.E. (1994) Tourism and sustainable development. In: Theobald, W. (ed.) *Global Tourism: The Next Decade*. Butterworth-Heinemann, Oxford, pp. 274–290.

Opschoor, J.B. and Vos, H.B. (1989) *Economic Instruments for Environmental Protection*. Organization for Economic Cooperation and Development, Paris.

Opschoor, J.B., Savornin Lohman, A.F. de and Vos, H.B. (1994) *Managing the Environment: The Role of Economic Instruments*. Organization for Economic Cooperation and Development, Paris.

Organization for Economic Cooperation and Development (1991a) *Environmental Policy: How to Apply Economic Instruments*. OECD, Paris.

Organization for Economic Cooperation and Development (1991b) *Environmental Indicators: A Preliminary Set*. OECD, Paris.

Organization for Economic Cooperation and Development (1991c) *Environmental Indicators: Progress Report*. OECD, Paris.

Organization for Economic Cooperation and Development (1993) *Core Set of Indicators for Environmental Performance Reviews: A Synthesis Report by the*

Group on the State of the Environment. Environmental Monographs No. 83, OECD, Paris.

O'Riordan, T. (1995a) Environmental science on the move. In: O'Riordan, T. (ed.) *Environmental Science for Environmental Management.* Longman, London, pp. 1–11.

O'Riordan, T. (1995b) *Extending Science for Global Environmental Change.* The 1995 Norma Wilkinson Memorial Lecture, University of Reading, Reading, 7 November.

Pearce, D.W. (1991) Towards the sustainable economy: environment and economics. *Royal Bank of Scotland Review* 172 (December), 3–15.

Pearce, D.W. and Markandya, A. (1989) *Environmental Policy Benefits: Monetary Evaluation.* Organization for Economic Cooperation and Development, Paris.

Pearce, D.W., Markandya, A. and Barbier, E. (1989) *Blueprint for a Green Economy.* Earthscan, London.

Pearce, D.W., Barbier, E. and Markandya, A. (1990) *Sustainable Development: Economics and Environment in the Third World.* Edward Elgar, Aldershot.

Pearce, F. (1993) How green is your golf? *New Scientist*, 25 September, pp. 30–35.

Peattie, K. (1995) *Environmental Marketing Management.* Pitman, London.

Pigram, J. (1990) Sustainable tourism: policy considerations. *Journal of Tourism Studies* 1(2), 2–9.

Royal Commission on Environmental Pollution (1988) *Twelfth Report: Best Practicable Environmental Option.* HMSO, London.

Silbertson, A. (1993) Economics and the Royal Commission on Environmental Pollution. *National Westminster Bank Quarterly Review*, February, pp. 29–39.

Smith, V.L. and Eadington, W.R. (eds) (1992) *Tourism Alternatives: Potentials and Problems in the Development of Tourism.* University of Pennsylvania Press, Philadelphia.

Stabler, M.J. (1995) Research in progress on the economic and social value of conservation. In: Burman, P., Pickard, R. and Taylor, S. (eds) *The Economics of Architectural Conservation.* Institute of Advanced Architectural Studies, University of York, pp. 33–50.

Stabler, M.J. and Goodall, B. (1996) Environmental auditing in planning for sustainable island tourism. In: Briguglio, L., Archer, B., Jafari, J. and Wall, G. (eds) *Sustainable Tourism in Islands and Small States: Issues and Policies.* Pinter, London, pp. 170–196.

Stabler, M.J. and Goodall, B. (1997) Environmental awareness, action and performance in the tourism industry: a case study of the hospitality sector in Guernsey, *Tourism Management* 18(1), 19–33.

Stankey, G.H., McCool, S.F. and Stokes, G.L. (1984) Limits of acceptable change: a new framework for managing the Bob Marshall Wilderness complex. *Western Wildlands* 103(3), 33–37.

Stanners, D. and Bourdeau, P. (1995) *Europe's Environment: The Dobris Assessment.* Office for Official Publications of the European Communities for the European Environment Agency, Luxembourg.

Steele, P. (1995) Ecotourism: an economic analysis. *Journal of Sustainable Tourism* 3(1), 29–44.

Sykes, L. (1996) Small is beautiful. *The Geographical Magazine* 68(3), 36–38.

Tolba, M.K. and El-Kholy, O.A. (1992) *The World Environment 1972–1992: Two*

Decades of Challenge. Chapman and Hall for the United Nations Environment Programme, London.

Tourism Canada (1990) *An Action Strategy for Sustainable Tourism Development: Globe '90.* Tourism Canada, Ottawa.

Troyer, W. (1992) *The Green Partnership Guide.* Canadian Pacific Hotels and Resorts, Toronto.

Turner, K., Pearce, D.W. and Bateman, I. (1994) *Environmental Economics*, Harvester Wheatsheaf, Hemel Hempstead.

UK Government (1994) *Sustainable Development: The UK Strategy.* Cmnd. 2426, HMSO, London.

United Nations Conference on Environment and Development (1992) *Agenda 21: A Guide to the United Nations Conference on Environment and Development.* UN Publications Service, Geneva.

United Nations Environment Programme: Industry and Environment (1995) *Environmental Codes of Conduct for Tourism.* UNEP, Paris.

United Nations Environment Programme/United Nations Joint Group of Experts on the Scientific Aspects of Marine Pollution (1986) *Environmental Capacity: An Approach to Marine Pollution Prevention.* UNEP Regional Seas Reports and Studies, No. 86, UNEP, Geneva.

Weiszaecker, E.U. von (1989) *Erdpolitik: Oekologische Realpolitik an der Schwelle zum Jahrhundert der Umwelt.* Wissenschaftliche Buchgesellschaft, Darmstadt.

Wheeller, B. (1994) Eco-tourism, sustainable tourism and the environment – a symbiotic, symbolic or shambolic relationship. In: Seaton, A.V. *et al.* (eds) *Tourism: The State of the Art.* John Wiley, Chichester, pp. 647–654.

Wight, P. (1994a) The greening of the hospitality industry: economic and environmental good sense. In: Seaton, A.V. *et al.* (eds) *Tourism: The State of the Art.* John Wiley, Chichester, pp. 665–674.

Wight, P.A. (1994b) Limits of acceptable change: a recreational-tourism tool in cumulative effects assessment. Paper presented at *Cumulative Effects in Canada: from Concept to Practice*, National Conference of the Alberta Society of Professional Biologists and the Canadian Society of Environmental Biologists, Calgary, Canada, 13–14 April.

Winpenny, J.T. (1991) *Values for the Environment: A Guide to Economic Appraisal.* HMSO, for Overseas Development Institute, London.

Winpenny, J.T. (1995) *The Economic Appraisal for Environmental Projects and Policies: A Practical Guide.* Organization for Economic Cooperation and Development, Paris.

World Commission on Environment and Development (1987) *Our Common Future.* Oxford University Press, Oxford.

World Travel and Tourism Council (1991) *WTTC Policy: Environmental Principles.* WTTC, Brussels.

World Tourism Organization (1995) Sustainable tourism development. Background paper by WTO Secretariat at the WTO Asian Tourism Conference, Technical Session, Islamabad, Pakistan, 13 January. WTO, Madrid, pp. 1–11.

World Travel and Tourism Council (1994) *Green Globe: An Invitation to Join.* WTTC, Brussels.

The Effectiveness of Sustainable Tourism Policies in Scotland

20

R. MacLellan

The Scottish Hotel School, University of Strathclyde, Curren Building, 94 Cathedral Street, Glasgow G4 0LG, UK

Introduction

In this chapter the author will attempt to address the question of whether sustainability is the best approach to common problems relating to tourism and the environment by examining some issues relating to sustainable tourism policies, in particular, definition and measurement difficulties. Examples will focus primarily on those initiated by government in the form of the Scottish Office and its agencies, in particular Scottish Natural Heritage and the Scottish Tourist Board.

A number of questions will be raised in this chapter.

1. Can an understanding and common agreement on the meaning of sustainable tourism and sustainability be reached and is this important?
2. Can the policies pre-dating the adoption of sustainable tourism or alternative policies to sustainable tourism be identified in order to make an objective judgement on levels of success?
3. Is it possible to measure changes (positive or negative) resulting from the adoption of sustainable tourism policies by the official agencies in Scotland, given the relatively short duration and wide scope of these policies?

In attempting to address these questions the chapter will touch on two critical issues in the examination of sustainable tourism. The first is the political and policy making context, much ignored in the literature on sustainable tourism (Hall, 1994). The second is the issue of monitoring and measuring the effectiveness of sustainable tourism policies. Thus the policies initiated by the

Scottish Office in the late 1980s, under the banner of sustainability, will be evaluated in terms of their effectiveness in achieving sustainable tourism objectives. The discussion will use two illustrative case studies: at national level the reorganization of conservation agencies and the debate over land use designations, in particular national parks; and at the local level the Tourism Management Programmes (TMP) set up under the Tourism and the Environment Initiative (Tourism and Environment Task Force, 1993a).

Tourism and Sustainable Development

The issues relating to sustainable tourism and sustainable development have been debated in the literature extensively. However, there are still a lack of agreement and clarity over definitions and a lack of empirical data on implementation and measuring effectiveness.

> Although a great deal has been written about sustainable development much of it is still highly theoretical or speculative and it is not yet clear how it can be put into practice.
>
> (Scottish Natural Heritage, 1993, p. 6)

This has been recognized by a number of authors, where the initial euphoric reception and adoption of sustainable principles has now been tempered by more sceptical critiques resulting from closer examination of the practicalities of implementation. The basic simplicity of the original definition of sustainable development in the Brundland Report (World Commission on Environment and Development, 1987), with its almost universal acceptance by governments, of all political shades, industry and the general public, has now spawned a vast range of refinements, applications and policies. These are often only remotely linked to the original definition, often contradict each other and are frequently impossible to apply in practice or measure effectively. It is therefore little wonder that the creeping scepticism in the sustainable development debate has occurred.

The confusion is compounded when sustainable development principles are applied to an activity as misunderstood as tourism. Thus one swings from the optimistic interpretation of sustainable tourism, for example Lane (1994), who views sustainable tourism as a triangular relationship between local people and their environment, the visitors and the tourism industry which minimizes environmental and cultural damage, optimizes visitor satisfactions and maximizes long term economic growth, to the more pessimistic views of Wheeller, who criticizes the confusion caused by the sustainable tourism debate, which he feels has provided no answers,

> just a never-ending series of laughable codes of ethics: codes of ethics for travellers; codes of ethics for tourists, for government and for tourism businesses. Codes for all – or more likely, codeine for all.
>
> (Wheeller, 1994, p. 651).

The breadth of acceptance of sustainable tourism is based less on real consensus and more on confusion, misunderstanding or pure exploitation for promotional purposes.

> The inherent vagueness of sustainability is its great weakness. At present it is being used by both industry and the conservation movement to legitimise and justify their existing activities and policies although, in many instances, they are mutually exclusive.
>
> (McKercher, 1993, p. 131).

Butler attempts to bring some clarity to the debate, pointing out the need to draw a distinction between sustainable development in the context of tourism, where tourism remains viable over an indefinite period and does not degrade or alter the environment, and sustainable tourism, which is thought of as 'tourism which is in a form which can maintain its viability in an area for an indefinite period of time' (Butler, 1993, p. 29). Thus, in Butler's working definition, sustainable tourism makes reference to environmental impacts only where degradation affects the viability of the industry (Butler, 1993).

As the brief discussion above indicates, many argue that sustainable tourism as a concept is fundamentally misguided. However, it is in the nature of the sustainable tourism paradigm that its worth can only be judged in the long term. For the purposes of this chapter, the lack of clarity and the contradictions in the relationship between sustainable development and tourism may be viewed as contextual background to sustainable tourism policies implemented in Scotland, for many of the difficulties in examining the success rate of these policies derives from the above confusion. Nevertheless, most would agree that there is some common ground in definitional terms and positive attributes to the sustainable tourism paradigm. As Bramwell and Lane point out:

> One must ask what the alternatives are to developing more sustainable tourism - presumably either to stand back and do nothing or else to criticise without offering any realistic, practical way forward.
>
> (Bramwell and Lane, 1993, p. 3).

Rather than attempt a universally acceptable definition of sustainable development or sustainable tourism, it is more helpful to narrow the focus, as the breadth of debate surrounding sustainability is one of the key dangers. Even within the confines of sustainability and tourism the field of study has formed a variety of discrete subsections, focusing on specific disciplines, sectors, locations or issues (MacLellan, 1994a, p. 636).

In fact, this breadth is a by product of the philosophy of sustainable development and is one part of the definition which has common agreement: namely, the concept of holistic planning and strategy making. Issues previously seen as separate, such as ozone depletion, waste production, water pollution, soil erosion, social and cultural impacts and landscape degradation, are now considered together in the sustainable tourism debate. This integration of issues has been represented in a variety of models by agencies and authors (World Tourism Organization, 1993; Lane, 1994; Muller, 1994) and the complexity and difficulties in implementing these explains, perhaps, the state we find ourselves in at present. To some extent the measure of sustainability has shifted to measuring the effectiveness of organizational systems rather than their outcomes. Nelson, for example, views the most pressing research need in respect to tourism and sustainability as 'studies of how people have worked and can work together in tourism and other industries or activities in a pluralist and sharing manner' (Nelson, 1993, p. 260). Thus the level of cooperation, agreement and common purpose is measured rather than the degree to which this leads to forms of sustainable development. This will be considered later in this chapter in the context of measuring the effectiveness of sustainable tourism in Scotland.

For the remainder of this chapter the focus will be on public policy responses to sustainable tourism, in particular in Scotland and the related issue of how to measure the effectiveness of these policies: What has been achieved in Scotland? Are the so called 'sustainable tourism' policies more or less effective in achieving sustainable development than the ones they replaced? How can this be evaluated?

Public Policy and Sustainable Tourism

The above confusion in the debate over sustainable development has led to some very broad interpretations (or abuses) of the term by governments. The universal agreement at Rio 1992 on the principles of sustainable development (United Nations Conference on Environment and Development, 1993) by governments of all political shades should have sounded warning bells. Not only do all politicians, but now all large corporations, claim to be pursuing pro-environmental policies. Waldstein (1991) terms this 'limousine environmentalism', which has become common in the USA, where lipservice masks a lack of positive action. This reaction of governments to pressures to adopt sustainable development policies may be put in the context of modern industrial societies' ability to resist change. Dovers and Handmer (1993) point out that, in many cases, sustainable tourism policies give the appearance of significant change in attitude while in reality they make little impact on underlying trends and institutional structures. In other words, governments have become adept at devising tactics which produce 'changes at the margins' or 'fine tuning', rather than making fundamental policy shifts. This aspect of sustainable tourism policies is often ignored, which is hardly surprising, given the lack of attention given to the political dimensions of tourism in general (Hall, 1994).

When national tourism policies are examined it becomes obvious they remain overwhelmingly geared to the generation of economic growth. Since the introduction of the Development of Tourism Act 1969, successive UK governments have viewed tourism in terms of economic growth, balance of payments, employment and regional balance. Where other non-economic policy objectives are included they tend to be given a lower priority. As Hall points out:

> ... in the case of Australian tourism policy, several commentators have argued that economic goals are given a far higher priority than social and environmental concerns in state and national governments' tourism policy agenda.
>
> (Hall, 1994, p. 114)

A second misleading area in governments' attitude to sustainable tourism policies is their emphasis on the 'holistic' approach, mentioned earlier in this chapter. Clearly there is much of value in including all interest groups in formulating sustainable tourism policies and much has been written on this subject, in particular relating to local community involvement (Murphy, 1995). However, there are clearly some pitfalls in placing the issue of cooperation, joint working and pluralism too high on the policy-making agenda.

First, it should be pointed out that this aspect is only one of a range of sustainability indicators. Cooperation and local community involvement does not necessarily lead to sustainable development. There are many instances of misguided policies which have widespread support, in particular from local groups, which lead to unsustainable tourism, for example the Aviemore Centre, in the Cairngorm Mountains of Scotland.

Secondly, including all interest groups and issues in the tourism planning process does not always mean equal representation. The uneven distribution of power in a community, in private sector decision making and in national politics should be apparent to all. In the context of sustainable tourism policies, why should this be any different? 'In tourism planning and policy making it is inequality rather than equality that is the order of the day' (Hall and Jenkins, 1995, p. 77). Participation in the decision-making process does not necessarily influence policy outcomes. Haywood argues that in many cases, public participation in tourism planning may be more a form of placation than a means of giving power to communities to form their own decisions: '... the public participants will be seen as a token body, and the participation as a hollow exercise' (Haywood, 1988, p. 109). In this way, interest groups may be marginalized or placated. It could be argued that the conservation movement in Scotland has been consulted in the plans for Loch Lomond and the Cairngorms, more than ever before. In this way, they have been brought into the 'sustainable tourism' fold (in a vague sense), their power base has been undermined by a government that appears to be 'green', and the plans continue to place the greatest priority on the commercial aspects of sustainable tourism.

The third concern over public policy emphasis on the holistic aspects of sustainable tourism is that, in most cases, this is addressing an organizational issue in tourism rather than an environmental one. The inefficient, fragmented nature of the tourism industry has long been recognized and has been particularly problematic in Scotland. However, this has been addressed through the reorganization of institutional structures in Scottish tourism and the formulation of a new strategic plan (Scottish Tourist Board, 1994). Sustainable tourism policies may be part of this strategy but should go beyond this and be seen as having objectives other than purely organizational. The ability of sustainable tourism policies to bring interest groups together is only one indication of their success. The problem is that in many cases (examined later in the Scottish context) this is given the highest priority and is the only aspect currently with sufficient measurable evidence.

Sustainable Tourism and Government Policy in Scotland

The official agencies reporting to the Scottish Office may be viewed as the key policy-making organizations in Scotland relating to tourism and the environment. Although local authorities play a critical role in the tourism planning process, their powers have been reduced in contrast to the powers of the quangos in Scotland. The 1990s have seen the inclusion of sustainable development terminology and indications of the philosophy of sustainability in policy documents and initiatives emanating from statutory agencies, in particular the Scottish Tourist Board, Scottish Natural Heritage (Countryside Commission for Scotland and Nature Conservatory Council for Scotland to 1992), Scottish Enterprise and Highlands and Islands Enterprise, Historic Scotland and the Industry Department and Environment Department of the Scottish Office itself. In many instances these sustainable development policies have been in the form of joint initiatives established by voluntary groupings. The Scottish Tourism Coordinating Group (STCG) includes all the above agencies and was set up with, as part of its remit, the promotion of sustainable tourism in Scotland.

The inclusion of sustainable terminology in statutory agency literature and the adoption of sustainable tourism principles came at an interesting, some would argue opportune, time for the politicians planning tourism in Scotland. A number of difficulties were apparent in the management of tourism activities in the more fragile natural environments of Scotland. Long standing disputes existed over the development of tourist facilities in remote areas, involving a diverse range of interest groups and particularly the statutory agencies, which frequently adopted opposing views. The classic conflicts between conservation and development agencies were further complicated in the Scottish context by ambiguous access laws, powerful landowner groups and increased tourism demand with specific infra-structural requirements, such as skiing.

The introduction of sustainable policies also came at the end of a lengthy period of consultation into the management of mountain areas in Scotland (Countryside Commission for Scotland, 1990), which finally recommended the establishment of four national parks. This issue will be examined in greater detail later in the chapter. A final strand, which has further muddied the waters into which sustainable tourism policies dropped, has been the process of reorganizing tourist board and local authority structures over much of the early 1990s, resulting in the Area Tourist Boards network being reduced from 36 to 14 in April 1996. This was in response to the perceived inefficiencies of the previous systems and the relatively poor performance of the Scottish tourism sector in general. The reorganization of local goven-ment, where some elements of Agenda 21 were adopted, including the development of sustainability indicators, further added to the confusion.

Therefore the performance of sustainable tourism initiatives in Scotland must be judged by bearing in mind the complexity of organizational evolution and the political context. It is hardly surprising that it is difficult to discern a clear sustainable development (or sustainable tourism) policy for Scotland, as the initiatives have evolved and shifted in response to specific issues and in line with specific agency objectives. Focusing on two strands may illustrate this point and shed some light on the effectiveness of sustainable tourism policies. First, the work of the Scottish Tourism Coordinating Group (STCG) and related initiatives and, second, the work of Scottish Natural Heritage (SNH) are examined.

Scottish tourism sustainable initiatives

In response to criticism of the factional, fragmented nature of the Scottish tourism organizational structure, the Scottish Office established the STCG. The emphasis was on cooperation, partnership and joint initiatives, which matched the holistic side of sustainable development principles. In 1992, as part of their remit to promote sustainable tourism, the Scottish Tourist Board (STB), on behalf of the group, published their first report, *Tourism and the Scottish Environment: A Sustainable Partnership* (Scottish Tourism Co-ordinating Group, 1992). The report, in presentational terms, was clearly designed to promote, to a wide audience, the message of sustainable tourism. The benefits of tourism to Scotland were presented in economic terms followed by a detailed review of the environmental costs of tourism in Scotland. The solution to reducing these costs was presented as 'the adoption of sustainable tourism aims and priorities' (p. 22). These were to be identified and implemented through the establishment of a 'Tourism Management Initiative' (TMI) linking national priorities with local projects, which should follow the objectives of sustainable tourism. However, the first practical difficulty arose in the provision of funding for these projects. As Hughes points out:

> ...much of the initial deliberations were premised on the need for additional financial support from central government to provide the incentive for local participation in this initiative.
>
> (Hughes, 1995, p. 54).

This financial support was not forthcoming and the coordinating group reconvened to form the Tourism and Environment Task Force (TETF) late in 1992, with two slightly watered down objectives: (i) to promote an increased awareness of the environment within the tourism industry; and (ii) to draw up guidelines, based on the philosophy of sustainability, to be implemented in projects at local level by tourism representatives from the public and private sectors. The first objective led to the production of the report *Going Green: Guidelines for the Scottish Tourism Industry* in June 1993 (Tourism and Environment Task Force, 1993b). Like many similar sustainable tourism guidelines, this report is open to criticism for being public relations led, responding to media attacks on tourism, rather than voicing environmental concerns. However, unlike the earlier English Tourist Board (1991) report *The Green Light: a Guide to Sustainable Tourism*, described in scathing terms by Ashworth (1992, p. 326) Is this better than nothing or worse than useless?', *Going Green* does at least recognize tourism can harm the environment: 'The environment is Scotland's major asset. It must be treated, developed and promoted in a sympathetic and sustainable manner' (Tourism and Environment Task Force, 1993b, p. 7).

In addition, the report offers some practical guidance for industry, unlike many previous sustainable tourism guidelines which lost credibility by suggesting idealistic, naive proposals or 'platitudinous points, which sounded like a converted and over-virtuous boy scout promise' (Ashcroft, 1993, p. 146).

It is a pity these practical guidelines tended to emphasize developing 'the green product' and creating 'a green image' in response to the increase in 'the green market'. Although some progress has been made in raising awareness of green issues in mainstream tourism sectors, in particular large hotels and transport operators, and the development of niche green products, which may be environmentally friendly, for example dolphin watching, it is questionable whether many of these measures necessarily lead to more environmentally friendly, sustainable, forms of tourism. In addition, there is no evidence, to the author's knowledge, of attempts to measure or monitor the effectiveness of these guidelines.

The second objective of the task force led to more substantial developments. The consultative process on local sustainable tourism projects resulted in the publication of the third report (confusingly also in June 1993): *Tourism and the Scottish Environment: Tourism Management Initiative.*

Guidelines for the Development of Tourism Management Programmes (Tourism and Environment Task Force, 1993a).

The report offered examples of appropriate approaches to sustainable development but concentrated primarily on the stages necessary to establish a 'Tourism Management Programme' (TMP). TMPs are viewed as 'a way of piloting sustainable tourism to benefit the visitor, the place and the host community' (Bryden, 1995, p. 4). Their essential characteristics relate to a multifaceted planning approach, which includes local preparation and implementation, a defined geographical area and management through partnership arrangements. The performance of individual TMPs, to date, will be discussed later in this chapter.

As Hughes (1995, p. 54) points out, guidelines on sustainable tourism had been reduced to organizational procedures 'more administrative in their thrust than substantive'. In addition, TMPs receive no additional financial support from central government. TMPs are funded by reprioritizing the existing budgets of partners, primarily relevant statutory agencies, local authorities and, to a limited extent, the private sector.

Scottish Natural Heritage

Although the responsibilities of SNH go far beyond tourism it is never-theless the key Scottish agency for managing tourism activities in the countryside. More importantly it is an example of a government agency's attempt to put sustainable development into practice.

> Given our broad remit and the fact that our founding legislation requires us to consider the sustainability of developments in Scotland, SNH is potentially in a very influential position.
>
> (Scottish Natural Heritage, 1993, p. 23).

The question is, have they used, or been given the powers to use, this influence effectively?

The timing of the creation of SNH is worthy of note. The Scottish Office had a number of related issues to resolve at this time. Focusing on the three most relevant may provide an insight into the gestation period of the setting up of SNH. First, the divisions between the two key Scottish conservation agencies Countryside Commission for Scotland (CCS) and Nature Con-servancy Council for Scotland (NCCS) had become acute: 'nature was being protected from people, rather than for people' was how some viewed the role of the NCCS (Arnott, 1993, p. 73). Secondly, the CCS had completed a lengthy consultation exercise on the management of mountain areas, culminating in the recommendation that four national parks be created (Countryside Commission for Scotland, 1990). This was a highly con-troversial issue in the Scottish context, which will be examined in more detail

later. Finally, as previously discussed, tourism organizations were undergoing a period of restructuring, linked to the promotion of sustainable
tourism, led by the STCG. In this way the creation of SNH could satisfy a
number of Scottish Office objectives under the flexible term, or its flexible
interpretation of the term, 'sustainable development'. The emphasis, again, is
on the organizational aspects of bringing interdependent bodies together,
resolving conflicting issues and forming partnerships rather than addressing
substantial matters such as environmental protection.

Although only four years old, SNH has come under considerable
criticism regarding its structure and its success rate in implementing
sustainable development policies. In terms of SNH structure, the merging of
two such disparate bodies as the CCS and NCCS, as Arnott (1993, p. 73) puts
it, 'slowly bridging the division between the scientific and the aesthetic by a
more holistic view', success seems to have been partial. However, simply
joining the organizations has not resolved the underlying tensions in their
predecessors' objectives, such as those between recreation and access (CCS)
and land management (NCCS), which are as prevalent as ever. The
experimental SNH model, it seems, is not one the government wishes to
adopt south of the border. The Department of the Environment ruled out the
establishment of a similar body for England:

> ... after initial signs that the same principles would be applied in England,
> the evidence is that the Government was not sufficiently persuaded of
> the success of SNH to spend money setting up a similar body.
>
> (Herald, 1994, p. 20)

One of the key criticisms of SNH has been the lack of clear policies on issues
such as access reforms, the management of recreational activities and
developing tourism facilities in fragile environments. This lack of leadership
may be linked to their interpretation of sustainable development principles
which emphasize cooperation, consultation and consensus.

> The main inherent weakness of the organisation is that it has to rely on the
> voluntary principle, which is simply not effective enough to deal with
> increasingly urgent problems ...
>
> (Herald, 1994, p. 20)

SNH is strong on sustainable development philosophical objectives but weak
on statutory powers. Great emphasis has been given to statements such as
'enabling enjoyment of the natural heritage in a manner which is sustainable'
and pride is (or was) taken in the fact that 'our founding legislation enshrines
the word sustainable for the first time in UK legislation' (Scottish Natural
Heritage, 1992, p. 3).

Before examining two examples where 'sustainable development' policies suffer from lack of statutory powers, it is worth re-emphasizing the
importance of the political context. On policy-making, four political and
economic considerations have arguably been much more influential than any
sustainable tourism philosophies.

1. Policy making from the Scottish Office in the 1990s has been driven by the search for cuts in public sector spending (Bryden, 1995).
2. The current government has a philosophical aversion to legislation which controls or restricts the operation of the free market.
3. The large estate owners, partly through their influence on the board of SNH, continue to wield a powerful influence on land management decisions in Scotland.
4. Economic development remains the foremost consideration in the sustainable tourism policies of the key agencies in Scotland, whereas exploiting environmental assets is encouraged: 'A growing interest in the environment, natural history and remote places has given rise to a demand for so called "green holidays". Scotland is well placed to serve this demand' (Scottish Tourist Board, 1994, p. 31).

The following two examples illustrate the importance of political context in addition to the difficulty in applying sustainable tourism policies in practice and the even greater difficulty in accurately measuring the effectiveness of these policies.

National parks and sustainable tourism policies in Scotland

A crucial test of the sustainable development policies of a country is how it deals with the protection of its most valued environmental assets. In Scotland, the management of upland areas has been the focus of much conflict, in particular in locations under intensive tourism pressures such as the Cairngorms and Loch Lomond area. It is the resolution of these conflicts that has put the policies of the Scottish Office and the role of SNH under the microscope. As leading countryside managers noted:

> The areas in question (Loch Lomond and Cairngorms) are the finest pieces of Scotland's natural heritage which deserve the best stewardship and protection the nation can provide. If decisive action for their proper stewardship and protection is not taken soon, their sustainability must remain in doubt ...
>
> (Edwards et al., 1993, p. 150)

The debate over the creation of national parks goes back to the last century and has been well documented by a number of authors, although the reasons for their rejection remain shrouded in the mists of past politics (Moir, 1991; MacLellan, 1993). National parks, in their various forms, are by no means perfect as mechanisms for managing valued natural environments. Nevertheless, they were recommended by the statutory agency responsible, after lengthy consultation, as the best means to resolve the, at times, acrimonious conflicts in Scottish mountain areas (Countryside Commission for Scotland, 1990). The recommendation that four national parks be established in Scotland had, according to opinion polls, overwhelming public support:

> In the Scottish Office survey of December 1991 when asked if they thought that some parts of Scotland should be set aside as National Parks so that they are protected from development, 90% said 'yes'.
> (Arnott, 1993, p. 69)

The Scottish Office response to the CCS report was complicated by a range of counter proposals and initiatives and it is here that policies with sustainable terminology begin to appear. Not only did this period mark the launch of the STCG initiatives, as described earlier in this chapter, it also saw the Scottish Minister for the Environment announce three new related policy objectives: (i) the intention to seek World Heritage Site listing for the Cairngorms was announced; (ii) that a new working party be set up to look into the management of the Cairngorms (chaired by Magnus Magnusson, now chairman of SNH); and (iii) a new conservation designation, Natural Heritage Areas, was launched. The final complication came with the creation of SNH in 1991. Not only did this effectively countermand the recommendation to designate national parks but also the Act creating SNH did not include enabling legislation to deal with Scotland's mountain areas, as suggested by the CCS in 1990 (Barron, 1992).

The link between rejecting national parks and sustainable principles is seen in the literature produced by the individual working parties, SNH and the Scottish Office. The philosophy of sustainable development was adapted to mean less controls and regulations and more cooperation, voluntary agreements and partnership boards.

> The explosion of worthy publications must be examined in the context of an absence of government action in resolving how to adequately protect fragile mountain environments. These laudable sentiments on sustainable tourism do not offer solutions to the problems identified in the CCS Mountain Areas report.
> (MacLellan, 1994b, p. 12)

It is ironic that the same government (albeit a different department) reiterated, in the strongest terms, its commitment to national parks in England and Wales, viewing them as 'models for the sustainable management of the wider countryside' (Scottish Office, 1993, p. 175).

It seemed that sustainable tourism rhetoric provided the government with an opportunity to provide a 'green gloss' to their policies whilst giving priority to underlying political and economic objectives, in particular, cuts in public spending. As in the past, safeguarding the Scottish environment was compromised by a government unwilling to enter into a conflict with landowning interests or make major financial commitments (Scott, 1992).

The sustainable development credentials of these policies must be judged against international criticism, such as a World Conservation Union report which condemned Scotland for 'operating one of the weakest management arrangements for vulnerable areas in Europe' (Edwards, 1993, p. 6). The weakness of the current management arrangements have been highlighted recently in the Cairngorms, where the partnership board lacks the powers and finance to withstand pressures by commercial interests to build a funicular railway up the face of Cairngorm. It is depressing to see the same interest groups, including statutory agencies, clash over the same issues as they did 15 years ago in the classic conflict over proposed developments in Lurcher's Gulley. 'The fundamental issues are the same – protection of fragile mountain habitat and the rare species that depend on it, versus economic expansion' (Buie, 1996, p. 15).

Rather than resolving these issues through a national park authority, with strong planning powers and funding from the public purse, reliance is to be placed on the 'voluntary principle', 'a partnership board' and 'sustainable development guidelines', all of which do nothing to clarify or resolve an already complex and acute situation.

Tourism management programmes

The background to setting up the TMPs in Scotland under the TETF was referred to earlier. The more manageable scale of TMPs, as they are locally prepared and implemented, should, in theory, provide more hard evidence on which to judge the performance of sustainable tourism policies. In addition, at the outset, their plans are structured to include monitoring and measurement procedures to

> enable the Task Force to establish to what extent the joint objectives of the Initiative are being met, i.e. the promotion of tourism and the conservation of the environment on which it depends.
>
> (Tourism and Enviroment Task Force, 1993a, p. 7)

There are currently around ten TMPs operating in Scotland with several more being proposed. Therefore, to this extent the national initiative seems successful. Although they are in an early stage of their evolution, with the longest-running being around three years old, for example in St Andrews, some monitoring of the progress of TMPs has already been attempted. However, embarking on premature monitoring of something as long term as sustainable tourism is not the only problem encountered to date. The second fundamental obstacle to devising monitoring procedures for TMPs lies in the breadth, variety and number of their objectives. As each TMP is based on the holistic approach to planning, cooperation and local partnerships are essential. This often involves agencies with almost incompatible statutory objectives and local interest groups with priorities poles apart. The establishment of these TMPs should be viewed as a positive indicator of sustainable tourism policies; yet it creates inherent difficulties in setting clear objectives with measurable outcomes and leads to tensions within TMP boards. At worst, monitoring would be determined by the TMP member with most power (normally funding), which could lead to effectiveness being measured in the one-dimensional term of commercial performance. The third monitoring difficulty identified is the lack of tested methodologies for assessing programmes as diverse as these. Nelson, considering the monitoring of tourism in the context of sustainable development, points out:

> the problem is what and how to monitor, in an efficient and effective way, with only general concepts or criteria, such as economic health, diversity, productivity, maintenance of essential processes and equity in mind. What data and criteria should be used, by whom, according to what scheme and why?
>
> (Nelson, 1993, p. 261)

A full examination of monitoring procedures is outside the scope of this chapter. However, the TMPs in Scotland illustrate some practical difficulties which must be overcome if sustainable tourism policies are to be effectively implemented.

Giving two, perhaps extreme, examples is apposite. One focuses on the physical environment, the other on economic indicators, they illustrate some practical difficulties. The Trossachs TMP commissioned a study to propose 'a monitoring system to address the issue of sustainability in tourism development in the Trossachs' (Dargie et al., 1994, p. 1). The result was a detailed, comprehensive report recommending baseline surveys, a review of key sites, detailed monitoring procedures based on the principle of 'limits of acceptable change' and recommendations to enlarge the programme to include wider social and economic impacts. The recommendations of this report, excellent in an academic sense, are unlikely to be implemented in full on the grounds of being very time-consuming, costly and perhaps restrictive of further tourism development. Indications are that they will be replaced by a 'simpler' system. Although the proposed Trossachs monitoring system is valuable as an ideal for which to strive, in evaluating visitor impacts on natural environments, it fails to address the commercial and political realities of operating TMPs. The danger here is that the cruder but cheaper monitoring may not pick up changes until after irreparable damage has been done.

The St Andrews and Pitlochry TMPs illustrate two schemes based more on the built heritage and with even more pressing time and budgetary constraints built into their operations. Both have made good progress in terms of coordinating tourism activities and initiating environmental improvement schemes. However, their funding is limited and, with little long term security, they are under pressure to evaluate their performance (or justify their existence) on an almost annual basis. Rather than providing meaningful indicators such as physical environmental, economic or social change over time, they are forced into producing lists of activities initiated in the short term. Gathering reliable benchmark data from which to measure change in the future takes time and funding. One-off visitor or community surveys offer little information of the changes which are occurring over time.

The short term economic pressures on the operation of the TMPs may lead to the environmental protection element being lost in an attempt to show improvements in economic performance. Therefore, monitoring becomes reduced to measuring 'the feel good factor' of the business community or ad hoc visitor surveys, notoriously fickle and unreliable, even as short-term indicators of tourism performance.

However, once identified, these short-term pressures may be guarded against as there are many positive indicators of TMP performance. For example, Hughes (1995, p. 56) identified the notion of synergy as a key contribution of TMPs where multi-agency partnerships allowed 'capital rich agencies to coordinate with revenue spenders to design mutually supportive programmes'. They have also provided a means through which conservation (SNH) and development agencies (Local Enterprise Companies (LECs)) can possibly work together with the private sector, voluntary sector and communities to achieve environmental improvements, for example in the Skye and Lochalsh Footpath Scheme. The national TMP has also had some success in raising general awareness of the need to protect the environment for and from tourism. Whether this heightened awareness translates into positive action is another question.

It is a pity that these positive aspects of TMPs are being undermined by some political manipulation of sustainable tourism ideals and by short termism. The supporters of these initiatives, at local and national level, need to be patient and commit themselves to longer term investments if hard reliable evidence of the effectiveness of sustainable tourism policies is ever to emerge.

Conclusions

Measuring both positive and negative changes resulting from the adoption of sustainable tourism policies by official agencies has been problematic given their short duration and breadth. Some indications at present seem positive but impossible to quantify, such as the success of awareness-raising campaigns at national or local levels and the progress made in fostering cooperation and partnerships in decision-making affecting tourism in the Scottish environment. In reality, Scotland is still in the early stages of spreading the sustainable tourism message and implementing local pilot projects. One of the critical concerns, at present, is that this message remains confused and contradictory and that some local projects (TMPs) are showing signs of losing direction.

At national level, the adoption of sustainable tourism policies must be viewed in the wider political context of current public spending cuts, a desire to reduce regulations and the existence of powerful landowning interests. Accordingly one should look critically at the motivations of the Scottish Office. Have some sustainable policies been counterproductive – for example, where pursuit of sustainable (viable) tourism equals weaker environmental protection or where conservation is only valued if it results in more tourists or where sustainable tourism is simply viewed as a business 'feel good factor'?

Whether deliberate or not, the Scottish Office have gone some way, through promoting sustainable tourism initiatives, towards deflecting media criticism and the environmental lobby, whilst implementing spending cuts and weakening conservation measures. It could be argued that without sustainable policies to hide behind, the Scottish Office would have been more likely to give in to pressures for the establishment of national parks, with all the management and financial benefits afforded those in England and Wales. Although national parks are perhaps not a perfect solution, it is surely a tried and tested one, with more clarity of purpose than embryonic, experimental sustainable tourism ideals. Therefore, clarifying sustainable tourism objectives and providing hard data on the effectiveness of sustainable tourism policies will go some way towards avoiding criticism of political obfuscation in future.

References

Arnott, J. (1993) Issues facing Scotland. In: Fladmark, J.M. (ed.) *Heritage*. Donhead Publishing, London, pp. 63–76.

Ashcroft, P. (1993) One hundred and one dull machinations. *Journal of Sustainable Tourism* 1(2), 146.

Ashworth, G.J. (1992) Planning for sustainable tourism. *Town Planning Review* 63(3), 325–329.

Barron, S. (1992) Conservation agency for Scotland. *The Planner* April, 10–11.

Bramwell, B. and Lane, B. (1993) Sustainable tourism: an evolving global approach. *Journal of Sustainable Tourism* 1(1), 1–5.

Bryden, D. (1995) Tourism and the environment – maintaining the balance. In: *Proceedings of the Association for the Protection of Rural Scotland (APRS) Conference – Tourism and Recreation: A Sustainable Approach for Rural Areas*. APRS, Edinburgh, pp. 1–6.

Buie, E. (1996) Making tracks on Cairngorms. *Herald*, 30th January, p. 15.

Butler, R.W. (1993) Tourism – an evolutionary perspective. In: Nelson, J.G. *et al.* (eds) *Tourism and Sustainable Development: Monitoring, Planning, Managing*. University of Waterloo, Ontario, pp. 27–43.

Countryside Commission for Scotland (1990) *The Mountain Areas of Scotland*. CCS, Edinburgh.

Dargie, T., Aitken, R. and Tantram, D. (1994) *Trossachs Tourism Management Programme: Environmental Monitoring*. Highlands and Islands Enterprise, Inverness.

Dovers, S.R. and Handmer, J.W. (1993) Contradictions in sustainability. *Environmental Conservation* 20(3), 217–222.

Edwards, R. (1993) Tourists could destroy loch, report warns. *Scotland on Sunday*, 11th July, p. 6.

Edwards, T., Pennington, N. and Starrett, M. (1993) The Scottish parks system: a strategy for conservation and enjoyment. In: Fladmark, J.M. (ed.) *Heritage*. Donhead Publishing, London, pp. 141–151.

English Tourist Board (1991) *The Green Light: a Guide to Sustainable Tourism*. ETB, London.

Hall, C.M. (1994) *Tourism and Politics, Power and Place*. Belhaven Press, London.

Hall, C.M. and Jenkins, J.M. (1995) *Tourism and Public Policy*. Routledge, London.

Haywood, K.M. (1988) Responsible and responsive tourism planning in the community. *Tourism Management* 9, 105–118.

Herald (1994) Environmentally ineffective. *Herald*, 11th October, p. 20.

Hughes, G. (1995) The cultural construction of sustainable tourism. *Tourism Management* 16, 49–59.

Lane, B. (1994) Sustainable rural tourism strategies: a tool for development and conservation. *Journal of Sustainable Tourism* 2(1 & 2), 102–111.

McKercher, R. (1993) The unrecognised threat to tourism: Can tourism survive sustainability? *Tourism Management* 14, 131–136.

MacLellan, L.R. (1993) Tourism and Scottish mountain environments: values and public policy. In: *Proceedings of Conference: Values and the Environment, University of Surrey*, September 1993.

MacLellan, L.R. (1994a) Tourism and the environment: introduction. In: Seaton, A.V. *et al.* (eds) *Tourism: The State of the Art*. Wiley and Sons, Chichester, pp. 635–637.

MacLellan, L.R. (1994b) The tourism/environment debate: sustainable realism, guideline idealism, eco tourism distractions and public policy cynicism. In: *Proceedings of Sustainable Tourism Think Tank Conference*, Portugal, November 1994. CEMP, University of Aberdeen.

Moir, J. (1991) National parks: north of the border. *Planning Outlook* 34(2), 61–67.

Muller, H. (1994) The thorny path to sustainable tourism development. *Journal of Sustainable Tourism* 2(3), 131–136.

Murphy, P.E. (1995) *Tourism: A Community Approach*. Methuen, London.

Nelson, J.G. (1993) Are tourism growth and sustainability objectives compatible? Civics, assessment, informed choice. In: Nelson, J.G. *et al.* (eds) *Tourism and Sustainable Development: Monitoring, Planning, Managing*. University of Waterloo, Ontario.

Scott, M. (1992) What future for the Cairngorms? *Ecos* 13(2), 16–23.

Scottish Natural Heritage (1992) *Working with Scotland's People to Care for our Natural Heritage*. Mission Statement and promotional brochure, SNH, Edinburgh.

Scottish Natural Heritage (1993) *Sustainable Development and the Natural Heritage. The Scottish Natural Heritage Approach*. SNH, Perth.

Scottish Office (1993) *Common Sense and Sustainability: A Partnership for the Cairngorms*. The Report of the Cairngorms Working Party, March, HMSO, London.

Scottish Tourist Board (1994) *Scottish Tourism Strategic Plan*. STB, Edinburgh.

Scottish Tourism Coordinating Group (1992) *Tourism and the Scottish Environment: A Sustainable Partnership*. Scottish Tourist Board, Edinburgh.

Tourism and Environment Task Force (1993a) *Tourism and the Scottish Environment, Tourism Management Initiative. Guidelines for the Development of Tourism Management Programmes*. Scottish Tourist Board, Edinburgh.

Tourism and Environment Task Force (1993b) *Going Green: Guidelines for the Scottish Tourism Industry*. Scottish Tourist Board, Edinburgh.

United Nations Conference on Environment and Development (1993) *The Earth Summit, Rio 1992*. Graham and Trustman, New York.

Waldstein, F.A. (1991) Environmental policy and politics. In: Davies, P.J. and Waldstein, F.A. (eds) *Political Issues in America*. Manchester University Press, Manchester.

Wheeller, B. (1994) Ego tourism, sustainable tourism and the environment – a symbiotic, symbolic or shambolic relationship? In: Seaton, A.V. *et al.* (eds) *Tourism: The State of the Art*. Wiley and Sons, Chichester, pp. 647–654.

World Commission on Environment and Development (1987) *Our Common Future*. Oxford University Press, Oxford.

World Tourism Organization (1993) *Sustainable Tourism Development: Guide for Local Planners*. WTO, Madrid.

The Community Show: A Mythology of Resident Responsive Tourism

21

G. Taylor and D. Davis

Department of Hospitality and Tourism, University of Central Lancashire, Preston PR1 2HE, UK

Introduction

Within the last ten years, the concept of community involvement in tourism development has moved discussion of the social environment nearer to the centre of the sustainability debate. In a decade of enterprise and self interest Murphy (1985) argues for the communal voice, for tourism planning to be part of the social consciousness of the destination. More recently, Getz and Jamal (1994) refer to the environment–community symbiosis and Brent Ritchie (1993) is convinced that resident responsive tourism is 'the watchword of tomorrow'. Even earlier, in one of his 'guidelines to aspiring destinations', D'Amore (1985) sees the involvement of local people as a necessity if they are to maintain their lifestyle and control the pace of change. But what if this lifestyle is becoming less communal? The desire for settlement, for a life where we are at ease with things, may be stronger than ever but in many areas of the world the 20th century has seen communities become dislocated and destroyed. The ideology of communitarianism and the rhetoric of community empowerment may appear persuasive but what if 'the community' is no more than an association characterized by individualism and competitiveness? Surely the creation of some touristic Utopia, sensitive to and driven by local desires, depends on a fundamentally cohesive community.

The notion of a shared vision of some sustainable future may be highly romanticized but what of the political correctness of community participation? As far as Lane (1995, p. 324) is concerned, he adds a further spin, 'the community route to development is academically and politically correct'.

The political lexicon contains a growing number of references to partnership and subsidiarity and the idea of involving and empowering the community has appeared in countless local plans and regional strategy documents in all parts of the democratic world. In Alberta (Canada) some 400 communities (out of 427) completed the Community Tourism Programme. Each community formed a Tourism Action Committee and, following a set of instructions contained in a manual, produced, in the 20 hours (maximum) allowed for the task, a Tourism Action Plan (Lane, 1995). While even the most committed supporter of spreading the tourist load might think this is going a little too far (and after all this the net result was local disappointment), the political agenda is all too transparent.

If the relationship between tourism development and the community dynamic is problematic and there are suspicions regarding the motivation for intervention, what of the ethics of local involvement? The Code of Ethics for the Canadian tourism industry calls for services and facilities that will contribute to 'community identity ... and the quality of life of [the] residents' (D'Amore, 1992, p. 261). The idea of a partnership with local residents is more implied than stated in the Code of Ethics but the responsibility of the developer to the community is clearly articulated. The problem here is that the residents of destination areas are 'seen increasingly as the nucleus of the tourism product' (Simmons, 1994, p. 98). For Murphy (1985) there is both the recognition of tourism as a community resource and the expectation that local people are part of 'the community show' (p. 169) and 'must act as hosts whether they are directly involved or not with the industry' (p. 138). Residents, however willing they might be, have to play a part in shaping future tourism development and at the same time be part of the attraction. If they are successful, they will not only help to increase visitor numbers but somehow control the level of impact; they will simultaneously become part of the problem and part of the solution.

The aim of this chapter, then, is to question some of the commonly held beliefs about resident responsiveness, to examine the apparent 'correctness' of community involvement, and to consider the role of the resident as perhaps an essential part of the tourism product.

Backstage

It goes without saying that resident attitudes to tourism development will vary both within a community and between one community and another. In Hong Kong, residents 'generally favour the growth of tourism,' (Mok *et al.*, 1991, p. 292) while a longitudinal study (over six years) in a Rocky Mountain community (Silver Valley) describes initial feeling towards tourism development to be 'very positive', but finds later 'that support has diminished over time' (Johnson *et al.*, 1994, p. 629).

In a recent study of local attitudes in cities in the United States and United Kingdom, Madrigal (1994) identifies three types of community clusters, tourism 'realists', 'haters', and 'lovers'. Only 13% of the total sample, the 'lovers', are convinced that the benefits of tourism outweigh any negative aspects while nearly one third (31%), the 'haters', felt that the reverse was true. Prentice (1993) argues, not surprisingly, that those who benefit most from tourism are more likely to favour its development, a view supported by a survey of 199 households in Fiji, where residents both depend on tourism and seek its expansion (King *et al.*, 1993). Research also points to the unequal sharing of economic benefits within a community. Madrigal (1994) notes that the political organization of a community is often 'dominated by individuals benefiting either directly or indirectly from a specific development alternative' (p. 87), and goes on to quote from the research of Canan and Hennessy (1989), who acknowledge that 'tourism development usually benefits only a small proportion of local residents' (p. 98). Ryan and Montgomery (1994, p. 368) observe that among the residents of Bakewell, a small town in Derbyshire, UK, there is a grudging recognition of inequality, which they summarize as 'others get the benefits, not me'.

Differences in opinion regarding the type and extent of development within a community may relate to the degree of economic dependence, though factors such as length of residence and native-born status can influence resident attitudes to a greater or lesser degree (Madrigal, 1994; Ryan and Montgomery, 1994). Discussion of the community dynamic is further complicated by the fact that those who appear to gain most, the tourism entrepreneurs within a community, may not actually be part of that community. They may be 'off-comers', strangers who 'import qualities which do not and can not stem from the group itself' (Simmel, 1950, p. 402), or they may be in some way marginal, perhaps better equipped to profit from tourist enterprises. As Nunez observes (1978, p. 269), when a local community is discovered, 'this is the arena for the marginal individual to appear as a leading performer'.

The problem for those who support both tourism development and resident involvement is to find some way to prevent a conflict of interests between groups, and individuals, who may be differentially affected by tourism. Pearce (1995) divides the community into two groups. There are those who have no direct involvement in tourism (the broad host community) and those who work in and plan tourism (the tourism community professionals). If Pearce (1995, p. 151) is to be believed, a mixture of 'persuasive communication' and involving local people in planning will provide 'a rich understanding of each party's position [which] is a necessary pre-requisite to resolving planning conflicts'. For Getz and Jamal (1994) the notion of collaboration is central to tourism planning, though their sample of 'community stakeholders' in Canmore (Alberta, population 5700 in 1991) appears somewhat biased in favour of the 'tourism professionals' described by Pearce (1995) above. In an attempt to isolate key local concerns, Getz and Jamal (1994) sought the views of five businesses, six local government members, one regional government official, three environmental groups, two local consultants and six local residents, five of whom were involved with a local resident organization. From their analysis of interviews and from data gathered elsewhere, Getz and Jamal (1994) were able both to support the collaborative process, and to provide a realistic appraisal of the problems of its implementation. In their discussion of implications for the future study of community-based collaboration, two fundamental difficulties emerge. Can planning conflicts be resolved 'where diametrically opposed viewpoints on growth and development exist', and to what extent does the 'visible minority that appears at most public meetings represent the "silent" majority'? (Getz and Jamal, 1994, p. 171). Madrigal (1994) is convinced that only the 'lovers' and 'haters' of increased development 'would feel strongly enough to appear in public forums', and that the community cluster he labels 'realists' may actually represent this silent majority. The 'realists' recognize that tourism brings benefits to the community as well as costs such as increased traffic and litter. Madrigal (1994) believes that the 'realists' are more likely to support an increase in tourist numbers if the positive aspects of development are reinforced by the internal marketing efforts of local government.

To suggest that disparate opinions between groups and individuals can be aligned toward some communal vision fails to recognize that tourism development is fundamentally different from other kinds of economic development. The economic benefits of tourism may be unevenly distributed but the costs, the intrusion, congestion and rising prices will impact on the lives of the silent majority as much as on those who support and those who hate tourism. It might be more appropriate to consider the role of tourism in widening, rather than narrowing, community differences. It might even be that a previously cohesive community becomes factionalized as tourist numbers increase.

It is also questionable whether the silent majority can be encouraged to participate in decision making by 'persuasive communication' (Pearce, 1995), 'internal marketing' (Madrigal, 1994), or by the collaboration process (Getz and Jamal, 1994). If, as Madrigal (1994) suggests, the silent majority are in fact 'realists', perhaps they are realists because they recognize the contradiction between the political rhetoric of participation and the 'fundamental inequalities found in [Western] societies' (Prentice, 1993, p. 219). Even where formal mechanisms for participation are in place, as in Malaga, where the local government is reviving traditional festivals for local people as well as visitors, the perception of residents is that 'there is much dressing up for the tourists' (Barke and Newton, 1995, p. 132). Emphasizing the heritage of Malaga in its promotion might appear to be more 'appropriate' but the effect could be to increase the suspicion of local people that it is their 'way of life', their community, that has become the attraction. The meaning that politicians give to 'community' is framed by notions of cooperation and power. For local people, the view of the common good might be nearer to that expressed by several of Canmore's more recent residents ... 'not in my backyard' (Getz and Jamal, 1994, p. 165).

In spite of all this, Brent Ritchie (1993) believes that a destination vision can be 'crafted' by community members who will define the broad parameters within which tourism development should take place. According to Brent Ritchie (1993) this process of 'visioning' will provide a strategy which is 'dynamic and evolving' and, like the corporate culture rhetoric from which it emerges, emphasizes the importance of harmony, integration and empowerment. In the 'vision statement' for the Lancashire Tourism Strategy (Lancashire County Council, 1995) the words 'balance', 'residents' and 'visitors' in close proximity create the illusion that some sort of shared vision is possible. The contemplation of the future, whether it involves residents or not, is less important than trying to understand what such 'visions', or strategies, mean to individuals. The mission statement for Canmore (Alberta) refers to development which is 'sensitive to community aspirations' and contributes to the 'quality of life of the residents' (Getz and Jamal, 1994, p. 164). What is not clear is how an increase in visitor numbers can improve the quality of life of the owners of second homes in Canmore, where some 25% of homes are owned by non-residents. Indeed, as Getz and Jamal (1994) report, even newly arrived residents 'want to preserve the lifestyle for which they came to [the] town'. There is a fundamental tension between the indigenous members of a community, the 'off-comers', and the owners of second homes, and since it is axiomatic that these three groups will be represented in many destination areas it is something of an understatement to say that the task of reaching a consensus is a challenge (Brent Ritchie, 1993). What seems often to be overlooked by the political interventionists is that it is individuals who have aspirations and not communities.

Murphy (1985) argues that not only should the wishes of local people shape future development but the community's heritage and culture will provide a more individualistic tourist product (p. 151). It is perhaps ironic that it is often the search for distinctiveness and the attempt to mark a place as 'worth a visit' that creates the stereotype. Cohen (1993) identifies two ways in which the community gives meaning to its members. There is the sense they have of its perception by strangers and there is a highly personalized view, 'refracted through all the complexities of their lives and experience' (p. 74). Where the selection of heritage and the shaping of hospitable behaviour is informed by a perception of 'what the visitor wants', then what is being presented is the community's public face, 'where internal variety disappears or coalesces into a simple statement' (Cohen, 1993, p. 74). It is political and economic interests that will define the community's public face and any vision it might have for the future.

The Community Show

For Murphy 'the twin foundations of the [tourism] industry are its destination attractions and hospitality' (1985, p. 12). In many communities, both rural and urban, the perception of attractiveness will involve heritage and culture, though, as Ashworth (1994) points out, tourist heritage will be selected according to the expectation of the visitor, expectations which may be substantially different from those of the resident. The debate about culture and its consumption is not new and much has been written both before, and especially since, the question posed by MacCannell (1992) at the start of his analysis of the 'Locke case' in 1982, 'should a community be a commodity?' (p. 172). For local residents, though, the question may be something of an irrelevancy. If, as stated in the recent report, *Sustainable Rural Tourism* (Countryside Commission, 1995), the local community does not necessarily have a veto (p. 31), then residents would automatically become part of the commodity and be seen as participants in what is often regarded as a key community asset, the 'warmth' of its welcome. The community show requires a friendly community and attempts are often made to include community members in the 'atmosphere' of destinations in an effort to 'tangibilize' the idea of hospitality (Taylor, 1995, p. 488). When the Mayor of Eilat (Israel) writes that 'the residents are friendly and warm people who greet and accept visitors, guests and holidaymakers in an open-hearted way' and goes on to suggest that visitors 'walk round our neighbourhoods' and 'get to know the residents of Eilat' (Kadosh, 1994, p. 1), he seeks to define the community in a way that can surely have no legitimacy. In setting the community's public face towards the reception of visitors, the Mayor is making the community a symbol for hospitality where the showering of local people with clichés – 'warm', 'friendly', 'welcoming', 'smiling', 'romantic' – becomes more than advertising hyperbole. How can there be communal endorsement for such messages, which (often blatantly) suggest the price of the holiday includes domestic as well as commercial hospitality?

The Blue Guide to Crete (Cameron, 1988, p. 37), notes that 'people in rural areas are friendly and helpful out of a tradition of hospitality to strangers' and, while this might appear to be another case of increasing visitor expectation for unsolicited hospitality, there is no doubt that the kind of behaviour described is part of the social fabric in many parts of the world. In Crete, and indeed in the rest of Greece, traditional hospitable behaviour derives from the concept of *philoxenia*, where the stranger is received as a friend. Interviews with residents of a Cretan village (Taylor, 1996) confirmed a local respect for the tradition, and, though there were suggestions of a weakening influence, it was interesting to note that, when confronted by such unexpected friendliness, some visitors appeared suspicious. As one resident remarked, 'You can see it in their faces: why are they doing this?'

Such hospitality is part of a belief system involving gift exchange and reciprocity. It is a non-monetary mode of exchange, yet it enters the commercial environment because if a place is to function as a tourist destination the services of local people are required. Indeed, it is the local inhabitant, as receptionist or waiter, who is often central to the process of delivering service to the visitor. In his analysis of the tourism product, Smith (1994) distinguishes between hospitality, 'the expression of welcome', and service, which is 'the technically-competent performance of a task'. If 'hospitality' here means the enhancement of service by providing the 'something extra' that customers now expect (Smith, 1994, p. 588), then the responsiveness of local people will inevitably be framed by the conventions of commercial rather than domestic hospitality. The promise of staff who will 'pamper you morning, noon and night' (Sandals advertisement, *Daily Mail*, 30th January 1995) defines in its starkest form what is expected of the service provider.

The local person working in the hospitality industry will not only be expected to provide friendliness 'on demand' s/he will also be seen as representative of that locality. When a hotel manager in Crete states that the visitors expect a Greek atmosphere and even the traditional hospitality of the Cretan (Taylor, 1996, p. 472), then what is provided, and the way it is provided, will need to suggest a local distinctiveness. As Urry (1990, p. 68) notes, though, 'the "service" partly consists of a process of production which is infused with particular social characteristics'. In Crete, such characteristics are bound up in local traditions of hospitality and will conceivably be carried to the hotel as part of the cultural baggage of the employee. Several hotel managers referred to 'the Greek way' or 'the traditional way' to treat a guest based on the concept of *philoxenia*. As one manager put it, '*philoxenia* is the big hospitality we have inside of us', and when asked how this translates to the service arena it was the 'friendliness' of the employees that was not surprisingly recognized as a valuable natural asset (Taylor, 1996, p. 477).

When Grecotel, the Greek hotel management company based in Crete, promise that their staff 'with their warm and natural way will make you feel at home' (*Grecotel News* 1995), they both draw on a local conception of hospitality, and imply that this 'natural' friendliness can be brought under management control. In attempting to energize the spirit of *philoxenia* within the commercial environment of the luxury hotel, Grecotel appear to be achieving a degree of success (see Taylor, 1996, for a more detailed account). They are perceived as a good company to work for, staff in operational areas as well as managers talked of their regard for the clients, and the clients commented on the friendliness of staff and on the care that was clearly demonstrated. As one chambermaid remarked, 'it makes me feel good to see the clients happy, that's the Cretan way' (Taylor, 1996, p. 477). The motto of Grecotel is 'The Greek Hotel Family', and in their ability to retain staff (turnover less than 1%) the company can demonstrate that this is no mere rhetorical device. One function of a family, though, is to provide a measure of control and it is interesting to note the concern of some managers that customers may find Cretan friendliness overwhelming. As one manager explained, for some of his customers the Greek way 'is not so familiar ... if we come so open-hearted to them they look strange' (Taylor, 1996, p. 478). There are echoes here from the local village where visitors were surprised by the reception they received. In the hotel the degree of friendliness cannot be left to chance. As one corporate executive noted, 'it is better to be more friendly than cool but there is a limit'. It is the staff seminars that 'help [the staff] to see where the limit is' (Taylor, 1996, p. 478).

It is interesting to note the contradiction between some hotel managers in Crete, who see a danger in too much friendliness, with that of hotel managers in Warsaw, who reported that the hardest thing to do is to get the employees to smile (Nickson and Taylor, 1994). In Israel, where one hotel manager stated that 'the Israelis are rough, less polite,' the hotel company Isrotel have introduced an incentive scheme where visitors can nominate an employee who has provided friendly service (Taylor and Davis, 1995). The name chosen for this scheme, 'miles of smiles', in some ways symbolizes the dilemma facing hospitality managers, who may not have the benefit of some heightened local sensitivity towards hospitable behaviour.

In a study of tourism in Pefkochori (northern Greece) Wickens (1994) describes the search for Greek authenticity and the 'real atmosphere' of a Greek village, of a Greek taverna (p. 821). Staged performances, she notes, of Greek traditional dances are mass produced for consumption by the tourist (p. 822). In the same way *philoxenia* can be borrowed from the communal milieu to add value to the commercial hospitality product. Grecotel may be able to provide a working environment where friendly behaviour is both supported and encouraged but what of other hospitality operations both in Greece and elsewhere? Those who pay for the community show expect to be made welcome and, in the dismal language of the customer care manual, 'the customer is king'.

Conclusions

For the most part, the literature concerning community involvement, or more precisely the involvement of community members, has not taken sufficient account of the relationship between the resident and the tourism product. The political correctness in struggling to involve local people in the search for a sustainable, more equitable future has diverted attention from the ways in which they may be exploited. The theatrical metaphor in which 'stakeholders' become 'cast-members' assumes a greater significance where the traditions and heritage of local residents become key attractions in the community show. As MacCannell (1992, p. 175) notes, if, as a local representative of an 'ethnic attraction', it is your own house that the tourists come to see, 'they can only be a source of inconvenience and potential embarrassment'. In the face of such intrusion it may well be that attempts to increase tourist numbers will widen any differences that exist. As Cohen (1993, p. 74) observes, 'the community's public face is symbolically simple, in its private mode differentiation, variety and complexity proliferate'. Ioannides' (1995, p. 590) description of a small-scale, sustainable development in Cyprus sees the problem through the eyes of the residents, who felt that 'agrotourism would stereotype the inhabitants as backward peasants living like their forebears'; yet it is often this stereotype, this public face, that the tourist seeks.

That the community and its residents are essential components of the tourism product is perhaps best illustrated in relation to the provision of hospitality. In many ways, this may represent the saddest exploitation of resident responsiveness. Within the commercial environment local codes of hospitable behaviour can be adjusted by the demands of the market, but what of the village where the search for the 'friendly welcome' may be fuelled by a mixture of nostalgia and the promises of advertising copywriters? The expectation of hospitality will make a commodity out of a tradition and there is a price to pay. Lewis (1984, p. 177) in his vivid evocation of social life on the Costa Brava in the 1950s sees only resignation. When a fisherman turned waiter confides, 'You can safely say we've sold ourselves,' there is an acceptance that the community is finally on stage, that the show has begun.

References

Ashworth, G.J. (1994) Let's sell our heritage to tourists, some queries from the Maritimes. In *Proceedings of the Conference, Tourism in Canada*. 1994 Conference for Canadian Studies, Canadian Embassy, London.

Barke, M. and Newton, M. (1995) Promoting sustainable tourism in an urban context: recent developments in Malaga City, Andalucia. *Journal of Sustainable Tourism* 3(3), 115–134

Brent Ritchie, J. (1993) Crafting a destination vision – putting the concept of resident-responsive tourism into practice. *Tourism Management* 14, 379–389.

Canan, P. and Hennessy, M. (1989) The growth machine, tourism and the selling of culture. *Sociological Perspectives* 32, 227–243.

Cameron, P. (1988) *Blue Guide Crete*. A. and C. Black, London.

Cohen, A.P. (1993) *The Symbolic Construction of Community*. Routledge, London.

Countryside Commission (1995) *Sustainable Rural Tourism*. Countryside Commission, Cheltenham.

D'Amore, L. (1985) Social and cultural strategies. In: Murphy, P.E. *Tourism: A Community Approach*. Methuen, London.

D'Amore, L. (1992) Promoting sustainable tourism – the Canadian approach. *Tourism Management* 13, 258–262.

Dogan, H.Z. (1989) Forms of adjustment: sociocultural impacts of tourism. *Annals of Tourism Research* 16, 216–236.

Getz, D. and Jamal, T.B. (1994) The environment–community symbiosis: a case for collaborative tourism planning. *Journal of Sustainable Tourism* 3, 152–173.

Ioannides, D. (1995) A flawed implementation of sustainable tourism: the experience of Akamas, Cyprus. *Tourism Management* 16, 583–592.

Johnson, J.D., Snepenger, D.J. and Akis, S. (1994) Residents' perceptions of tourism development. *Annals of Tourism Research* 21, 629-642.

Kadosh, G. (1994) *All You Need to Know About Eilat, 1994–95.* Golden Guide Advertising, Eilat, Israel.

King, B., Pizam, A. and Milman, A. (1993) Social impacts of tourism, host perceptions. *Annals of Tourism Research* 20, 650–665.

Lancashire County Council (1995) *A Tourism Strategy for Lancashire.*

Lane, B. (1995) In: Taylor, G., Tourism in Canada. *Tourism Management* 16, 323–325.

Lewis, N. (1984) *Voices of the Old Sea.* Penguin, Harmondsworth.

MacCannell, D. (1992) *Empty Meeting Grounds.* Routledge, London.

Madrigal, R. (1994) Resident's perceptions and the role of government. *Annals of Tourism Research* 22(1), 86–102.

Mok, C., Slater, B., and Cheung, V. (1991) Residents' attitudes towards tourism in Hong Kong. *International Journal of Hospitality Management* 10(3), 289–293.

Murphy, P.E. (1985) *Tourism: A Community Approach.* Methuen, London.

Nickson, D. and Taylor, G. (1994) Getting the service right? Hotel management in Warsaw. In: *Proceedings of Council for Hospitality Management, Fourth Annual Conference,* Norwich.

Nunez, T. (1978) Touristic studies in anthropological perspective. In: Smith, V. (ed.) *Hosts and Guests: The Anthropology of Tourism.* Blackwell, Oxford.

Pearce, P.L. (1995) From culture shock and culture arrogance to culture exchange: ideas towards sustainable socio-cultural tourism. *Journal of Sustainable Tourism* 3(3), 143–154.

Prentice, R. (1993) Community-driven tourism planning and residents' preferences. *Tourism Management* 14(3), 218–227.

Ryan, C. and Montgomery, D. (1994) The attitudes of Bakewell residents to tourism and issues in community responsive tourism. *Tourism Management* 15(5), 358–369.

Simmel, G. (1950) *The Sociology of Georg Simmel.* trans Wolff, K.H. Free Press, New York.

Simmons, D.G. (1994) Community participation in tourism planning. *Tourism Management* 15(2), 98–108.

Smith, S.L.J. (1994) The tourism product. *Annals of Tourism Research* 21(3), 582–595.

Taylor, G. (1995) The community approach: does it really work? *Tourism Management* 16(7), 487–489.

Taylor, G. (1996) 'Put on a happy face': culture, identity and performance in the service role. In: Robinson, M., Evans, N. and Callaghan, P. (eds) *Managing Cultural Resources for the Tourist.* The Centre for Travel and Tourism, University of Northumbria, Business Education Publishers Limited, Sunderland.

Taylor, G. and Davis, D. (1995) Interview material from hotel managers. Eilat, Israel (unpublished).

Urry, J. (1990) *The Tourist Gaze*. Sage, London.

Wickens, E. (1994) Consumption of the authentic: the hedonistic tourist in Greece. In: Seaton, A.V. *et al.* (eds) *Tourism: The State of the Art*. Wiley, Chichester.

Local Environmental Protection Initiatives in Crete

E. Prinianaki-Tzorakoleftheraki

Technological Educational Institute of Heraklion, Department of Tourism Industries, Stavromenos, Heraklion 71500, Crete, Greece

Introduction

European and classical Mediterranean tourism is at a watershed, with a number of European, governmental, and regional initiatives over the past few years, all pointing to the need for environmentally sensitive tourism development. A recurring theme in all these initiatives has been the emphasis placed on the role of the stakeholders concerned and on the role of education and training in increasing responsibility for sustainable tourism development.

Whilst tourism is Europe's largest service activity and is, for many countries, including Greece, of vital economic, social, and political significance, it has not always strived for sustainability. Tourism development in many areas has not considered either local inhabitants or the environment in sufficient measure. In fact, certain destinations have appeared to concentrate more on profit and employment creation and less on possible negative consequences of mass tourism. This could also be the case in Crete, where tourism is increasingly being used as an economic development tool.

Recognizing the problems that exist, a change in attitude seems to be occurring in the government of the Cretan region, the tourism industry and society. This chapter examines the relationship between tourism and the environment in Crete, which, in turn, includes the trends and events which are moving towards achieving and sustaining further tourism development. A shift from marketing to the planning of an environmental–cultural based product is suggested, while the role education can play is pinpointed.

The Tourism Industry in Crete and its Environment

The Greek tourism industry is a major contributor to the national economy. The most recent estimates by the Ministry of Tourism suggest that it is a US $6.5 billion a year industry. It employs 280,000 people in all tourism-related sectors, representing 7.6% of the country's workforce. In Crete, Rhodes, Corfu and other major tourism destinations, it is the top source of employment (Soteriadis, 1994).

Although Greece has been the focus of small-scale historical and cultural tourism for many years, mass tourism is a more recent phenomenon, as Greece was seen as an attractive, unspoiled and competitively priced alternative to increasingly congested south Mediterranean resorts.

Statistical data provide information that 33,300 tourists passed the borders in 1950, 1,600,000 twenty years later in 1970, 9,300,000 another two decades later in 1990, to reach 11,301,722 arrivals in 1994.

Crete is the primary and most dynamic tourism destination in Greece, contributing 18% of available bedspaces, 21% of total bednights, 25% of air arrivals and 42% of the country's tourism exchange influx (Editorial, 1995a). It enjoys the enviable position in the Mediterranean of being a gateway to three continents. It is the largest island of Greece and the fifth largest of the Mediterranean islands. The surface area of the island is 8258 square km, it is 260 km long, ranging in width between 12 and 60 km. The extent of the coastline reaches approximately 1046 km, 155 km of which consists of sandy beaches. Mainly mountainous, its geophysical layout is determined by the three important mountain ranges (Lefka Ori, Psiloritis and Dikti) which cross the island, rearing up to heights of 2500 metres.

Crete has a landscape of exceptional scenic beauty, full of variety and contrast. The physical assets of this island include numerous bays, wild gorges, and more than 1000 caves. On the island there are more than 400 different flower and plant species, of which, apart from the typical Mediterranean flora, there are 130 species unique to Crete. However, Crete's advantages and strong points are not only based on its natural environment but also on a distinctive cultural dimension, supported by a wealth of archaeological and historical features and traditional villages. Crete was the centre of Minoan culture (2600–1100 BC) and there are many remains of their palaces and tombs. Also, the Roman and Byzantine buildings, together with the castles of the medieval Venetian period, comprise an extraordinary variety of historical and cultural interest. The island and its 545,000 inhabitants enjoy a typical Mediterranean climate. Early travellers commented on the healthy climate, reputedly the mildest in Europe (Papadokostaki, 1993).

The incoming tourist flows to Crete are dominated by only a few origin markets. The vast majority of foreign visitors come from Europe, accounting for 98% in 1994, the German, British, and the Scandinavian markets alone contributing 71% (Editorial, 1995a). This concentration in relatively few markets leaves Crete extremely vulnerable to sudden changes in trends and events in its main market countries.

Although the climate is suitable for holidaymaking from March through to November, as in most EU countries, the demand in Crete is heavily concentrated in the summer months. In the last ten years 37% of the average arrivals were in July and August, while the period May to September attracted 74% of total arrivals. This over-concentration of tourism in the high season, and the corresponding under-utilization of capital and human resources in the low season, represents one of the greatest problems tourism faces in Crete (Papoutsis, 1996).

The unbalanced geographical distribution of the tourism activity is yet another area of concern for Cretan tourism. The greatest concentration of tourism development is on the northern coast of the island. Despite these obvious drawbacks, Crete, whose possibilities for industrial development are limited, has used tourism as a means of economic development. For over three decades, the island's image as a paradise of sun and sea has been aggressively pushed by travel agents and government. As a result, Crete became a popular destination for packaged tourism. At present, package-holiday tourism accounts for 90%, a market that, as related evidence suggests, has become saturated and highly competitive (Papoutsis, 1996), thus creating an additional concern. However, compared to other European countries and regions, the Greek and Cretan tourism industry is still relatively small but is growing and is capable of adapting to new demands and changing tourists' tastes.

The Environmental Issue - a Worldwide and a Local Affair

Environmental awareness is high all over the world. Environmentally friendly products are very popular and political power is increasingly being given to environmentally friendly parties, while in the tourism industry there is a growing recognition that travellers place a great importance on the quality of the environment. The tourist today is not the same, even compared to five years ago. The interests, values and expertise of most people have been enriched. Knowledge and experience have given them new horizons, their economic and social levels have risen, tourism is perceived as a priority while at the same time personal options and values are more quality centred (Editorial, 1996a). Within this context, quality also refers to environmental factors and considerations due to deepening concern for the environment. The tourists of the present are increasingly looking for safety, including environmental safety, they are increasingly critical about the role of business in the environment, while the tourists of the future will be even more concerned. In fact, predictions suggest that, after the year 2000, viability and further development of tourism will be dialectically conditioned by the environmental quality (Costa, 1995; Kue-Sung and Singh, 1995; Paravoliasakis, 1996).

In Crete, with the growth and increased importance of tourism, coupled with a recent decline in tourism arrivals, the issue of tourism in relation to the environment is being addressed more frequently as it is increasingly thought not to meet the needs of all concerned. Evidence suggests that:

1. tourism businesses want government to invest in order to increase their business prospects, declaring that environment protection is a priority.
2. the local communities are often afraid of unbalanced development affecting the environment and their quality of life.
3. the representatives of the government and the political parties give assurances that they are striving for sustainability, calling for cooperation by all concerned (Editorial, 1995b).

A central theme of debate over the last few years concerns the relationship between tourism and the environment in the island. This discussion centres upon several issues, namely:

- the negative effects of tourism on the environment.
- the extent to which tourism will further develop and its potential impact on the environment.
- the stakeholders' responsibility.
- the importance of education in sustainable tourism development.

These issues will be briefly considered in turn. However, any attempt to look at the negative impacts of tourism on the environment is hampered by the diversity of interest and understanding of tourism by concerned groups. Nevertheless, the growth of tourism has, in recent years, at least in part reflected a view that this is an important industry with a need for sustainable planning and development. However, several bodies have voiced concerns about the economic dependency, environmental degradation, limited energy resources and insufficiency of infrastructure. For instance, Crete faces an acute energy problem, while the airport of Heraklion, which receives over 80% of the total arrivals to the island, has exceeded its carrying capacity. A project for its extension is approved, increasing therefore the debate about the extensive noise pollution for the inhabitants in the nearby area (Editorial, 1995c, 1996b).

At the same time, the national government, the regional authorities, and the local tourism community recognize that the major issues, identified earlier, are becoming more acute in the Cretan tourism industry:

- unbalanced geographical distribution of the tourism activity.
- intense seasonality.
- economic dependency.
- overuse of the natural resources.
- low competitiveness.

Regarding tourism's further development, Cretans increasingly voice concern about the uncontrolled expansion of tourism. The basis for this concern is that tourism development may, in the near future, exceed the carrying capacities of certain regions in the island, with unforeseen negative impacts on the environment, indicating that new developments need to be preceded by an environmental impact assessment.

To quote from the minutes of the recent Tourism and Environment Convention held at Heraklion, 17–19 March 1995:

> tourism is a vital part of our economy with great hopes placed on it, provided that the environment is protected. But, environmental protection and awareness not only in words but in actions, (is) not a sensitivitism but a prerequisite for the future.
>
> (Editorial, 1994)

With regard to stakeholders' responsibility in relation to past and future tourism development, all seem to agree that the government's role is crucial for it creates the framework within which tourism is developed. At the same time, it has been extensively supported that the environment's protection and the related balanced development are a great national obligation of every member of society, as well as a serious responsibility of the government.

The responsibility of the government goes without saying. Yet what is the merit of private sector's responsibility? This is infrequently debated, as if government is the only player in the tourism arena and as if developers, for instance, have one and the same responsibility every member of society has. However, it has been increasingly recognized that laws in force and published guidelines aiming at minimizing the effects of tourism on the natural environment are potentially inadequate if there is not:

- sufficient human care.
- sufficient development control.
- adequate collaboration between the private and the public sectors.

The planners and managers of the public and private sectors need to understand tourism issues so the future of Cretan tourism as a long-term activity will remain sustainable. However, it is necessary that new forms of tourism be evaluated first in relation to resource capabilities and then a decision made as to whether development is appropriate and on what scale. Public benefit should predominate over individual interest.

Striving for Sustainability

A major problem affecting the tourism industry is that tourism development and tourism expansion, sooner or later, run up against the law of diminishing returns, because the tourism industry as a whole makes extensive use of resources that scarcely increase. These resources are the cultural and natural environments. As a result, dilemmas such as tourism or environment appear from time to time. A different but related dilemma posed for tourism development in relation to the environment refers to whether they disqualify each other or can coexist.

Certainly, tourism without an environment cannot exist. The physical, cultural and structural environment comprise the core of tourism, the reason for its existence. However, there are cases where uncontrolled development and rash tourism actions have altered and damaged the environment minimizing tourism potential. This is exactly where the problem lies; not in tourism *per se*, but in uncontrolled tourism development.

Sustainable tourism management and development, 'αειφορος ανά–πτυξη' in Greek, is defined as:

a philosophy of tourism management and development that strives for a viable tourism that is based on and enhanced by the culture and the environment in the present time and always.

It is also defined as an approach intended to harmonize relationships between tourists, residents, the industry and the environment (Editorial, 1994).

The first and second Peripheral Conferences in 1994 and 1995 respectively, held in Crete, pinpointed the need for sustainability: 'tourism development must match the needs and aspirations of international tourism with the peculiarities of the host areas today and in the future', they declared. The representative of the Greek National Tourism Organization referred to the need for improving the quality of the Cretan tourism product with strategies such as: the development of alternative types of tourism; taking steps towards environmental protection; infrastructure relative to tourism needs; promotion of the famous monuments and cultural treasures of Crete and association of tourism development with the cultural heritage; and improvement of tourism and hospitality education and training (Editorial, 1995b, 1996b).

Although a legal framework in relation to the environmental protection has been in place from the early stages of tourism development, it seems that the philosophy of sustainability has only recently served as a basis for tourism planning and development. Whilst tourism was growing and expanding over the last two decades, concern was also growing about the need to preserve the unique qualities of Crete and to protect its physical and cultural diversity.

It seems that it is now understood that the long-term viability of the tourism industry in Crete depends on maintaining and enhancing these assets and on solving current problems. A number of new governmental laws and policies, aiming to respond to these concerns, are now in force. In parallel, the region of Crete has taken a number of initiatives. The idea is to protect fragile ecosystems, to develop new forms of tourism, to direct funds away from main tourism destinations and conventional tourism developments and to enhance Crete's tourism image. It is also recognized that management of existing operations in a manner which is sensitive to the principles of sound environmental management can play a role. The question that arises is how can one organize and persuade the small and medium-sized companies involved in the tourism industry to take environmental management seriously? Because tourism and hospitality are characterized by small and medium-sized operations, on an individual basis, many of them feel that they are too small to have any effect.

A Paradigm of Responsible Environmental Management in Crete

Grecotel, the biggest hotel chain in Greece, states:

> Due to prevailing circumstances, the participation and contribution of a tourism business in activities related to the protection of the environment, whether physical, cultural, historical or folklore, is considered a necessity. Tourism, culture, tradition and environment, go hand in hand and so called sustainable development respects equally whatever each of the above terms mean and everything that joins the human being with its environment.
>
> (Grecotel, 1995)

It is very important for Greece to present companies which operate under this perspective as it is for Greek tourism to have factual evidence and examples of environmental and cultural friendly practice and philosophy. Grecotel has indeed given such proof, by financing the excavation of ancient Eleftherna in Rethymno, as well as with its actual contribution to publicizing the findings. Grecotel has also employed a tourism policy with respect to environmental management in its hotels. Taking into account general concerns about environment sensitivism, Grecotel engaged in a responsible management policy in an effort both to add value to its product and to improve its image. One could argue that the efforts of a single company, even of Grecotel, are not enough, unless responsible environmental practices are put in place by all concerned. Certainly, individual efforts are not enough, but they are of crucial importance in giving an example. In effect, Grecotel has already been quoted as a good example, and it is, in many views, a leader in the Greek, if not in the international tourism market.

Crete's Regional Tourism Development Plan

Crete is not only sun, sand, sea. What Crete needs for long-term tourism viability, growth and development is a shift from marketing to the planning of an environmental–cultural based product in such a way that minimizes adverse impacts. Again, its physical beauty and diversity, its educational interest, its long history and traditions must come to the forefront to support and enhance the Cretan tourism product and to operate as a protective mechanism for these cultural and natural assets of the island. Fortunately, the Region of Crete is becoming more aware of the need to achieve sustainable development. Being such an important activity, tourism has to be managed and developed in a way that ensures its resources are not lost (Region of Crete, 1995). A tourism development practice for Crete is being initiated through Crete's Regional Tourism Development Plan, a project being funded by the government and the European Union.

The primary objectives of this strategy are: (i) the geographical and seasonal balanced distribution of the tourism activity; and (ii) the improvement of the quality offered, which in turn will enhance and upgrade the Cretan tourism.

The steps to be taken to fulfil this policy follows.

Environmental protection with particular emphasis on the relationship between tourism and culture In addition to strict planning control, new zoning laws and protected areas, specific projects for the preservation of cultural and natural heritage have been initiated. Village preservation projects and renovation of buildings are key areas of concern and action.

However, three new concepts have an important role to play here; *philoxenia*, *physi* and *philia*. The famous Greek hospitality – ΦΙΛΟΞΕΝΕΙΑ (*philoxenia*) – could be extended to include the rich cultural history and traditional lifestyle of Crete, which together with the traditional hospitable nature of its inhabitants could offer the traveller a real feeling for the local life. The extraordinary physical and cultural environment – ΦΥΣΗ (*physi*) – and the necessary friendly and sensitive attitude towards it could be the basis of improving the relationship between tourism and the environment and for tourism planning. Friendship – ΦΙΛΙΑ (*philia*) – within this framework, describes a friendly attitude towards both the tourist and the environment.

Extension of the tourism season At seven and a half months, Crete has the longest tourism period in the country. Yet further extension of the tourism season and a balanced geographical distribution of the tourism activity could increase the economic and social benefits without threatening the situation in the island. In fact, a series of initiatives which seek to promote winter tourism and small-scale development in southern and central Crete are now in force.

New types of tourism In line with the Cretan Region's programme objectives, tourism options that minimize adverse impact should be identified. This implies a departure from conventional mass tourism approaches, with emphasis being placed on new forms of tourism. The encouragement of cultural, conference, yachting, eco- and rural tourism, and the development of marinas and water spas are a few examples. A related project is the development of mountain paths and that of the marked path 'E4', which starts from Western Europe, continues through Greece and ends up in Crete.

Tourism infrastructure Mitigating current problems and developing new infrastructure investments are also priority areas within the regional development plan.

Promotion of tourism Crete is an established destination. However, it needs new markets; promotion costs are therefore being considered as an integrated activity in tourism development and not in isolation.

Two 'Ss', standing for 'sustainability' and 'strategy', predominate here. However, a third 'S', for synergy, is also required. Sustainability implies that environmental, natural and historical resources are being used and developed for the enjoyment of tourists but that these resources will be protected for the benefit of future tourists and residents. Strategy implies that there is strategic planning at regional and local level, supported by the government, for the coordination and implementation of sustainable development. Synergy suggests that the implementation of sustainable development is achieved with the cooperation of the public and private sectors, the tourists, the local people, the professionals, and society as a whole. Obviously, without strategy and synergy there will be no success in any effort towards sustainable growth and development and the necessary elements of *philoxenia*, *physi*, and *philia* cannot be supported and promoted. Still, the three 'Ss' here are more complicated and dynamic, involving struggles and requiring power, will and sensitivity by all concerned, not only the government. The role of education is predominant here.

Education and training, which focus on tourism-related agencies, are also included in the region's development plan. This area of activity is mainly devoted towards implementing seminars for tourism-related groups and producing various publications. This plan has been in action since 1994. Since then, a number of projects have been completed, several are in progress, and others are at the planning stage. The general attitude towards the plan is positive, although it is doubtful if it represents a definitive answer. The solution would be most be likely to be found in education.

Educating for Sustainability

It has been established that the tourism industry has reached a critical point in its development. After more than three decades of rapid growth, tourism is in desperate need of a sustainable approach for its further development. It has been argued that, unless responsible management practices are in place, the tourism industry can end up degrading the features on which its prosperity is based. Tourism managers, therefore, face the challenge of improving the effectiveness of their management of the relationship between tourism and the environment. This indicates the increasing need for well-educated and skilled managers, able to cope with the demands of the industry, which, in turn, requires effective tourism and hospitality education and training.

The departments of tourism industries of the technological educational institutes have a significant role to play here. Representing the leading tourism and hospitality management options available within the Greek tertiary-level educational system, they are in a position to meet an urgent need to reconceptualize tourism and hospitality education, to separate tourism studies from hospitality studies, and to provide a more innovative education than they are currently providing (Prinianaki-Tzorakoleftheraki, 1995).

A principle of education suggests that a pragmatic and scientific approach to curricular development should guarantee that the rate of education provision is greater than, or at least equal to, the rate of change. Tourism is undergoing a process of change and refocusing with environmental and sustainable issues being at the forefront, implying the need for extending the contextual boundaries and concerns of tourism. Tourism education can make a contribution to this process by broadening its scope of training to include ecological and sustainable development issues. Also, the study of the role various stakeholders play in tourism activity should be included and effectively combined with practical education in professional tourism curricula, while environment management issues seem to be an important and necessary component in sound hospitality management curricula.

In the author's view, the profitable and sustainable development of the tourism industry in Crete is going to depend to a great extent on the effectiveness and sustainability of tourism education and on the subsequent ability of tourism professionals to genuinely conceptualize and understand this complex industry and, importantly, to develop appropriate management systems to prevent potential environmental degradation.

Focusing on the education and training of tourism professionals is of major significance. However, the philosophy of sustainability–strategy–synergy also applies here, suggesting that education for the whole society is required. For this to be achieved it is proposed that:

1. environment and tourism subjects should be considered in the primary and secondary levels of the educational system in a more systematic way. These subjects should be included in the typical curricula.

2. continuing education models should be available for all employees of the tourism industry.

3. in addition, establishment of 'centres of environmental education' in all major towns, for the systematic and essential guidance and familiarization of the local and tourist populations could be beneficial. This has already been proposed for the town of Rethymno by its local authorities.

Politicians, developers, the tourism workforce, local and tourist populations will then have the resources, in terms of education and information, to show respect for the environment, to be more responsible, and more conscious of the need for environmental protection. They, together with the professionals, can guarantee the long-term sustainability of tourism in Crete.

References

Costa, J. (1995) International perspectives on travel and tourism development. *International Journal of Contemporary Hospitality Management* 7(7), 10–19.

Editorial (1994) Tourism and the environment. *Tourismos Ke Economia* March, 106–107.

Editorial (1995a) Crete – local point of Greek tourism. *Tourismos Ke Economia* May, 44–46.

Editorial (1995b) Citations from the 1st Peripheral Tourism Conference in Crete – 1994. *Tourismos Ke Economia* April, 138–139.

Editorial (1995c) Nea Dimocratia Conference: Tourism – problems and solutions. *Tourismos Ke Economia* November, 20–23.

Editorial (1996a) The Greek collection. *Travel World View* January, 7–8.

Editorial (1996b) Citations from the 2nd Peripheral Tourism Conference in Crete – 1995. *Tourismos Ke Economia* January, 68–69.

Grecotel (1995) Department of Environment and Culture: Grecotel's policy. Rethymnon, unpublished, pp. 1–2.

Kue-Sung, C. and Singh, A. (1995) Marketing resorts to 2000: review of trends in the USA. *Tourism Management* 16, 463–469.

Papadokostaki, A. (1993) Cretan tourism. *Cosmos ke Tourismos* March/April, 38–40.

Papoutsis, C. (1996) Tough competition in tourism. *Tolmi*, 29th March, p. A4.

Paravoliasakis, G. (1996) The future of Cretan tourism. *Tolmi*, 29th March, p. A11.

Prinianaki-Tzorakoleftheraki, E. (1995) Tourism and leisure education in Greece. In: Richards, G. (ed.) *European Tourism and Leisure Education: Trends and Prospects.* Tilburg University Press, Tilburg, pp. 93–124.

Region of Crete (1995) *Regional Environment Policy.* Heraklion, Crete.

Soteriadis, M. (1994) *Tourism Policy in Greece.* Technological Educational Institute of Heraklion, Heraklion, Crete.

23

Tourism and the Environment in Madagascar

J. Parsler

*Department of Environmental Management,
University of Central Lancashire, Preston PR1 2HE, UK*

Introduction

The world is currently in the midst of a major human induced extinction spasm. Nowhere is this more apparent than on the island of Madagascar. It sits, along with several offshore islands, on an isolated piece of continental crust 500 km from the east coast of Africa in the Indian Ocean. With a land area of 587,000 km² it is the fourth largest island in the world. Madagascar is culturally unique, inhabited by descendants of Indonesian settlers that first arrived some 1500 years ago. Biologically, the island has acted both as a refuge for relict species and as a mini-continent, the site of several impressive bursts of speciation. The resulting biota is of truly global significance. Madagascar is both a biodiversity 'hotspot' (Myers, 1988) and one of only 12 'megadiversity' countries that together support up to 70% of the world's species of, for example, vertebrates and higher plants (Mittermeier and Werner, 1990).

Despite its remaining biological wealth the Republic of Madagascar is economically poor. Gross domestic product per person was just US $247 in 1992. The human population, which doubled to 11.5 million from 1960 to 1992, is growing at an estimated 3.1% per year.

Since 1993, a democratically elected government, working within a new constitution, has seen tourism as one of the few sectors of the economy with the potential for rapid growth. Visitor rates are increasing and the island is seen as having major potential for high volume beach tourism (Economist Intelligence Unit, 1994). For tourism to contribute in any meaningful way to sustainable development it must not only bring in more US dollars but also contribute to the well-being of local people and to conserving biodiversity. This chapter seeks to outline the scale of the potential loss of biodiversity in Madagascar and begins to assess the possible role future increases in tourism activity may have in conserving the country's biological wealth.

Species diversity, endemism and habitat loss

The terrestrial fauna and flora of Madagascar consist of some 12,000 species of flowering plant (85% endemic), an estimated 100,000 species of invertebrate (80%+ endemic), 360 species of reptile and amphibian (93–99% endemic), 256 species of bird (66% endemic) and over 100 species of mammal (80%+ endemic) (International Union for Conservation of Nature et al., 1987; Glaw and Vences, 1994). Most of the fauna and flora are restricted by habitat preference to forest cover (Jolly et al., 1984). The fate of this rich assemblage of species is thus virtually one and the same as the fate of Madagascar's forests. The island was almost completely forested before the spread of early settlers across Madagascar 1500 years ago. Now only 5% of the original vegetation remains (Myers, 1988) although between 25% (International Union for Conservation of Nature et al., 1987) and 32% of the island (Myers, 1988) retains some form of woody cover. The principal agent in the destruction and degradation of this habitat has been *tavy* or slash and burn agriculture by the country's mainly rural population (Jolly et al., 1984; International Union for Conservation of Nature et al., 1987; Myers, 1988; Jolly, 1989; Goodman, 1993; Larson, 1994). Between 1000 and 1500 km² are estimated to be lost every year (Myers, 1988).

Environmental issues

The National Environmental Action Plan (NEAP), which the government began implementing in 1991, identified and focused on three general problem areas: (i) loss of biological diversity; (ii) soil degradation; and (iii) urban and rural water and sanitation problems (World Bank et al., 1988). All three sets of issues are linked, for while the forests harbour biological diversity they also act as reservoirs of soil and fertility and as ideal water catchments. While soil degradation (Randrianasijaona, 1983) and water and sanitation problems are acknowledged to be major issues, this chapter concentrates on conservation of biological diversity.

Tourism

Tourism is not well developed on the island. Other than from Reunion and Mauritius, Madagascar is a distant destination, little promoted and expensive to reach. However, with a tropical climate, many fringing reefs and 5000 km of white sand beaches it 'offers considerable potential for the development of high-volume beach resort tourism' (Economist Intelligence Unit, 1994). Visitor numbers are growing, from 35,000 in 1988 to an estimated 65,000 in 1994, and are projected to reach 100,000 by the year 2000 (Economist Intelligence Unit, 1994). Most visitors combine some kind of beach holiday with nature tourism. Indeed, the country's flora and fauna are one of its outstanding unique attributes. Nowhere else can one view over 1000 species of orchid, see lemurs in their natural habitat, or see any of the other hundreds of thousands of endemic species.

There are three major 'centres' of tourism activity: (i) around the country's capital, Antananarivo; (ii) on the offshore island of Nosy Be to the north west; and (iii) on the offshore island of Ile St Marie to the east. Despite the modest size of the tourist sector, it makes a major contribution to export earnings (US $50 million in 1993, second only to vanilla exports). The government has targeted tourism as perhaps the only sector of the economy that is capable of rapid growth. Prompted by the World Bank, several measures are being taken to encourage investment and growth in the sector. These include simplifying property laws, providing tax incentives for investment and improving marketing.

Tourism and Biodiversity Conservation

The most effective way of conserving biodiversity in Madagascar would be to end forest destruction. However, the main cause of forest destruction and degradation, *tavy*, is carried out by thousands of individuals of a dispersed rural population. Inevitably, the sets of economic and cultural circumstances and motivations that lead to *tavy* are complex and are probably well beyond the reach of environmental policy (Larson, 1994). It should also be noted that the present system of protected areas for nature conservation covers only approximately 2% of the land area. Many of these are small (< 2000 ha) and isolated. Given the current knowledge of species/area relationships and the fate of small populations (World Conservation Monitoring Centre, 1992), even if these areas were given full protection, this would have to be regarded as inadequate to conserve all of Madagascar's present biodiversity.

Any positive benefits to biodiversity conservation that may arise from increased tourism activity are likely to be limited. However, if they were concentrated on to a few protected areas they could be significant for some species, providing that they are not outweighed by negative impacts.

Assuming that increased numbers of visitors are concentrated into beach resort areas, three scenarios (negative, neutral and positive) regarding the impact of increased tourism on the environment can be described.

1. Negative. Increased tourism and related infrastructure development along with increases in local populations hasten the decline of forest areas and concentrate 'nature tourism' demands on small, fragile local 'protected areas', causing further declines in biological resources.
2. Neutral. Increased tourism does not significantly affect the rural economy and the general decline in biological resources continues, unaffected by tourism developments.
3. Positive. Increased tourism does not significantly affect the rural economy but generates resources for strengthening protection for nearby or easily accessible protected areas. This includes the employment of local people in nature protection, therefore providing an economic motivation for conservation. Localized but significant improvements in conservation of biological resources are achieved.

The three scenarios are not mutually exclusive. The potential for each needs to be carefully researched both in a strategic context and as part of the environmental impact assessment that should accompany any major developments. However, a preliminary assessment of the impact of increased tourism on protected areas can be made by looking at present impacts.

Tourism and Protected Areas

Two of the many protected areas in Madagascar appear to be particularly appropriate for further research. The Réserve de Faune de Perinet-Analamazoatra Andasibe (Perinet) is the most widely visited, with a high degree of involvement by statutory and international conservation agencies, while the Réserve Naturelle Intégrale de Lokobe (Lokobe) is less visited, and has little involvement of statutory agencies but growing involvement by local people.

Réserve de Faune de Perinet-Analamazoatra Andasibe (Perinet)

With 8000 visitors a year, Perinet is the most visited reserve in Madagascar (Andasibe Association of Guides, 1995). It is less than four hours' taxi ride from the capital Antananarivo and harbours the largest and most vocal of the lemur species, the indri (*Indri indri*). The reserve is small (810 ha) and is made up of areas of both primary and secondary medium altitude tropical moist forest. Over the last two decades one family group of indri have become habituated to human presence and tourists are virtually guaranteed both hearing the dramatic territorial whooping calls of groups and close sightings.

The reserve is biologically rich, containing at least 89 species of bird, 25 species of mammal, 40 species of amphibian and reptile and an unknown number of species of invertebrate and plant. Many species, particularly the mammals and some birds, survive in populations that depend on habitat beyond the reserve boundary as well as within the reserve. *Tavy* is encroaching on the reserve from both the south and the east, and the reserve is known to be used for the collection of animals for the national and international pet trade (International Union for Conservation of Nature *et al.*, 1987). Staff consists solely of a Chef de Station Forestière and two labourers. The reserve is inadequately policed.

Tourist visits tend to be concentrated in one small area, in the home range of the habituated indri. There is circumstantial evidence that the current level of usage of this part of the site has caused significant habitat deterioration through, for example, the cutting of lianas and dislodging of epiphytes.

The village of Andasibe lies just beyond the perimeter of the reserve. The village and surrounding area has an estimated population of 10,000 inhabitants. Most are involved in agricultural activities. The three other employment opportunities are in a local graphite mine, which employs about 40 villagers, in independent employment as guides for tourists visiting the reserve and as workers in the few small hotels that have grown around the reserve.

The 'Association of Andasibe Guides' was formed in 1992 to rationalize the then chaotic system of guide provision. To act as a guide in the reserve an individual must now be a member of the Association. In 1995 the Association had 19 members. All but one were men. All were thought to be aged under 40. The majority of tourists arrive in organized groups of up to 12 and spend between an afternoon and two days in the reserve surroundings, entering the reserve for up to three hours at a time for day and night excursions. A good guide (one that is vigilant enough to find new tourist arrivals first and with good language skills) can expect to get three to four bookings a week for most of the year (outside the cyclone season February–March). The hourly rate for a guide is fixed by the Association at 20,000FMG or about £3.00 (1995).

Réserve Naturelle Intégrale de Lokobe (Lokobe)

Lokobe is situated on the off-shore island Nosy Be, the 'centre' of the currently limited beach tourism industry in Madagascar.

The reserve is small (740 ha) representing the largest remaining remnant of humid primary forest on the now primarily agricultural island. It is surrounded by a 150–200 ha 'buffer' zone of secondary forest although there is ample evidence of this being cleared for manioc, rice, vanilla, and coffee production. The reserve is particularly important as it may be the only remaining site for several species of reptile and amphibian (International Union for Conservation of Nature *et al.*, 1987). At least 42 species of bird inhabit the reserve along with four species of lemur, including the black lemur, *Lemur macaco macaco*, which is confined to Nosy Be, one other offshore island and a few mainland sites.

Threats to the reserve arise from habitat destruction and from poaching of lemurs for food and the pet trade. Lokobe has one member of staff who must police it and prevent any illegal activities. Access to the reserve is difficult by land other than from the nearby village of Ampasipohy (estimated population 2000). Given that the reserve officer is stationed in the town of Hell-Ville, a two-hour car and pirogue trip away, and his only transport is a bicycle, policing the reserve effectively is impossible.

Tourist visits to the buffer zone around the reserve are mostly through organized tours led by one person, an enterprising former resident of Ampasipohy. Since 1980, this has brought thousands of tourists (no precise record has been kept) into contact with both the village and the extremities of the reserve. The contact with the tourists and the revenue gained from the sale of handicrafts has apparently brought about a shift in the attitude of villagers towards the reserve, culminating in 1993 in a vote to ban the hunting of lemurs from around the village and in the reserve. As compensation for the curtailing of this activity, a village tax is imposed on visitors brought to the reserve of 2500FMG (£0.36). The money collected in this way has enabled a teacher to be employed at the local school for much of the year.

Discussion

Detailed research needs to be carried out to identify the impacts of tourist visits to protected areas in Madagascar and into best practice for managing those impacts. However, the above case studies do allow for some preliminary discussion.

It is probably true to say that all protected areas in Madagascar have been designated in an opportunistic fashion without a proper assessment of how or if objectives can be achieved. All suffer to various degrees from encroachment and all are likely to be too small to fully protect all the species within them. Given this, one goal of managing tourist involvement in protected areas should be to conserve habitats and species beyond the perimeter of protected areas in so-called buffer zones. As with Lokobe, where visitors rarely penetrate the reserve itself, even a partially forested buffer zone can be the site for most tourist activity. This situation minimizes the physical impact of visitors on the reserve and could also serve the function of raising visitor awareness of some land-use pressures.

It is logical to assume that an acceptance of the function of a reserve by local people is vital in determining its effectiveness. Both Perinet and Lokobe provide examples of reserves where the revenues from tourist visits have provided motivation for at least some local people to alter their behaviour, allowing for the greater protection of the reserves. In Perinet there has apparently been a lessening of direct opposition to the policing of the reserve over the last decade (Andasibe Association of Guides, 1995) and in Lokobe there is evidence from the behaviour of *Lemur macaco macaco* that hunting has ceased. The involvement of the local people in voting for a ban on *Lemur macaco macaco* hunting in return for modest funds for village based development is perhaps the clearest example of a positive relationship between tourist involvement and conservation (see also Wells and Brandon, 1992). The positive nature of this relationship probably results from several factors, which include: the degree of control local people have over tourist involvement with the reserve and their village; the direct involvement of local people in the operation, including the provision of transport and food; the absence of involvement of outside agencies; and the relatively small size of the local community. In short, the small local community has virtual monopolistic control over visits from tourists. As such, they can virtually guarantee a return from their concessions to conservation.

The operation at Perinet presents a slightly different picture. Firstly, a relatively limited potential exists for direct employment (as guides) in relation to a larger local population. Some 8000 visitors provide direct employment for just 19 people. If economic benefits from tourist visits are a prime motivation for the acceptance of a protected area, this may have important consequences. Secondly, important aspects of the operation are controlled by people from outside the area or even non-nationals. Perinet can be visited in a day trip from the capital. Most of the accommodation at the site is owned by one private company. Thus, local people are not in control of access to the reserve and even the guides are in a competitive situation.

In addition, it should be noted that, while hourly rates are very high for guides when compared to other employment, the time spent waiting to be hired can be considerable. If a guide is not waiting at the gate when a party arrives to visit the reserve, no work is obtained, but a guide waiting is not earning. This problem could be resolved with a pre-booking system, or even salaried guides, but this would require much greater organization, which may not be achieved with the present essentially competitive system. If guides could work only for time spent in the forest, the benefits of employment could be spread more widely in the community. Older married guides would be able to work and gain enough money to support spouses and families. This could further increase the influence of those with a positive stake in the reserve.

Despite some evidence that tourist visits can, or could, be a positive influence on a reserve, some practices at reserves and around tourist facilities bear a striking resemblance to a conservationist's nightmare. Of prime concern is the practice of keeping captive wildlife, for example crudely tethered or caged lemurs, or captured chameleons, for tourists to 'interact with'. From the developer's point of view, when there is no wildlife around to satisfy the tourist's curiosity, bringing endangered wildlife to the tourists may seem to be a good second best. However, to come all the way to Madagascar to stand at the bars of a cage and watch lemurs displaying unnatural behaviour is far from ideal 'ecotourism' and hopefully will not satisfy many visitors. In the context of small fragile sites, almost equal concern should be expressed about the routine and deliberate flushing, or disturbance, of animals in natural habitats for visitors to glimpse them. Continual disturbance of resting and feeding places is a good way of rapidly changing protected areas into sub-optimal areas for larger fauna.

Conclusions

The wildlife of Madagascar represents one of the world's most significant pools of biodiversity. This resource is continually being eroded, mainly through the actions of a dispersed, poor, rural population.

Tourism is a small but significant and growing element of the economy. Tourism is unlikely to yield enough resources or provide the opportunity for major improvements in the conservation status of much of the island's biological resource. However, under some circumstances, particularly when controlled by local people, increased tourist visits could be of significant benefit to the effective functioning of some protected areas. There is an urgent need for further research in this area in order to allow maximum benefit to be gained from the projected increases in visitor numbers.

References

Andasibe Association of Guides (1995) Group discussion, personal communication.
Economist Intelligence Unit (1994) Madagascar. *EIU International Tourism Reports* 1, 25–44.
Glaw, F. and Vences, M. (1994) *A Field Guide to the Amphibians and Reptiles of Madagascar*, 2nd edn. Private publication, Bonn.
Goodman, S.M. (1993) A reconnaissance of Ile Sainte Marie, Madagascar: the status of the forest, avifauna, lemurs and fruit bats. *Biological Conservation* 65, 205–212.
International Union for Conservation of Nature/United Nations Environment Programme/World Wide Fund for Nature (1987) *Madagascar, an Environmental Profile*. Edited by M.D. Jenkins. IUCN, Gland.
Jolly A. (1989) The Madagascar challenge: human needs and fragile ecosystems, In: Leonard, H.J. *et al.* (eds) *Environment and the Poor: Development Strategies for*

*a Common Agenda.*Overseas Development Council, Washington, DC.

Jolly A., Oberle, P. and Albignac, R. (eds) (1984) *Key Environments Madagascar.* Pergamon Press, Oxford.

Larson B.A. (1994) Changing the economics of environmental degradation in Madagascar: lessons from the National Environmental Action Plan Process. *World Development* 22(5), 671–689.

Mittermeier, R.A. and Werner, T.B. (1990) Wealth of plants and animals unites 'megadiversity' countries. *Tropicus* 4(1), 4–5.

Myers, N. (1988) Threatened biotas: 'hot-spots' in tropical forests. *The Environmentalist* 8, 187–208.

Randrianasijaona, P. (1983) The erosion of Madagascar. *Ambio* 12(6), 308–311.

Wells, M. and Brandon, K. (1992) *People and Parks – Linking Protected Area Management with Local Communities.* World Bank, World Wildlife Fund, US Agency for International Development, Washington, DC.

World Bank, US Agency for International Development, Coopération Suisse, UN Educational, Scientific, and Cultural Organisation, UN Development Programme, and the World Wildlife Fund (1988) *Madagascar: Plan d'Action Environmental, Vol. 1, Document de Synthèse Générale et Propositions d'Orientations; Vol. 2, Synthèses Spécifiques et Recommendations.* World Bank, Washington, DC.

World Conservation Monitoring Centre (1992) *Global Biodiversity, Status of the Earth's Living Resources.* Chapman and Hall, London.

The Sustainability of Ecotourism

24

R. Burton

School of Geography and Environmental Management, University of the West of England, Coldharbour Lane, Frenchay, Bristol BS16 1QY, UK

Introduction

Butler's (1991) model of tourist development suggests that, once a destination begins to become popular, commercial forces set in train a series of events that tend to increase the levels of visitor use, change the nature of the resource and change the type of tourist experience until the resource loses its tourist appeal. Thus the model suggests that tourism as a phenomenon is intrinsically unsustainable. Indeed, Butler suggests that, unless steps are taken to intervene in this cyclic process, then, because the resources are 'common', in Hardin's (1968) terms, they will inevitably become overused. Tourism can only become sustainable, in terms of both its economic returns and the survival of the resource on which it is based in the long run, if there is intervention in the destination development process.

Steele (1995), arguing from the viewpoint of economic theory, also concludes that restriction of open access to 'common' renewable tourist resources is the only way to ensure economic and environmental efficiency, that is, genuinely sustainable tourism. The most direct way to intervene in the destination development cycle and to control open access to 'common' resources is to restrict tourist numbers, thereby holding tourism to an environmentally and an economically sustainable level.

Butler (1991) accepts that intervention in the form of limiting tourist numbers is politically difficult in a free market situation. Grabowski (1994) also argues that the political scenarios that would make this outcome a possibility are indeed rare, although Burton (1994) and Sofield (1991) report on two specific situations when alliances of interest between central government and local communities have resulted in strong policies that effectively curb tourist numbers.

Butler's three other options for sustainable tourism – changing the tourist type, hardening the resource, or educating the actors in the development cycle to accept controls and reduce impacts – are much more widely acknowledged as feasible and politically acceptable. They are more widely adopted in practice. Ecotourism has been proposed as an alternative form of tourism that achieves two of these options, by changing the tourist type, and by education, and which also contributes positively to the conservation of the resource. Ecotourism is a subset of nature; based tourism; the latter, according to Valentine (1992), is 'primarily concerned with the direct enjoyment of some relatively undisturbed phenomenon of nature' (p. 108). The Commonwealth Department of Tourism (1994) in Australia's National Ecotourism Strategy defines ecotourism as 'nature based tourism that involves education and interpretation of the natural environment and is managed to be ecologically sustainable' (p. 3), where ecologically sustainable tourism involves an appropriate return to the local community and long term conservation of the resource.

Valentine's definition spells out the management implications more clearly:

> ecotourism is restricted to that kind of tourism which is:
> - based on relatively undisturbed natural areas.
> - non-damaging, nondegrading, ecologically sustainable.
> - a direct contributor to the continued protection and management of the natural areas used.
> - subject to an adequate and appropriate management regime.
>
> (Valentine, 1993, p. 108)

Wight (1993) suggests that ecotourism 'involves a spectrum of experiences' (p. 57) and can be a mix of adventure, traditional, nature and cultural tourism but it is the ethical overlay of the actor and operator that translates these activities into sustainable ecotourism.

Thus, Valentine puts more emphasis on the need for a clear management regime to enable ecotourism to meet the sustainable criteria, while Bottrill and Pearce (1995) adopt a more extreme definition, which maintains that ecotourism can only occur in legally designated and protected areas of conservation value. On the other hand, Wight implies that the attitudes, motivations and beliefs of the participants translate a range of tourist experiences into 'ecotourism'.

If all these demanding definitional criteria have to be met in full, then it is unlikely that many tour operations would qualify as ecotourism. In reality, there is more likely to be a continuum of nature-based tourism, from those at one extreme that meet none of the criteria or just do the minimum required, to those at the other extreme that voluntarily meet all the criteria. Ziffer (1989) has identified a spectrum of tour operators that could be equated with this continuum; this is elaborated in Table 24.1. However, the problem lies in defining the point along this continuum at which genuine ecotourism and sustainable practice begin.

Background to Study

This chapter reports on a study of two areas of Northern Australia of nature-based tour operators who reflect the full range of this spectrum, and will argue the case that:

Table 24.1. Ziffer's spectrum of ecotourism.

Nature-based tourism	Added subdivisions
1. Tour operators that sell nature	(a) Those who are unaware or uncaring about its impact. (b) Operators that are aware of impacts, do the minimum to abide by any management rules, who do *not* seek to educate or change tourists' attitudes, but may provide information.
2. Sensitive tour operators	(a) Aware of impacts, actively seek to educate tourists by providing information. (b) Actively seek to influence tourists' attitudes and behaviour. Support conservation, e.g. members of conservation groups. (c) Practise minimum impact tourism (over and above management requirements), e.g. pack out rubbish, rotate sites.
3. Donors (in that they give something back to the environment)	Act positively to improve the environment they use and restore damage, e.g. pack out other people's rubbish, participate in restoration schemes, voluntarily donate a proportion of trip costs to conservation or management of the resource, plant trees, support local community.
4. Doers	(a) Those who initiate conservation projects or research. (b) Those involved actively in influencing policy and management towards sustainable practices.

Ecotourism

1. the number of tour operators that genuinely and voluntarily meet all the criteria is likely to be very small and these businesses are of a particular type.
2. the market for this type of tourism is also very small, and unlikely to be big enough to support very many of these tourism enterprises.
3. the tour operators that meet most of the criteria for classification as ecotourism may actually initiate tourism development in new destinations and, rather than being agents of conservation, may possibly initiate the Butler cycle in hitherto pristine environments. Increased commercial competition may lead ecotour operators to open up hitherto unvisited locations, in an attempt to sustain the quality of their tourists experience. This response is likely to trigger off the destination development cycle in yet another location and so accelerate the spatial spread of tourism rather than contain it.
4. ecotourism operators are subject to the same commercial pressures as all tourism businesses who participate in the destination development process, and ecotourism may represent merely an early and ephemeral stage in the destination development cycle.
5. ecotourism may in fact depend on strong interventionist management policies in order to survive, and the benefits of ecotourism may only be assured if total tourist numbers are constrained.

The tour operators included in the study were limited to those offering land based walking or motorized (four-wheel drive/coaster/air/tour bus) day or extended camping and accommodated nature-based tours. The operators of built facilities and independent travellers were excluded.

Case Studies

Australia was chosen as the most appropriate location for the research because it is rightly acknowledged as a world leader in ecotourism (Jenner and Smith, 1992). Also its international tourism is very strongly based on its unspoilt natural resources, while its national tourism, ecotourism and sustainable development policies are relatively well developed (Commonwealth Department of Tourism, 1992, 1994).

Ecotourism

Two regions (the Top End of the Northern Territory and Far North Queensland, see Fig. 24.1) were chosen as study areas because nature-based tourism is a huge component of their tourist industries. In the Top End of Northern Territory, land-based tourism to the World Heritage Kakadu wetlands is a relatively recent phenomenon. Tourism grew very rapidly after 1985, when the film *Crocodile Dundee* was released and focused world attention on Kakadu. Tourism growth in the adjacent national parks, such as

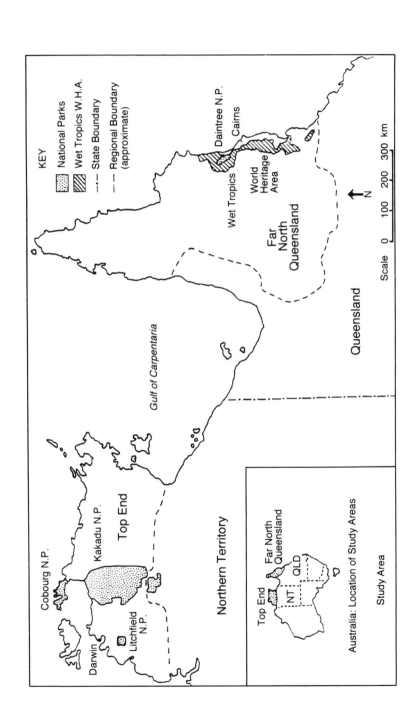

Fig. 24.1. Location of study areas.

Litchfield, is even more recent (see Fig. 24.2).

The tourism industry in Far North Queensland is much bigger and longer established (see Table 24.2). It is also primarily based on wildlife resources on both land in the tropical rainforest, the Wet Tropics World Heritage Area, and sea, the World Heritage Great Barrier Reef Marine Park (coral reefs). The study only included the land-based tours in the Wet Tropics.

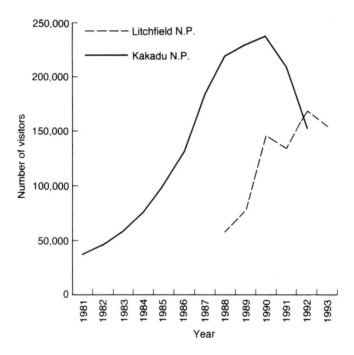

Fig. 24.2. Growth of tourism in Top End national parks 1981–1993.

Table 24.2. Growth of tourism in Top End and Far North Queensland 1986–1994. Number of visitors (international and domestic, in millions).

Year	Far North Queensland	Top End of Northern Territory
1994	1.184	
1993	–	
1992	1.852	0.524
1991	1.324	0.472
1990	1.531	0.495
1989	1.283	0.502
1988	1.235	0.492
1987	–	0.468
1986	0.867	

The two case study areas were chosen on the assumption that they represented destinations at different stages of the Butler cycle: the Top End being at an earlier stage, with a smaller but rapidly growing industry, than the more mature Queensland industry. Thus the purpose of the research was to study the development, management and tour operator behaviour at two different levels of tourism pressure and at two different stages of the Butler cycle.

In each location, a stratified sample of tour operators was selected and representatives of each interviewed. The sample included large and small companies, those offering day and extended tours, large and small group tours, up- and down-market products, and using one or several national parks. All provided nature-based tours, but with the variety of experiences outlined by Wight (1993). The interviews were in depth, comprised of structured open-ended questions, and generally lasted between one and three hours. Seventeen operators were interviewed in Northern Territory in 1992, and five key individuals that most closely met the criteria for ecotourism and/or were actively involved in the policy making and management process were re-interviewed in 1994. Ten operators were interviewed in Far North Queensland in 1995. These were selected to represent the same dimensions but concentrated more towards the 'ecotourism' end of the spectrum. Fewer interviews were completed in this year, partly because of time constraints but mainly as a result of there being a much more extensive research background, for example Jones (1993) and Card Miller (1994), in this location.

In both locations, as many of the tours as possible, time and finances allowing, were experienced as an unannounced participant observer.

A wide range of issues were discussed during the interviews, but key topics included:

- the operator's background and attitudes, and their role in education and tourist management.
- how they would or had responded to increases in tourist numbers and the amount of development in the sites they used on their tours.
- the growth of the market they were aiming at, and the growth (or otherwise) of their business.
- their attitudes to, and response to the current management practices and policies of the National Parks/World Heritage areas they used.

Classification of Tour Operators

The tour operators were first classified according to the expanded Ziffer (1989) spectrum (see Table 24.1) of nature-based (shallow green) tourism to (deep green) ecotourism. The results for each case study area are summarized in Tables 24.3 and 24.4. Predictably, further analysis suggests that the

Table 24.3. Top End classification of tour operators.

Operator continuum	Ecotourism ← Tourism operators									→ Nature-based tour operators							
	K	I	G	N	P	Q	E	F	J	A	B	C	D	H	L	M	O
Nature-based tourism																	
Selling nature, unaware/uncaring of impact										*	*						
Aware of impact																	
Abide by rules, do not seek to educate												*		*	*	*	*
Shallow green																	
Sensitive																	
Aware of impact																	
Educate						*		*									
Educate to alter attitude, and influence behaviour, member of conservation group	*	*	*	*	*	*	*	*	*								
Sensitive positive action																	
Minimum impact tourism, rotate sites	*	*	*				*						*				
Donors																	
Constructive action to improve environment, e.g. pack out others' rubbish, plant trees, donate % trip costs to conservation, support local initiatives	*							*									
Doers																	
Initiate conservation projects, involved in policy-making	*	*	*							*							
(Deep green): ecotourism																	

companies fulfilling more criteria for classification as 'ecotourism' were more likely to be:

- small ones employing up to three tour guides, although the Queensland companies tended to be slightly bigger.
- those offering small group tours (4–25 people per tour).
- relatively expensive tours (most over 98 Aus$ per day at date of survey).

- those run by people with either a professional, for example teaching, local government, consultant ecologist, background, or who were 'bushmen', defined as having experience of living and working/farming in the bush. Significantly, those who had trained and worked in the tourism industry or who had worked in any other exploitative industry, such as mining and water resources, were not involved.

In terms of Valentine's (1993) definition of ecotourism, only one company in the Northern Territory sample (K in Table 24.3) and one in the Queensland sample (no. 5 in Table 24.4) voluntarily met all the criteria. In terms of Ziffer's (1989) continuum, these clearly could be classified as 'doers' or 'donors'. If participating in policy-making was included then another two of the Northern Territory sample would qualify as 'doers' but not as (voluntary) 'donors'. In the Queensland sample, two of the 'deeper green' companies also employed guides who were very positive 'doers'.

Virtually all the companies in both case studies provided tourists with

Table 24.4. Wet Tropics (Queensland) classification of tour operators.

Operator continuum	Ecotourism			← Tour operators				→ Nature-based tour operators	
	3	5	6	1	2	4	7	8	9
Nature-based tourism									
Selling nature, unaware/ uncaring of impact								*	*
Aware of impact									
Abide by rules, do not seek to educate						*			
Shallow green									
Sensitive									
Aware of impact						*			
Educate				*	*		*		
Educate to alter attitude, and influence behaviour, member of conservation group	*	*	*						
Sensitive positive action									
Minimum impact tourism, rotate sites	*	*	*						
Donors									
Constructive action to improve environment, e.g. pack out others' rubbish, plant trees, donate % trip costs to conservation, support local initiatives	*	*							
Doers									
Initiate conservation projects, involved in policy-making	*	*							
(Deep green): ecotourism									

information, but significant numbers (see Tables 24.3 and 24.4) were adamant that they did not seek to educate or change visitors' attitudes. These tended to be the larger companies and included some of the long established local companies.

These data indicate that less than half of the sample voluntarily met any of the criteria for sustainable ecotourism, while the number meeting the most stringent definition was very small. The sample showed that particular types of business characteristics are associated with sustainable practices. Is the impression drawn from this quota sample, that few tour operators can be classified as genuine ecotourism providers, borne out by the findings of other surveys? Jones's (1993) survey of all 40 tour operators in the Daintree region of the Wet Tropics in Queensland yielded an 85% response rate. Her results (p. 44) showed that 32% of the respondents claimed to be 'donors' or 'doers' in that they said they 'participated in or sponsored some environmental rehabilitation or conservation projects'.

Weiler's (1992) national survey of Australian nature-based tour operators yielded 27 responses (a 46% response rate). This study showed that only 7% of operators aimed to be 'doers' or 'donors' in that they had environmental objectives that 'sought to raise funds or otherwise contribute to long-term research conservation activities' (p. 7). A Canadian study of nature-based tour operators (Bottrill and Pearce, 1995) indicated that, while the majority met 'reasonable' standards of environmental management, only five of the 22 companies would qualify as genuine ecotourism ventures. This evidence therefore leads to the conclusion that it is a minority of nature-based tour operations that qualify as ecotourism providers, and it may be a very small minority that voluntarily meet the most stringent criteria. The next section of this chapter addresses the question of whether ecotour operators are more likely to initiate the Butler cycle than contain it, and whether they are more likely to do so than the more mainstream nature based tour operators.

Tour Operator Behaviour in Response to Increasing Tourism

The two case-study regions of Northern Australia have both experienced a significant overall growth in tourism, but the level of tourism in Far North Queensland is two to three times that of the Top End of Northern Territory (see Table 24.2). The size of the industry is also much bigger, with an estimated 600 members of the Far North Queensland Tourism Promotion Bureau as opposed to approximately 122 members of the Darwin Regional Tourist Association (1994/95) and only 195 operators in the whole of the Northern Territory. The intensity of use of some of the national parks in Far North Queensland is much higher than that for Top End national parks (see Table 24.5). The detailed figures of growth of visitor numbers that are

Table 24.5. Intensity of use of study areas.

| | Top End | | Queensland | |
	Kakadu (19,757)	Litchfield (1461)	All Wet Tropics (9000)	Douglas Shire (Daintree) (1827)
Area (km²)				
1992/93 Visitor numbers	154,000	155,000	est. 1.4 million	240,000–331,500
Tourists per km²	Overall 7.7 (but to most intensively used zone est. 70–150)	106	est. 155	131–181

available for Top End national parks appear to show the classic progress of the Butler cycle: rapid growth, stabilization and decline, which was disrupted only by the atypical year of 1991 (see Fig. 24.2). In reality, the explanations of these profiles may be concerned more with differences in management policies than with the unfettered processes of the Butler cycle. It is not the purpose of this chapter to analyse the detailed reasons for this growth; suffice it to say that it has happened. The objective of this chapter is to analyse the ecotour operators' response to these changes and their role in spatially extending tourism to new pristine areas.

In the interviews the tour operators' response was explored by:

- discussing the extent to which *their* business was growing or not.
- their response in terms of attitudes to increased levels of tourism in particular places.
- any changes they had made to their tours.
- the success or otherwise of these changes (i.e. did they sell?).

The results are analysed and presented in terms of different patterns of response in relation to the shallow green–deep green continuum of operators, spanning the nature tour–ecotour spectrum. It was assumed that the deeper green the operator and the more of the ecotourism criteria met, the more likely it was that tours would be modified to avoid crowded areas subject to increasing use, and the more likely pristine environments would be sought. Tables 24.6 and 24.7 show that the genuine ecotour operators did all actively seek to avoid crowded locations, and sought pristine environments. In addition to the Top End tours (which included Kakadu), operator 'K' offered a variety of tours to varied remote wilderness locations where no other tour operators had yet penetrated. These received 'rave reviews' and the operators achieved various tourism and environmental awards. But the new tours did not sell consistently. The operator's response was to increase the frequency of the Top End tour that included Kakadu (which did sell regularly), even

Table 24.6. Top End tour operators' response to increasing tourism.

	Ecotours				← Tour operators →					Nature-based tourism							
	K	I	G	N	P	Q	E	F	J	A	B	C	D	H	L	M	O
Extent to which operator seeks to extend tourism																	
Actively avoids crowded place	*	*	*		*					*				*			
Avoids crowded times	*	*	*			*	*	*									
Offers new products, new pristine environments	*	*	*														
Business decreasing	*	*			*		*										
Business static			*	*						*	*	*	*		*	*	*
Business increasing						*			*				*			*	

Table 24.7. Wet Tropics tour operators' response to increased tourism.

	Ecotours ←			Tour operators				→ Nature-based tourism	
	3	5	6	1	2	4	7	8	9
Extent to which operator seeks to extend tourism									
Actively avoids crowded places	*	*	*	*		*			
Avoids crowded times	*	*	*	*	*	*			
Offers new products, new pristine environments					*				
Business decreasing		*	*	*	*	*			
Business static								*	*
Business increasing	*			*					

though it became increasingly difficult to avoid crowded sites. Operator 'I' had, up to 1992, run a regular high-cost tour to Litchfield Park that, with skilful timetabling and planning, provided a quality ecotourism experience. However, this was abandoned on two counts. First, it was increasingly difficult to provide the wilderness experience that justified the high price, which was the direct result of the managing authorities' policy of hardening, opening up and increasing the use of sites. Secondly, the tour began to under-recruit and run intermittently because of price competition from new and cheaper day tours. After attempting to sell very specialized scientific wildlife tours, the operator sought a new location for an extended wilderness ecotour. He found a suitable venue outside the national park system on private land to which he had exclusive access, thereby guaranteeing a pristine environ-ment, lack of crowds and lack of competition. This policy has sold tours, but not well enough for the business to survive on these alone so the company, reluctantly, now offers 'bread and butter' tours to Kakadu, which do sell regularly. The third Northern Territory operator, 'G', offers specialist bush walking tours that can still be located in the wilderness and pristine parts of the park, but he too is expanding his business outside the Kakadu area, as the suitable areas open to him in the park are progressively shrinking. The fourth ecotourism operator, 'N', in Northern Territory runs charter tours, so it is more difficult to apply the same analysis to his pattern of operations, as each tour is unique.

None of these four operators reported a significant increase in business in the Northern Territory Top End; rather they described it as static or decreasing, even though the overall visitor numbers to Top End of Northern Territory are still rising (see Table 24.2).

The Northern Territory group of 'shallow green' operators that met some of ecotourism criteria tended to show a different pattern of response to increasing levels of tourism. They generally avoided the peak times at the most heavily used sites rather than missing out these popular sites altogether. They tended to offer cheaper tours, with more 'fun' and less explicit education. The two operators offering 'budget tours' were the only ones which reported an increase in business. None had sought to expand their tours in other locations. The final group of operators, who did not meet the main criteria for ecotourism, did not avoid the congested sites or times. Some said they couldn't. All reported static or increasing business.

Thus, in the Northern Territory case study, it does appear to be true that those operators whose business most closely resembles the ideal form of ecotourism are most likely to respond to increases in visitor use by offering new tours in new pristine settings. This finding, that these new tours generally failed to sell well enough to support the business, which was then forced back to offering 'bread and butter' tours in the congested parks, was unexpected. How did the experience of the Queensland sample match that of the Northern Territory tour operators? The three 'deepest-green' tour operators all sought to avoid crowded locations but none had explicitly designed a new product as an alternative. These three operators had experienced sharply different business conditions. One reported a 10% increase in business, a second was cross-subsidizing his wilderness/wildlife ecotours with his more mainstream tours, which visited the busy spots, while the third company had been taken over and the operator now ran essentially the same tour but as a guide for another company. There was further anecdotal evidence of other 'genuine' ecotourism companies having been taken over or gone out of business but it was not possible to obtain interviews in these cases. One of the Queensland operators in the 'shallow green' category had, amongst other products, been offering a botanical rainforest walk but these were not selling, and again he was cross-subsidizing them and was on the verge of withdrawing that tour in order to (reluctantly) concentrate on the more mainstream tours. A consistent experience of all the Queensland nature-based tour operators was that the more expensive extended tours were suffering at the hands of cheaper/mass day tours from Cairns. They claimed that this was mainly due to the practices of the agents rather than as a response to demand.

Only one operator in the Queensland sample had attempted to offer a new tour to a less congested site, where only one other operator had a licence, outside the most crowded zones. This tour was still offered but the operator stated that with luck it would run 30 times a year in comparison to the daily demand for the Daintree tours. The other operators' response when asked if they would develop new tours to new sites was that there was nowhere else to go because the Wet Tropics had already reached its capacity to absorb tourism and there were no other opportunities to find new accessible pristine uncrowded environments.

To summarize the findings of both case studies, it is clear that the majority of the operators offering the most sustainable type of ecotourism were finding it difficult to run their tours profitably; indeed, some were struggling. There appeared to be more evidence of these problems in Queensland. This would fit the theoretical expectations if it is assumed that the Wet Tropics area has progressed further into the Butler cycle of destination development than the Top End.

Conclusions

Although it is impossible to draw conclusions that can be backed up by quantitative evidence from this type of research, the evidence put forward here does support the following conclusions, reinforced by the experience of Discovery Ecotours (Preece, 1995).

1. The market for the type of experience offered by the businesses offering the most sustainable ecotourism is very small and/or is much more price-conscious than theoreticians suppose. There are simply not enough tourists prepared to pay high prices for pristine environments to sustain such business enterprises.
2. The genuine ecotour operators may seek to extend tourism by offering better products in new pristine sites but they do not have the marketing resources effectively to sell these new tours, given that the existing market is so small. So, if the operators cannot get access to suitable sites in the areas promoted and marketed by umbrella organizations, they have either to change their product to a lower quality type of tour or go out of business.
3. Operators offering 'sustainable ecotourism' are therefore part of the cyclic process of destination development. They are not immune from the

commercial forces that drive the cycle. The fact that they offer environmentally sustainable products does not protect them from these economic forces. According to the Butler theory, they will inevitably be replaced and eventually squeezed out as the destination and type of visitor change.

If these general conclusions are valid, what are the implications of this for tourism and environmental management policies and practice? The managing authorities have several options.

They may try to reduce the ecotourism operators' competitive disadvantage by introducing management regimes that force all operators to become 'donors' by introducing the user-pays system. All operators, whether or not they follow sustainable practices such as recycling and minimum impact camping, would contribute funds to the management of the resource, which could be used for the rehabilitation and restoration of sites. This option does not necessarily increase sustainable practice. On its own, it merely provides funds to manage the consequences of unsustainable tourism.

Policies could be introduced to encourage, require or force all operators to adopt more sustainable practices. Systems of operator accreditation would fall into this category of management tool. But such schemes tend to set minimum standards of practice rather than to encourage best practice. These schemes may merely 'cut out the cowboys', though they should protect the better operators from a certain form of unfair competition.

More radical solutions are necessary if the most environmentally sustainable tour operations are themselves to be sustained and encouraged. If such ecotourism operations do provide a more desirable form of tourism than that provided by other operators, then the managing authorities need to provide contexts in which their operations can be viable. This inevitably means more direct intervention in the market forces that threaten to squeeze out such businesses. This can be done in various ways, all politically difficult, but some perhaps being slightly more politically acceptable than others.

This chapter would argue that the sustainable ecotour businesses identified in this research would be most likely to continue in business if they had exclusive access to suitable sites, where their own environmentally responsible action would guarantee a high quality/pristine environment to the tourist, and the lack of competition would guarantee the high quality wilderness type experience that such tourists are supposed to want.

This course of action would have three further beneficial consequences. It would: (i) remove the incentive for the ecotour operators to spread tourism to new untouched sites; (ii) meet Steele's (1995) main criterion for environmentally and economically sustainable tourism, by restricting open access; and (iii) be likely to reduce the environmental damage done by each tourist. The implementation of exclusive access is fraught with difficulties but management techniques are available to implement this.

1. Non-competition agreements. The operator is guaranteed exclusive use of a site for a specific number of years. This has occasionally occurred in Northern Territory. It is a technique that can probably only be politically justified when a development requiring high capital investment in a high risk situation is deemed desirable.

2. Concessions and competitive tendering for sites combined with operator licensing schemes. These have been used more frequently as a set of techniques in different combinations in both Northern Territory, for example in Kakadu, and Far North Queensland, for example by the Great Barrier Reef Marine Park Authority. The industry perception of the system is that although it is unpopular. However it is reasonably acceptable because it favours the big companies. One of the criteria for the assessment of tenders could be the extent to which the company voluntarily meets the criteria for ecotourism and the extent of its sustainable practices. This would give the smaller companies, who are more likely to be genuine sustainable ecotourism businesses, a better opportunity to win the tender.

These management techniques obviously need to be considered in the political context in which the managing authority finds itself, and some measures may be more acceptable than others. Creating 'monopoly' situations is politically risky and open to corruption and abuse. The managing authority would have to be seen to be impartial and would have to set up genuine site monitoring systems so that it could be demonstrated that the operators given exclusive rights were using the site sustainably, and that the blame for any environmental damage was laid at the right door. It is also particularly politically difficult for publicly funded national park authorities to justify the exclusion of the public from sites used by commercial operators, even though most of the tour operators, managers and rangers contacted in this study agreed that the behaviour of the independent tourists often caused more environmental problems in the parks than the visitors on guided tours.

The political situation each managing authority finds itself in is different. The Northern Territory state-run Litchfield National Park is being explicitly managed deliberately to increase visitor numbers, harden sites and encourage recreation succession, which, combined with proposals to create new remote ecotour/adventure tour parks, simply serve to accelerate the Butler cycle (Burton, 1994). On the other hand Kakadu, run under Commonwealth Government supervision, finds it politically acceptable and even necessary to limit visitor numbers and restrict visitor access, which has the net effect of concentrating most tourists into a diminishing area of the park. Both result in some tour operators, including the ecotour operators, seeking opportunities on private land outside the parks. The Wet Tropics Management Authority has inherited a situation of very intensive tourist development with a huge built-in growth potential in the form of licences already given but not yet taken up. Here tour operators are also being encouraged to seek access to privately owned land in and around the Wet Tropics area. The keener price competition here means that it is a less attractive option to the Queensland tour operators but it is welcomed by the authority as they perceive it as a means of relieving pressure on their area. In all three situations, the genuine ecotour operator is finding it hard to survive.

In current political climates of opinion, a growing role for the private sector is favoured but it is most unlikely to lead to an overall increase in sustainable resource use practice. This can only be guaranteed through strict management controls on the public land of the national parks. Policy makers and park managers will eventually have to 'bite the bullet' and intervene in the market forces that inexorably lead to the progression of the Butler cycle and to the non-sustainable use of natural resources. Steele (1995) argues from economic theory rather than empirical research but comes to the same conclusion. He argues that open access leads to environmental and economic inefficiency and thus to lower profits in the long term. He states that both controls on 'damage per tourist' and controls on total numbers of tourists are required in order to achieve economically and environmentally sustainable tourism.

In this chapter it is concluded that the genuine ecotour operators who voluntarily adopt sustainable practices are the most likely to ensure that damage per tourist is minimized, but that the survival of this type of operator in the marketplace depends on a management regime that controls total tourist numbers and limits open access. Market intervention may create the circumstances in which the genuine ecotourism businesses may themselves be sustainable.

References

Bottrill, C.G. and Pearce, D.G. (1995) Ecotourism: towards a key elements approach to operationalising the concept. *Journal of Sustainable Tourism* 3(1), 45–54.

Burton, R.C.J. (1994) Making sustainable tourism a reality: tourism management in the National Parks of Australia's Top End. In: *Expert Meeting on Sustainable Tourism and Leisure*. Department of Leisure Studies, Tilburg University, The Netherlands, December.

Butler, R.W. (1991) Tourism, environment and sustainable development. *Environmental Conservation* 18(3), 201–209.

Card Miller, M. (1994) Preaching to the converted? Environmental education and ecotourism. MEd, thesis, James Cook University of North Queensland, Australia.

Commonwealth Department of Tourism (1992) *Tourism, Australia's Passport to Growth: a National Tourism Strategy*. Australian Government Publishing Service, Canberra.

Commonwealth Department of Tourism (1994) *National Ecotourism Strategy*. Australian Government Publishing Service, Canberra.

Grabowski, P. (1994) Sustainable tourism development – no chance. In: *Expert Meeting on Sustainable Tourism and Leisure*. Department of Leisure Studies, Tilburg University, The Netherlands, December.

Hardin, G. (1968) The tragedy of the commons. *Science* 162, 1243–1248.

Jenner, P. and Smith, C. (1992) *The Tourism Industry and the Environment*. EIU Special Report No. 2453, London.

Jones, K. (1993) Ecotourism operator guidelines – a case study of the Cape Tribulation – Daintree Section of the Wet Tropics World Heritage Area. BA (Tourism) thesis, James Cook University of North Queensland, Australia.

Preece, N. (1995) Discovery ecotours – sustainable tourism in outback Australia. In: Harris, R. and Leiper, N. (eds) *Sustainable Tourism – an Australian Perspective*. Butterworth-Heinemann, Australia, pp. 140–144.

Sofield, T. (1991) Sustainable ethnic tourism in the South Pacific: some principles. *Journal of Tourism Studies* 2(1), 56–72.

Steele, P. (1995) Ecotourism: an economic analysis. *Journal of Sustainable Tourism* 3(1), 29–44.

Valentine, P.S. (1992) Nature based tourism. In: Weiler, B. and Hall, C. (eds) *Special Interest Tourism*. Belhaven, London, pp. 105–127.

Valentine, P.S. (1993) Ecotourism and nature conservation. A definition with some recent developments in Micronesia. *Tourism Management* 14, 107–115.

Weiler, B. (1992) Nature based tour operators – are they environmentally friendly or are they faking it? Paper to *First World Congress on Tourism on the Environment*, Belize, April/May.

Wight, P.S. (1993) Sustainable ecotourism: balancing economic, environmental and social goals within an ethical framework. *Journal of Tourism Studies* 4(2), 54–66.

Ziffer, K. (1989) *Ecotourism: the Uneasy Alliance*. Conservation International, Ernst and Young, Washington, DC.

Index